INTERMEDIATE ALGEBRA
and
ANALYTIC GEOMETRY
MADE SIMPLE ®

BY

WILLIAM R. GONDIN, Ph.D.

Associate Professor, The City College of N. Y.

with

practice exercises and answers

BY

BERNARD SOHMER, Ph.D.

Instructor, The City College of N. Y.

MADE SIMPLE BOOKS
®
DOUBLEDAY & COMPANY, INC.
GARDEN CITY, NEW YORK

ABOUT THIS BOOK

In an age when science and engineering make international headline news, he who would keep abreast of the times must have more than an elementary knowledge of mathematics. For mathematics is still "the first of the sciences." And as the key to all the other sciences, it has become the engineer's most indispensable tool.

Today, an ever-increasing number of people who never before studied the subject beyond elementary algebra and geometry, or who remember it only as an irksome "requirement" of their earlier "schooling," now find that they need some command of its more advanced techniques on the jobs by which they earn their living. Countless others have long since discovered that, without an understanding of what advanced mathematics is about, one cannot feel altogether comfortably at home in the culture our twentieth century has produced.

This book, together with its companion volume, *Advanced Algebra and Calculus* in the same *Made Simple* series, is designed to help meet such needs on a plan explained in detail in the introductory chapter which follows. But first a few words of more general preface are in order here.

The concern of these books is not with mathematics as a set of arbitrary rules to be followed blindly as an exercise in compliance with classroom "discipline." Rather, it is with *mathematics as a way of thinking — as an application, in a somewhat specialized subject matter, of ordinary commonsense and reasoning power.*

Of course, the subject does have a notorious reputation for being "abstruse." But one of the convictions upon which the writing of these books has been based is that the *seeming* difficulty of mathematics is due largely to the fact that too many students are introduced to it in such a way that they never get to apply their native wits to best advantage.

The core of the trouble appears to be that, *unlike more complex subject matters, the peculiarly simple subject matter of mathematics lends itself to abbreviated treatment.* It lends itself to abbreviated representation by the shorthand of mathematical symbols. And it lends itself to abbreviated exposition by the shortcut of formal demonstration. This is both its unique marvel and its main stumbling block for the uninitiated.

Once we understand the nature of the problems and ideas in a given branch of mathematics, we have the key to its systems of abbreviation. Then we can appreciate formal treatment for its beauty and economy . . . . perhaps the trial-and-error research of generations of mathematicians over the centuries, brilliantly concentrated in a few crisp lines of "proof."

But until we have grasped what these problems and ideas are all about, we lack the key to their codes of abbreviation. Then formalism may compound confusion with despair. And all too often, we may attempt the impossible task of frantically trying to "memorize" what we do not understand.

Especially for the benefit of readers who must study without the help of a teacher sensitive to this difficulty, these books therefore introduce each new topic by a consideration of its underlying ideas in terms of the problems out of which these ideas arise. Only after such matters have been carefully discussed is the pace later quickened by the more usual, formal methods of presentation. Even then, more space than usual is devoted to such questions as: *What kinds of answers may be expected from the described formal method of solving problems? Does the method always work? If not, why and under what conditions does it fail? And what interpretation are we to place upon the conditions of our problem in each case?*

For valuable suggestions on the treatment of several topics in these books I am indebted to Associate Dean Sherburne F. Barber of the City College School of Liberal Arts and Science. For aid in checking galleys of the first volume, *Intermediate Algebra and Analytic Geometry Made Simple*, I am indebted to Mr. Henry Jacobowitz.

I am also so indebted to my colleague, Dr. Bernard Sohmer, for most of the practice exercises, for preparing the answers to the exercises in the appendix, and for re-checking the entire work from typescript to page-proof, that I have insisted his name share its cover and title-page.

The original plan and execution of the text have been its author's sole responsibility, however, and any shortcomings of either must be charged to his failure to make the most of these gentlemen's better counsel.

            — WILLIAM R. GONDIN

TABLE OF CONTENTS

SECTION ONE — INTRODUCTION

CHAPTER I

SECTION TWO — INTERMEDIATE ALGEBRA

CHAPTER II

CHAPTER III

Chapter IV

Chapter V

Chapter VI

Chapter VII

Chapter VIII

TRIGONOMETRIC FUNCTIONS AND EQUATIONS 92

SECTION THREE — PLANE ANALYTIC GEOMETRY

Chapter IX

POINTS, DISTANCES, AND SLOPES 111

Chapter X

STRAIGHT LINES 125

INTERMEDIATE ALGEBRA
AND ANALYTIC GEOMETRY
MADE SIMPLE

CHAPTER I

PRELIMINARY REVIEW

The Purpose of These Books

Once the student of mathematics has a reasonably good command of elementary algebra and geometry, he stands at the threshold of a whole new world of mathematical discovery and experience. Perhaps the greater subtleties of *higher* mathematics are still beyond him. But here at hand, under the general heading of **advanced mathematics**, are some of the most important analytic tools of modern science.

With relatively little more time and effort, the student may now reach out and grasp those tools. The purpose of this book and of its companion volume, *Advanced Algebra and Calculus Made Simple*, is to help you, the reader, make the most of that opportunity.

But first of all, a few words of general explanation . . . Before you start a new enterprise, it is always a good idea to take a glance ahead, a glance around, and then a last glance back. That is the purpose of this present chapter.

What is *Advanced Mathematics?*

Although the term is sometimes used differently, *advanced mathematics* is most often understood to be the content of first courses in such subjects as *college algebra, analytic geometry, vector analysis, differential calculus,* and *integral calculus.*

Of these, calculus is by far the most powerful as a basic analytic tool of modern science. For that reason you will find the most interesting practical problems in the second of these two volumes where both differential calculus and integral calculus are introduced. For the same reason, the other mathematical subjects

in these two books are presented in such a way as best to lead up to their later applications in calculus.

Before you can concentrate on the essentially new ideas of calculus, however, you first need to understand the use of analytic methods in geometry. And before you can handle either calculus or analytic geometry effectively, you need more algebraic technique than the reader who has studied only elementary mathematics is likely to command.

This present volume therefore continues with a section on those relatively advanced topics in algebra which are applied directly in analytic geometry. And the next volume begins with a second section on algebra dealing with further topics required in calculus.

Together, these two algebra sections cover topics which are usually presented in a *college algebra* course for students who have not had intermediate and advanced algebra in high school. On the customary high school plan of division, however, Chapters II through V in this volume may be classed as *intermediate algebra;* and the remaining algebra chapters in both volumes may be classed as *advanced algebra.*

Advanced Approaches to Mathematics

In what ways do the above mentioned subjects differ from more elementary ones?

Most obvious is the fact that their "advanced techniques" are ways for solving more difficult kinds of problems, or for solving simple problems more efficiently. This is illustrated throughout these volumes, but first in chapters IV and VII below.

Less commonly realized is the fact that much of the power of advanced mathematical techniques derives from a difference of approach — a broader point of view concerning the nature of mathematical problems, with resulting deeper insights into connections between seemingly unrelated mathematical subjects. This is best illustrated by the section on analytic geometry, but first in Chapter III below.

An approach to mathematics may also be more advanced, however, only in that it is a more careful re-examination of ways of dealing with problems — including even very simple ones which the elementary student may feel he already fully understands. It may raise such questions, for instance, as: *Does a given method always work? If it does not work, under what conditions does it fail? Why does it fail? And what interpretation are we to place upon the conditions of our problem and the results of our mathematical operations in each case?* This is first illustrated in Chapter II.

Recommended Sequence of Study

The usual subjects of advanced mathematics are arranged in these two volumes to form a natural study sequence. Each section, as a unit, leads on to the next. So the reader may begin at any point which his previous preparation will permit, and proceed systematically thereafter.

However, the chapters in both of these volumes have been so arranged that they may also be studied in other sequences if the reader wishes. After finishing Chapter III in this book, for instance, you may proceed at once to the first two chapters on analytic geometry, and complete the rest of the algebra section later.

Since alternative sequences like these may be preferred by some students, the reader is informed in the text wherever a reasonable possibility of choice occurs. But if you have any doubt at all as to which way to proceed at such points, you are strongly advised to *elect the sequence which temporarily skips you ahead in the text.* The reason is this . . .

In these volumes the usual topics of advanced mathematics are organized in separate sections under the conventional headings of different "subjects" only because some school curricula still parcel them out that way, and because some readers may therefore feel that they should concentrate on one subject in one section at a time.

But many school curricula have also begun to cut across the traditional subject-matter lines in mathematics. The sound theory behind this practice is that, *in some respects, the concurrent study of two, or possibly even three, of the conventionally separated subjects is more practicable, and indeed easier, than the study of any one of them separately.*

For instance, the reader who has carefully covered Chapters II and III of the following algebra section should find the analytic geometry of points, distances, lines, and slopes, in Chapters IX and X, little more than clarifying exercises in the application of simple linear equations and elementary graphing technique with which he has just become familiar. And very similar relationships exist between several other parallel sets of chapters.

Consequently, even though you may start with a more limited objective in the use of these books, by taking recommended alternative choices of chapters in later sections, you may be able to accomplish that objective better. Meanwhile, for little, if any, additional cost in time and effort, you will also have acquired command of a more advanced subject matter as an extra-dividend, so to speak.

Recommended Pace of Study

The reader who begins his study with the first chapters in any major section of either of these two volumes may perhaps feel that the text's *pace* — the rate at which new material is introduced — is "somewhat slow." But the reader whose mathematical preparation encourages him to begin his study in the middle of a section may perhaps have the opposite feeling that its pace is "a bit fast."

This is explained by another important feature of the deliberate plan of these books. As a new subject is introduced in each major section, careful consideration is given to basic underlying ideas which should be clearly understood before they are likely to be lost to view in a mass of technical details. But once the student has grasped these fundamentals, he is thereafter able to cover the remaining material much more quickly.

A sound study practice, then, is this: *If the pace of the text seems to you to be too slow at any point, make sure you are really getting its main ideas before you begin to skip ahead in the same section. But, if the pace of the text seems to you to be too fast at any point, look back to earlier chapters which contain explanations that will enable you to cover the later material more rapidly.*

Recommended Use of Practice Exercises

The best test of your understanding of mathematics is your ability to apply it in solving problems. For that reason, practice exercises are given at regular intervals throughout the text, with answers at the end of the book.

Moreover, the solutions of many of the problems in these exercises are applied again later, either in examples illustrating more advanced techniques or in further problems.

It is a good idea, therefore, to *do these exercises regularly. Check your results with the answers at the end of the book. Make sure that you can do most of the problems before you go on further. And keep your work organized systematically in a notebook so that you can refer back to it later.*

Assumed Background of the Reader

In this book and its companion volume, *Advanced Algebra and Calculus Made Simple*, the reader is assumed already to be familiar with the following —

Elementary Topics	(Chapters in Mathematics Made Simple)
Signed numbers and algebraic expressions	(VII)
Algebraic formulas and equations	(VIII)
Roots, powers, and exponents	(IX & X)
Factoring	(XI)
Common logarithms	(XII)
Number series	(XIII)
Elementary geometry	(XIV)
Trigonometry for angles between 0° and 90°	(XV)

NOTE: The *Mathematics Made Simple* references in parentheses here are to chapters of the book by that title in the same publisher's *Made Simple* series. Further references to this source are made with the abbreviation, "MMS".

Whenever you are in doubt about such elementary subjects, you should of course review them before attempting to go on in either of the two present volumes. In the long run, that will save both time and effort. And it will help to assure a sounder foundation for all further study.

Note on Tables of Formulas

There is a summary of main points at the end of most chapters in this book. Some of these summaries contain **tables of formulas** which you will find very useful, not only for your first study of the text, but also for reference purposes later.

The meaning and application of all tabulated formulas in later chapters is explained at length. However, this present chapter concludes with a table of formulas which are only briefly described because they summarize definitions, principles, and other elementary mathematical relationships with which you are assumed already to be familiar. Many are referred to at relevant points in the text below. But you are advised to *scan the entire table now as a final review of your previous study of elementary mathematics.*

The first few formulas on this table may seem so obvious as hardly to be worth stating. For instance, the fifth — reading: "$xy = yx$" — means only that "the ordinary quantities of elementary mathematics may be multiplied together in any order." As an example, $3(7) = 21$ and $7(3) = 21$; hence it makes no difference whether we multiply 3 and 7 together as 3 times 7 or as 7 times 3. But you will see later on in vector analysis (*Advanced Algebra and Calculus Made Simple*), that this is not true for all kinds of quantities, since the vector product of two vector quantities depends upon the order in which these quantities are multiplied.

In some of these formulas, the value, 0, is excluded for one of the quantities by a parenthetical *inequality* in the form, $x \neq 0$, to be read: "x not equal to 0." Reasons for this special treatment of the value, 0, are explained later in the text.

Note that the formulas in this table are written with the designator-letter R (Review). This is for convenience in distinguishing them from other series of formulas tabulated, and referred to, in the text.

REVIEW TABLE OF ELEMENTARY FORMULAS

R-Series Formulas
(Fundamental Operations)

R1: $x + y = y + x$.

Quantities may be added in any order.

R2: $x + (y + z) = (x + y) + z = x + y + z$.

. . . or in any grouping.

R3: $x + (-y) = x - y$.

Adding $-y$ to x is the same as subtracting y from x.

R4: $x - (-y) = x + y$.

Subtracting $-y$ from x is the same as adding y to x.

R5: $xy = yx$.

Quantities may be multiplied in any order.

R6: $x(yz) = (xy)z$.

. . . or in any grouping.

R7: $x(-y) = -xy$.

The product of a negative quantity and a positive quantity is negative.

R8: $(-x)(-y) = xy$.

The product of two negative quantities is positive.

(Operations Involving Parentheses)

R9: $a(x + y) = ax + ay$.

All terms within a parenthesis are to be multiplied by a coefficient outside the parenthesis.

R10: $-(x - y) = -x - (-y)$
           $= -x + y$.

A minus sign preceding a parenthesis applies to all terms within the parenthesis.

R11: $x - [a(y - z)]$
$= x - [ay - az]$
$= x - ay + az$,
      or
$x - [a(y - z)]$
$= x - a(y - z)$
$= x - ay + az$.

Sets of parentheses, or brackets, are to be removed "from the inside out" or "from the outside in."

(Operations With Fractions)

R12: $\dfrac{x}{y} = x\left(\dfrac{1}{y}\right)$, $(y \neq 0)$.

Dividing by y is the same as multiplying by $1/y$, provided $y \neq 0$.

R13: $\dfrac{x}{y} = \dfrac{nx}{ny} = \dfrac{\dfrac{x}{n}}{\dfrac{y}{n}}$, $(n \neq 0)$.

The numerator and denominator of a fraction may both be multiplied or divided by the same quantity, n, without changing the fraction's value, provided $n \neq 0$.

R14: $\dfrac{\frac{1}{x}}{y} = \dfrac{y}{x}$.

Dividing by a fraction is the same as multiplying by the reciprocal of that fraction.

R15: $\dfrac{x}{a} + \dfrac{y}{a} = \dfrac{x + y}{a}$.

The sum of two fractions with a common denominator is the sum of the numerators over the common denominator.

R16: $\dfrac{x}{a} + \dfrac{y}{b} = \dfrac{bx}{ab} + \dfrac{ay}{ab} = \dfrac{bx + ay}{ab}$.

Fractions with different denominators must be reduced to a common denominator before being added.

R17: $\left(\dfrac{x}{a}\right)\left(\dfrac{y}{b}\right) = \dfrac{xy}{ab}$.

The product of two fractions is the product of their numerators over the product of their denominators.

(Operations With Exponents)

R18: $(xy)^n = x^n y^n$.

R19: $\left(\dfrac{x}{y}\right)^n = \dfrac{x^n}{y^n}$.

An exponent outside a parenthesis is understood to apply to each factor within the parenthesis.

R20: $x^m x^n = x^{m+n}$.

The product of two powers of the same base quantity is the same base quantity with an exponent equal to the sum of the original exponents.

R21: $\dfrac{x^m}{x^n} = x^{m-n}$.

The quotient of two such powers is the same base quantity with an exponent equal to the exponent of the numerator minus the exponent of the denominator.

R22: $x^1 = x$.

Quantities written without exponents are understood to have the exponent, 1.

R23: $x^0 = 1, \quad (x \neq 0)$.

Any quantity, except 0, with the exponent, 0, equals 1.

R24: $x^{-n} = \dfrac{1}{x^n}$.

A base quantity with a negative exponent is equal to the same quantity in the denominator with the sign of its exponent changed.

R25: $(x^m)^n = x^{mn}$.

The n'th power of the m'th power of a quantity is the mn'th power of that quantity.

R26: $x^{\frac{1}{n}} = \sqrt[n]{x}$.

R27: $x^{\frac{m}{n}} = \left(x^{\frac{1}{n}}\right)^m = \left(x^m\right)^{\frac{1}{n}}$
$\qquad = (\sqrt[n]{x})^m = \sqrt[n]{x^m}$.

The numerator of a fractional exponent indicates a power, and the denominator a root, of the base quantity.

(Special Factors)

R28: $ax + ay = a(x + y)$.

R29: $x^2 - y^2 = (x - y)(x + y)$.

R30: $x^2 + 2xy + y^2 = (x + y)^2$.

R31: $x^3 - y^3 = (x - y)(x^2 + xy + y^2)$.

R32: $x^3 + y^3 = (x + y)(x^2 - xy + y^2)$.

(Operations With Logarithms)

R33: $\log xy = \log x + \log y$.

The logarithm of a product is the sum of the logarithms of the product's factors.

R34: $\log x/y = \log x - \log y$.

The logarithm of a quotient is the logarithm of the dividend minus the logarithm of the divisor.

R35: $\log x^n = n \log x$.

The logarithm of the n'th power of a quantity is n times the logarithm of that quantity.

R36: $\log \sqrt[n]{x} = \dfrac{\log x}{n}$.

The logarithm of the n'th root of a quantity is the log of that quantity, divided by n.

(Formulas From Elementary Geometry)

R37: $A = bh$.

Area, A, of a parallelogram with base, b, and altitude, h.

R38: $A = b^2$.

Area, A, of a square with side, b.

R39: $A = \frac{1}{2}bh$.

Area, A, of a triangle with base, b, and altitude, h.

R40: $c^2 = a^2 + b^2$, or
$c = \sqrt{a^2 + b^2}$.

The Pythagorean theorem: the square of the hypotenuse, c, of a right triangle is equal to the sum of the squares of the other two sides, a and b.

R41: $C = 2\pi r$.

The circumference, C, of a circle with radius, r.

R42: $A = \pi r^2$.

Area, A, of the same.

R43: $V = abc$.

Volume, V, of a rectangular solid with edges, a, b, c.

R44: $V = b^3$.

Volume, V, of a cube with edge, b.

R45: $V = Bh$.

Volume, V, of a parallelopiped or cylinder with base, B, and altitude, h.

R46: $V = \frac{1}{3}Bh$.

Volume, V, of a pyramid or cone with base, B, and altitude, h.

R47: $A = 4\pi r^2$.

Area, A, of a sphere with radius, r.

R48: $V = \frac{4}{3}\pi r^3$.

Volume, V, of the same.

(Formulas From Trigonometry)

R49: $\sin x = 1/\csc x$, or $\csc x = 1/\sin x$.

R50: $\cos x = 1/\sec x$, or $\sec x = 1/\cos x$.

R51: $\tan x = 1/\cot x$, or $\cot x = 1/\tan x$.

R52: $\tan x = \sin x/\cos x$.

R53: $\sin^2 x + \cos^2 x = 1$.

R54: $a/\sin A = b/\sin B = c/\sin C$.

The law of sines for triangle, ABC with side, a, opposite angle A, etc.

R55: $a^2 = b^2 + c^2 - 2bc \cos A$.

The law of cosines for the same triangle.

(Axioms of Equality)

If $x = y$ and $a = b$, then —

1: $x + a = y + b$.

2: $x - a = y - b$.

3: $ax = by$.

4: $\dfrac{x}{a} = \dfrac{y}{b}$,

$(a = b \neq 0)$

Equal quantities may be added to, subtracted from, multiplied into, or divided into, both sides of an equation without destroying the equality, provided that no divisor is zero.

5: $x^n = y^n$.

6: $\sqrt[n]{x} = \sqrt[n]{y}$.

The same powers or roots of both sides of an equation are equal.

Practice Exercise No. 1

A. Find the value of each side of the equation in formulas R1 through R21 and in formula R25 when $x = 3$, $y = 2$, $z = 4$, $a = 5$, $b = 6$, $n = 2$, $m = 3$.

B. Derive formula R22 by letting $n = m - 1$ in formula 21.

C. Derive formula R23 by letting $n = m$ in formula 21.

D. Derive formula R24 by letting $m = 0$ in formula 21 and applying formula R23.

E. Find the value of each side of the equation in formulas R26 and R27 when $x = 4$, $m = 3$, $n = 2$.

F. Repeat E with $x = 27$, $m = 2$, $n = 3$.

G. Continue A above for formulas R24 through R32.

H. Continue G for formulas R33 through R36, but with $x = 10{,}000$; $y = 1{,}000$; $n = 2$.

I. Find A, c, C, or V in formulas R37 through R48 when $a = 3$, $b = 4$, $h = 5$, $c = 5$, $h = 6$, $r = 9$, $B = 7$.

J. Find the value of each side of the equation in formulas R40, and R49 through R55, when $x = A$ $= 30°$, $B = 60°$, $C = 90°$, $a = 1$, etc., as in the right triangle of Figure 1.

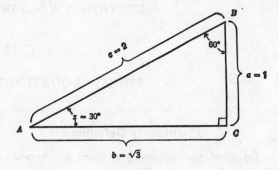

Fig. 1

SECTION TWO — INTERMEDIATE ALGEBRA

CHAPTER II

LINEAR EQUATIONS IN TWO UNKNOWNS

Preliminary Definitions

By what methods may a given mathematical problem be solved? How many, and what kind of solutions, may the problem have? Is the problem really solvable?

Answers to questions like these depend largely upon the **types of equations** involved. For that reason, different types of equations are considered here in separate chapters, and each type is first carefully described.

In elementary algebra, for instance, the reader should already have dealt with equations like

(1) $3x = 5$, (2) $x - 2y = -8$, (3) $-x + 3y - 2z = 14$.

These are classed as **first degree equations** — also called **linear equations** — because *their unknown quantities appear no more than once in each term* and *only with the understood exponent, 1,* (Formula R22). An equation like $4xy = 7$ is therefore *not* linear because its unknown quantities, x and y, appear *twice* in the term, $4xy$. And an equation like $3x^2 - 4y = 1$ is *not* linear because its unknown quantity, x, appears in the term, $3x^2$, in the *second degree*. The latter are both second degree equations, discussed in Chapters IV and V below.

For many mathematical purposes it is also important to identify equations according to **the number of unknown quantities** which they contain.

For instance, the first equation above is classed as a linear equation **in one unknown**. Knowledge of how to treat this type of equation is here taken for granted.

The third equation above is classed as a linear equation **in three unknowns**. These are discussed in Chapters VI and VII.

This present chapter is mainly concerned with **linear equations in two unknowns**, illustrated by the second example above.

A set of equations which are to be considered together in connection with the same problem is called **a system of simultaneous equations**; or, for short, **a system of equations**.

A **solution** of an equation, or of a system of equations, is *a set of values for the unknown quantities which satisfy the mathematical conditions expressed by the equation or equations.* When only one such set of values for the unknowns is possible, the solution is said to be a **unique solution**. When all possible solutions for a system of equations have been found, the system is said to have been **solved simultaneously**.

Examples for systems of simultaneous linear equations in two unknowns follow throughout the rest of this chapter. Examples for other types of equations follow in later chapters.

Systems of Two Equations

From elementary algebra you should already know how a system of two linear equations in two unknowns may be solved simultaneously. To review —

EXAMPLE 1: Solve simultaneously – –

$x - 2y = -8$,   (Equation #1)
$x + y = 7$.   (Equation #2)

SOLUTION (By the method of comparison):

$-3y = -15$ (Subtracting #2 from #1, by axiom 2, page 16)

$y = 5$ (Dividing by $-3 = -3$, axiom 4)

$x + 5 = 7$ (Substituting $y = 5$ in #2)

$x = 2$ (Transposing 5; that is: subtracting $5 = 5$ from the above equation by axiom 2)

Answer: $x = 2$, $y = 5$.

SOLUTION (By the method of substitution):

$$x = 2y - 8 \quad \text{(Transposing } 2y \text{ in \#1)}$$
$$2y - 8 + y = 7 \quad \text{(Substitution for } x \text{ in \#2)}$$
$$3y = 15 \quad \text{(Transposing, etc.)}$$
$$y = 5 \quad \text{(Dividing by } 3 = 3)$$
$$x = 2(5) - 8 \quad \text{(Substituting } y = 5 \text{ in}$$
$$x = 2y - 8)$$
$$x = 2 \quad \text{(Removing parentheses, etc.)}$$

Answer: $x = 2, y = 5$

How do we know that the result arrived at by the above methods is really a solution of the system?

The *arithmetic method* of verifying your work is **check by substitution**. This tests whether the numerical values for the unknowns in the supposed solution actually satisfy the mathematical conditions expressed in the original equation or system of equations.

In the above case, for instance, where the supposed solution is $x = 2, y = 5$, we get by this method —

$$x - 2y = 2 - 2(5) = 2 - 10 = -8 \; \checkmark \quad \text{(Substitution in Equation \#1)}$$
$$x + y = 2 + 5 = 7 \; \checkmark \quad \text{(Substitution in Equation \#2)}$$

Thus $x = 2, y = 5$, is seen to be a true solution of the given equations, and is therefore said to **check**.

But is the result which checks as above the unique (only) solution of the system?

Since this is an important practical question in some applications of equations, we answer it here by the following algebraic reasoning: Suppose there is some other solution for the system. Let the amounts by which this other solution's values for x and y differ from the above be m and n respectively. The supposed other solution will then be $x = 2 + m, y = 5 + n$. Checking these values in the original equations as above, we get —

$$2 + m - 2(5 + n) = -8 \quad \text{(Substitution in \#1)}$$
$$2 + m - 10 - 2n = -8 \quad \text{(Removing parentheses)}$$
$$m - 2n = 0 \quad \text{(Transposing, obtaining new equation \#1a in } m \text{ and } n)$$

$$2 + m + 5 + n = 7 \quad \text{(Substitution in \#2)}$$
$$m + n = 0 \quad \text{(Transposing, obtaining new equation \#2a in } m \text{ and } n)$$
$$-3n = 0 \quad \text{(Subtracting \#2a from \#1a)}$$
$$n = 0 \quad \text{(Dividing by } -3 = -3)$$
$$m - 2(0) = 0 \quad \text{(Substitution in \#2a)}$$
$$m = 0 \quad \text{(Removing parentheses, etc.)}$$

Answer: $m = n = 0$.

But m and n are the amounts by which we supposed any other solution to differ from the solution $x = 2, y = 5$. Hence, the latter must be the only possible solution — in other words, the unique solution — of the original equations.

Practice Exercise No. 2

Solve the following systems of simultaneous equations. Check all solutions by substitution. Verify by the above method that the solutions checked for the first five problems are unique.

(1) $\quad x - 2y = -8$
$\quad\;\; 2x - y = -1$

(2) $\quad x + y = 7$
$\quad\;\; 2x - y = -1$

(3) $\quad x - 2y = -8$
$\quad\;\; 2x - y = -10$

(4) $\quad x + y = 7$
$\quad\;\; 2x - y = -10$

(5) $\quad -\frac{1}{2}x + y = -4$
$\quad\;\; 2x - y = -10$

(6) $\quad 2x + y = 7$
$\quad\;\; 3x - y = 3$

(7) $\quad 2x - 5y = 6$
$\quad\;\; 4x + y = 1$

(8) $\quad x + y = 15$
$\quad\;\; x - y = -1$

(9) $\quad x + 2y = -1$
$\quad\;\; 2x + 5y = -7$

(10) $\quad 2x + 3y = -9$
$\quad\;\;\; 3x + 4y = -10$

Determinate and Indeterminate Systems

In mathematics, as in daily life, some problems have definite answers but others do not. The main difference between the two situations is that, in mathematics it is easier to tell which problems are which, and why.

When a system of equations, or a verbally stated problem leading to such a system, has *a definite number of definite solutions*, the system or problem is said to be **determinate**.

Otherwise the system or problem is said to be **indeterminate**.

For instance, the system of equations in Example 1 above is *determinate* because it has the *unique solution*, $x = 2$, $y = 5$. All the problems in Practice Exercise No. 2 above are also determinate because they likewise have unique solutions. These examples illustrate the fact that, although different kinds of systems may have different definite numbers of definite solutions —

A determinate system of two linear equations in two unknowns has a unique solution consisting of a single pair of definite values for the unknown quantities.

Hence, *when we have a verbally stated problem leading to such a system, we may expect to find for our answer a single pair of definite values for the unknown quantities.*

EXAMPLE 1A: Liquid begins to flow *into* an empty tank at a rate which raises its level $\frac{1}{2}$ foot per minute. A second tank is already filled to a level of 6 feet, but 3 minutes later liquid begins to flow *from* it at a rate which lowers its level 1 foot per minute. After how many minutes from the beginning of flow in the first tank will the level in both tanks be the same, and what will both levels then be?

SOLUTION: Let x be the number of minutes of flow into the first tank. Then the level in this tank after x minutes will be

$$y = \tfrac{1}{2}x. \qquad \text{(Equation \#A1)}$$

Since flow from the second tank begins 3 minutes later, the time of its flow will be $x - 3$. And since its level is 6 feet to begin with, its level after x minutes will be

$$
\begin{aligned}
y &= 6 - (x - 3) &&\text{(Equation \#A2)}\\
&= -x + 9. &&\text{(Removing parentheses, etc.)}
\end{aligned}
$$

Hence,

$$
\begin{aligned}
\tfrac{1}{2}x &= -x + 9 &&\text{(Substitution in \#A2)}\\
\tfrac{3}{2}x &= 9 &&\text{(Transposing } -x)\\
x &= \tfrac{2}{3}(9) = 6 &&\text{(Multiplying by } \tfrac{2}{3} = \tfrac{2}{3})\\
y &= \tfrac{1}{2}(6) = 3 &&\text{(Substitution \#A1)}
\end{aligned}
$$

Answer: In 6 minutes the level in both tanks will be 3 feet.

Check:
$$
\begin{aligned}
3 &= \tfrac{1}{2}(6) = 3 \ \checkmark &&\text{(Substitution in \#A1)}\\
3 &= -6 + 9 = 3 \ \checkmark &&\text{(Substitution in \#A2)}
\end{aligned}
$$

Examples in elementary textbook exercises are usually all determinate like those above. Such examples produce comforting (definite) answers by standard methods for the simple reason that they have been hand-picked to work out that way. This is so beginning students will not be confused or discouraged by exceptions to the mathematical rules which they are studying for the first time.

But in the practical work of engineering and scientific research, are the equations with which you have to deal always determinate? Do standard methods always produce such fondly-hoped-for solutions as those in the above examples? Or, if they fail, when and why do they fail? And what do we then learn about the conditions of the problems out of which our equations arise?

These are extremely practical questions which the more advanced student must learn to answer, beginning here with the simple case of linear equations in two unknowns.

Consistent and Inconsistent Equations

To test the notion that standard methods of solving equations must always work, consider —

EXAMPLE 2: Solve simultaneously,

$$
\begin{aligned}
x - 2y &= -8 &&\text{(Equation \#1)}\\
-3x + 6y &= -6 &&\text{(Equation \#3)}
\end{aligned}
$$

ATTEMPTED SOLUTION (By comparison):

$$
\begin{aligned}
3x - 6y &= -24 &&\text{(Multiplying \#1 by } 3 = 3)\\
0 &= -30! &&\text{(Adding to \#3)}
\end{aligned}
$$

Since this is not a solution, but a *mathematical contradiction*, try —

ATTEMPTED SOLUTION (By substitution):

$$
\begin{aligned}
x &= 2y - 8 &&\text{(Transposing } -2y \text{ in \#1)}\\
-3(2y - 8) + 6y &= -6 &&\text{(Substitution in \#3)}\\
-6y + 24 + 6y &= -6 &&\text{(Removing parentheses)}\\
0 &= -30! &&\text{(Transposing, etc.)}
\end{aligned}
$$

Here again there is no solution, but only the same mathematical contradiction!

What does this mathematically absurd result, $0 = -30$, *mean?* The system of equations in Example 2 is of the same general type as that in Example 1. The methods of solution are also the same, following equally valid steps. *Do the "laws of mathematics" then sometimes break down and fail us?*

To see that this is not really so, note that in attempting to solve the equations of Example 2 simultaneously we are assuming that the equations do have a common solution. All we have shown by the mathematical steps of the above attempted solutions is that,

IF $x - 2y = -8$,   AND

IF $-3x + 6y = -6$ SIMULTANEOUSLY,

THEN $0 = -30!$

But this is the logic of *reduction to the absurd* — the kind of reasoning we appeal to when we make such a statement as: "*If* Jane was a major in World War II, *and if* she is still only 29 years old, *then* I am Napoleon's grandmother!" What the absurdity of our conclusion tells us is simply that the two IF's with which we started cannot both be true at the same time — simultaneously, that is. In other words, *we learn from such a result that the system of equations can have no simultaneous solution.*

In a case like this, where a system of two equations is *indeterminate* in that it has *no possible simultaneous solution,* the two *equations* are said to be **inconsistent** or **incompatible.**

Likewise, when a verbally stated problem leads to such a pair of equations, then the *problem* is said to be **indeterminate** because its *conditions* are inconsistent or incompatible. This means that the problem cannot be solved because its two conditions cannot both be true of the same physical situation. And, as in Example 2, the typical sign of such an inconsistency is a resulting mathematical contradiction of the form: some non-zero number, N, $= 0!$

EXAMPLE 2A: Two minutes after the beginning of flow *into* the first tank in Example 1A above, liquid begins to flow *into* a third tank at a rate which will raise its level $\frac{1}{2}$ foot per minute. After how many minutes will both levels be equal, and what will the levels then be?

ATTEMPTED SOLUTION: As before, the level in the first tank after x minutes will be

$y = \frac{1}{2}x.$ (Equation #A1)

And the level in the third tank after x minutes will be

$y = \frac{1}{2}(x - 2)$ (Equation #A3)
$\quad = \frac{1}{2}x - 1$ (Removing parentheses)

Hence,

$\frac{1}{2}x = \frac{1}{2}x - 1$ (Substitution)
$0 = -1!$ (Transposing $\frac{1}{2}x$)

Answer: The resulting contradiction shows that this problem must be indeterminate with no possible solution because its conditions are inconsistent. Actually, the level in the third tank can never equal that in the first tank because the latter starts rising 2 minutes later and rises at the same rate.

In contrast to the equations in Examples 2 and 2A, *equations* like those in Examples 1 and 1A, which have a common solution, are said to be **consistent** or **compatible.** And when the conditions of a verbally stated problem lead to such equations, as in Example 1A, then these *conditions* are said to be **consistent** or **compatible.**

Practice Exercise No. 3

Solve all the determinate problems among the following. Identify by attempted solution, as above, all problems which are indeterminate because their equations, or verbally stated conditions, are inconsistent.

(1) $3x - 5y = 2$ (4) $5x + 6y = 7$
$\quad\;\; x + 2y = 3$ $\quad\;\; 2y + 3x = 8$

(2) $2x - 7y = 1$ (5) $6x + 8y = 5$
$\quad\;\; 3x + 5y = -6$ $\quad\;\; 3x + 4y = 3$

(3) $2x - 3y = 2$ (6) $4x + 5y = 7$
$\quad\;\; 6x - 9y = 7$ $\quad\;\; 3x - 6y = 2$

(7) A man desires to fence an area. He wants to make it twice as long as wide and he has 300 feet of fencing. How big an area can he enclose?

(8) Two tanks A and B are being filled with water. Tank B is filled at 100 gal. per hour. Tank A starts 30 minutes later and fills at 200 gal. per hour. When do they have the same volume? How late could A have started so that they meet before they contain 800 gal.?

Independent and Dependent Equations

Is a system of two consistent equations in two unknowns always determinate?

For an answer to this question, and further to dispel the notion that standard methods must always produce elementary-textbook results, consider next —

EXAMPLE 3: Solve simultaneously

$$x - 2y = -8 \qquad \text{(Equation \#1)}$$
$$-\tfrac{1}{2}x + y = 4 \qquad \text{(Equation \#1', to be read: "number one-prime")}$$

ATTEMPTED SOLUTION (By substitution):

$$x = 2y - 8 \qquad \text{(Transposing in \#1)}$$
$$-\tfrac{1}{2}(2y - 8) + y = 4 \qquad \text{(Substitution in \#1')}$$
$$-y + 4 + y = 4 \qquad \text{(Removing parentheses)}$$
$$-y + y = 4 - 4 \qquad \text{(Transposing 4)}$$
$$0 = 0! \qquad \text{(Adding similar terms)}$$

Zero equals zero! Far from being a solution, this is a mere *truism* — a statement which, although true, gives us only uselessly trivial information. So let us try —

ATTEMPTED SOLUTION (By comparison):

$$-x + 2y = 8 \qquad \text{(Multiplying \#1' by } 2 = 2)$$
$$0 = 0! \qquad \text{(Adding to \#1)}$$

Again the same trivial result!

What does this mathematically trivial result, $0 = 0$, mean? Even if the "laws of mathematics" do not break down, are they sometimes useless for solving mathematical problems?

The fact that our result this time is not contradictory tells us that our equations are consistent. But it is also possible for equations to be *too consistent* — so consistent, indeed, that even though they may appear to be different, for all mathematical purposes they are actually the same.

Note in Example 3 that if you multiply each side of equation #1 by $-\tfrac{1}{2}$ you obtain equation #1', and that if you multiply each side of equation #1' by -2 you obtain equation #1. This means that equations #1 and #1' are really *the same equation written in two different ways*. Hence, every possible solution of the equation in one of its forms must also be a solution of the equation in its other form. Moreover, there is *an unlimited number of such possible solutions* — sometimes called "an *infinite* number."

For instance, we can write equation #1 in the transposed form, $x = 2y - 8$, as above. Then, for each possible value of y there will be a corresponding value of x. As an example, if we let $y = 10$, then $x = 2(10) - 8 = 20 - 8 = 12$. And this solution of equation #1, $x = 12$, $y = 10$, is also a solution of equation #1' as may be shown by the following substitution:

$$-\tfrac{1}{2}x + y = -\tfrac{1}{2}(12) + 10 = -6 + 10 = 4. \; \checkmark$$

In a case like this, where one equation can be derived from another by multiplication of all its terms by a constant, the two *equations* are said to be **dependent or equivalent**. And we see from Example 3 that a *system of two dependent linear equations in two unknowns is indeterminate in that it has an unlimited (infinite) number of possible solutions.*

When such equations express the conditions of a verbally stated problem, then the *problem* is said to be **indeterminate** because its *conditions* are **dependent** or **equivalent**. *This means that the two supposedly different conditions of the problem are really only one mathematical condition stated in two different ways.*

In other words, dependent or equivalent equations are those which are so compatible that they have all possible solutions in common. But inconsistent or incompatible equations are those which are so independent that they have no possible solutions in common. Hence —

For a system of two linear equations in two unknowns to be determinate, its equations must be both consistent (compatible) and independent (non-equivalent).

This means that **for a problem leading to such equations to be determinate, its conditions must be both consistent (compatible) and independent (non-equivalent).**

EXAMPLE 3A: Two minutes after liquid begins flowing into the first tank in Example 1A, liquid begins flowing into a fourth tank at a rate which raises its level $\frac{1}{2}$ foot per minute. The level of the liquid in this fourth tank, however, is 1 foot to begin with. After how many minutes will both levels be equal, and what will the levels then be?

SOLUTION: As before, the level in the first tank after x minutes will be

$$y = \tfrac{1}{2}x. \qquad \text{(Equation \#A1)}$$

And the level in the fourth tank after x minutes will be

$$y = \tfrac{1}{2}(x - 2) + 1 \qquad \text{(Equation \#A1')}$$
$$= \tfrac{1}{2}x - 1 + 1 = \tfrac{1}{2}x. \qquad \text{(Removing parentheses, etc.)}$$

Hence,

$$\tfrac{1}{2}x = \tfrac{1}{2}x \qquad \text{(Substitution)}$$
$$0 = 0! \qquad \text{(Transposing } \tfrac{1}{2}x)$$

Answer: The resulting mathematical truism, $0 = 0$, shows that the problem is indeterminate because its conditions are equivalent. Beginning with the flow of liquid into the fourth tank at the end of the first two minutes, the two levels are equal and increase at the same rate. Hence, indefinitely (infinitely) many possible answers may be found by substituting values for x in either equation (providing for the purposes of this *physical* problem, that $x = 2$ or greater). For instance, when $x = 2$ min., $y = \frac{1}{2}(2) = 1$ ft.; when $x = 3$ min., $y = \frac{1}{2}(3) = 1\frac{1}{2}$ ft.; when $x = 10$ min., $y = \frac{1}{2}(10) = 5$ ft., etc.

Practice Exercise No. 4

Solve all the determinate problems among the following. Identify by attempted solution as above, all problems which are indeterminate because their equations or verbally stated conditions are dependent. Find the possible solutions of the latter when $x = 0, 1, \ldots, 10$.

(1) $3x - 5y = 7$
   $6x - 10y = 14$

(2) $3x - 5y = 3$
   $6x - 10y = 5$

(3) $3x - 5y = 0$
   $6x + 10y = 14$

(4) $2x + 6y = 2$
   $x + 3y = 7$

(5) An airplane flew 400 miles in 2 hours with a headwind. If it flew half as fast it would have flown only 250 miles. What was the speed of the wind?

(6) An airplane flew 250 miles in 2 hours with a tailwind. If both the airplane's speed and tailwind are doubled, it flies 500 miles in 2 hours. What is the speed of the tailwind?

Defective and Redundant Systems

Thus far we have considered only systems of *two* linear equations in the *same* number of unknown quantities.

A system which has *fewer equations than unknown quantities* is called a **defective system.** Obviously, *the only possible defective system in two unknowns is the special case of a single equation.* And we have already seen from the cases of equations equivalent to a single equation in Examples 3 and 3A above that *such a system is indeterminate in that it has an unlimited (infinite) number of possible solutions.*

On the other hand, a system which has *more equations than unknown quantities* is called a **redundant system.** Thus a redundant system in two unknowns is one containing three or more equations.

A redundant system of linear equations in two unknowns has a unique solution in the special case where each sub-pair of equations separately has the same solution. This special case is most quickly identified as follows: *First,* solve any determinate sub-pair of equations for its unique solution. *Then,* substitute these values of the unknowns in the remaining equations to determine whether they check there too.

EXAMPLE 4: Solve simultaneously,

$$x - 2y = -8 \qquad \text{(Equation \#1)}$$
$$x + y = 7 \qquad \text{(Equation \#2)}$$
$$2x - y = -1 \qquad \text{(Equation \#4)}$$

SOLUTION: The unique solution of the sub-pair of equations, \#1 and \#2, has already been found to be
$$x = 2, \quad y = 5. \qquad \text{(Example 1 above)}$$

Moreover:

$2(2) - 5 = 4 - 5 = -1$, Check.   (Substitution in #4)

Hence $x = 2$, $y = 5$, is the *unique* solution of the redundant system.   (Since #1 and #2 can have no other solution)

The above illustrated method may apply even when one or more sub-pairs of the equations are dependent.

EXAMPLE 5: Solve simultaneously,

$$x - 2y = -8 \qquad \text{(Equation \#1)}$$
$$-\tfrac{1}{2}x + y = 4 \qquad \text{(Equation \#1')}$$
$$x + y = 7 \qquad \text{(Equation \#2)}$$

SOLUTION: When we attempt to solve simultaneously equations #1 and #1' we obtain the result, $0 = 0$, as in Example 3 above. However, when we treat equations #1 and #2 simultaneously we obtain the unique solution, $x = 2$, $y = 5$, as in Example 1 above. Moreover, this solution must also check in equation #1' which is equivalent to #1. Hence $x = 2$, $y = 5$, is *the unique solution of the redundant system*, even though the sub-pair of equivalent equations, #1 and #1', also has (indefinitely many) other sub-solutions which are not solutions of the entire system.

However, redundant systems of equations are not exceptions to the **general rule** that *no system of equations has a solution if any sub-pair of its equations is inconsistent*. This follows from the fact that any simultaneous solution of the system must also be a simultaneous solution of the inconsistent sub-pair of equations. But this is impossible by the definition of inconsistency between two equations (page 21). Hence, in applications of the above illustrated method, the finding $N = 0$ in solving any sub-pair of equations is at once a warning that the entire system has no possible solution.

EXAMPLE 6: Solve simultaneously,

$$x - 2y = -8 \qquad \text{(Equation \#1)}$$
$$x + y = 7 \qquad \text{(Equation \#2)}$$
$$-3x + 6y = -6 \qquad \text{(Equation \#3)}$$

SOLUTION: As before, we already know that the unique solution of equations #1 and #2 is

$$x = 2, \quad y = 5. \qquad \text{(Example 1, above)}$$

But this solution does not check in equation #3, since

$-3(2) + 6(5) = -6 + 30 = 24$, which is not $= -6$.

Moreover, the attempt to solve simultaneously the sub-pair of equations, #1 and #3 produces the result, $N = 0$, as in Example 2 above. Hence we are doubly assured that the redundant system can have no simultaneous solution.

Moreover, in *a redundant system of linear equations in two unknowns*, each sub-pair of equations will in general have a different solution, if any. Hence, even though each sub-pair of equations in such a system is separately independent and consistent, *the redundant system as a whole will, in general, be indeterminate with no possible simultaneous solution*.

EXAMPLE 7: Solve simultaneously,

$$x - 2y = -8 \qquad \text{(Equation \#1)}$$
$$x + y = 7 \qquad \text{(Equation \#2)}$$
$$2x - y = -10 \qquad \text{(Equation \#5)}$$

ATTEMPTED SOLUTION: Again we know that the unique solution of equations #1 and #2 is

$$x = 2, \quad y = 5. \qquad \text{(Example 1, above)}$$

We may similarly find that the unique solution of equations #1 and #5 is

$$x = -4, \ y = -2; \qquad \text{(Steps as above)}$$

and that the unique solution of equations #2 and #5 is

$$x = -1, \quad y = 8. \qquad \text{(Steps as above)}$$

But since all of these solutions of sub-pairs of equations are unique, none will check in the remaining equation. Hence the redundant system as a whole has no possible simultaneous solution.

We see from these examples that **to be determinate a system of linear equations in two unknowns must contain two, and** *in general* **only two, independent, consistent equations.** The *exception to the general rule* is the special case in which each sub-pair of equations in a redundant system happens to have the same solution.

Consequently, **to be determinate a verbally stated problem leading to linear equations in two unknowns must specify two, and** *in general* **only two, independent, consistent condi-**

tions. The *exception to the general rule* is the special case of a redundant set of conditions which leads to the exceptional type of redundant system of equations mentioned above.

EXAMPLE 4A: Four and one half minutes after liquid begins to flow into the first tank in Example 1A above, liquid begins to flow into a fifth empty tank at a rate which raises its level 2 feet per minute. After how many minutes will the level in the fifth tank be equal to that in the first two tanks, and what will the levels then be?

SOLUTION: As before, the levels in the first two tanks are given respectively by the equations,

$y = \frac{1}{2}x,$ (Equation #A1)
$y = -x + 9.$ (Equation #A2)

The unique solution of these equations we already know to be

$x = 6, \quad y = 3$ (Example 1A, above)

The level in the fifth tank is given by the equation,

$y = 2(x - \frac{9}{2}) = 2x - 9.$ (Equation #A4)

And the solution, $x = 6$, $y = 3$, checks in this equation as follows:

$3 = 2(6) - 9 = 12 - 9 = 3.$

(Substitution in #A4)

Hence this is one of the above described *exceptional* cases. The unique solution of the redundant system is $x = 6$, $y = 3$, which means that after 6 minutes the level in all three tanks will be 3 feet.

EXAMPLE 5A: Six minutes after liquid begins to flow into the first tank in Example 1A, liquid begins to flow into a sixth empty tank at a rate which raises its level 2 feet per minute. After how many minutes will the level in the sixth tank be equal to that in the first two tanks, and what will the levels then be?

SOLUTION: This time the levels in the three tanks are given respectively by the equations:

$y = \frac{1}{2}x$ (Equation #A1)
$y = -x + 9$ (Equation #A2)
$y = 2(x - 6) = 2x - 12$ (Equation #A5)

As before, the unique solution of equations #A1 and #A2 is

$x = 6, \quad y = 3$ (Example 1A, above)

But these values do not satisfy equation #A5:

$2(6) - 12 = 12 - 12 = 0$ which is not $= 3$.

Hence this must be one of the more typical cases in which each sub-pair of equations has a separate solution, if any. Indeed, equations #A1 and #A5 may be found to have the unique solution, $x = 8$, $y = 4$; and equations #A2 and #A5 may be found to have the different unique solution, $x = 7$, $y = 2$. From this we must conclude that the first and second tanks will have the same level of 3 feet at the end of 6 minutes, the second and sixth tanks will have the same level of 2 feet at the end of 7 minutes, and the first and sixth tanks will have the same level of 4 feet at the end of 8 minutes; but all three tanks will never have the same level at the same time.

Practice Exercise No. 5

Solve all the determinate problems among the following redundant examples. Identify any sub-pairs of equations or conditions which have separate unique solutions, which are inconsistent, or which are dependent. In terms of these findings, explain why certain of the problems are indeterminate.

1.(a) $5x + 2y = 7$
  (b) $3x - y = 4$
  (c) $6x - 2y = 8$

2.(a) $10x - 3y = 7$
  (b) $3x + 2y = 6$
  (c) $6x + 4y = 12$

3.(a) $3y - 2x = 7$
  (b) $2y + 5x = 3$
  (c) $4x - 3y = 2$

4.(a) $\frac{1}{2}x + \frac{3}{4}y = 7$
  (b) $x + \frac{3}{2}y = 14$
  (c) $3x - 7y = 2$

5.(a) $3x + 2y = 7$
  (b) $4x - 5y = -6$
  (c) $7x - 3y = 1$

6.(a) $14x + 42y = 6$
  (b) $7x + 21y = 3$
  (c) $2x + 6y = \frac{6}{7}$

7.(a) $2x + 3y = 8$
  (b) $3x + 2y = 7$
  (c) $2x + 3y = 3$

8.(a) $2x + y = 5$
  (b) $3x - y = 0$
  (c) $5x - 2y = -1$

9.(a) $2x - 3y = 4$
  (b) $6x - 9y = 12$
  (c) $8x - 12y = 3$

10.(a) $3x + y = 5$
   (b) $2x + 3y = 1$
   (c) $5x + 8y = 2$

Summary

A **determinate mathematical problem** is one which has *a definite number of definite solutions.* Elementary students are usually given only such problems to solve.

An **indeterminate mathematical problem** is one which has either *no possible solution* or else *an unlimited (infinite) number of possible so-*

lutions which are equally suitable. More advanced students must learn to recognize and interpret these too when they occur.

A **determinate** system of two linear equations in two unknowns has a **unique solution** consisting of a single pair of possible values for the unknown quantities.

A system of linear equations in two unknowns is **indeterminate**, however, in the following cases:

(1) If it contains two **inconsistent** (**incompatible**) equations. Then there are *no possible solutions*, and standard elementary methods result only in the *mathematical contradiction, N = 0*, when an attempt is made to solve two such equations simultaneously.

(2) If it consists only of **dependent** (**equivalent**) equations. Then there are *an unlimited (infinite) number of possible solutions*, and elementary methods result only in the *mathematical truism, 0 = 0*, when an attempt is made to solve two such equations simultaneously.

(3) If the system is **defective**. Then there are an *unlimited (infinite) number of possible solutions*.

(4) In general, if the system is **redundant**. Then there are *no possible solutions* unless, by exception, each sub-pair of equations happens to have the same solution.

Consequently, to be **determinate** a system of linear equations in two unknowns must be (1) independent, (2) consistent, (3) nondefective, and (4) in general non-redundant.

This means that, **to be determinate a verbally stated problem leading to linear equations in two unknowns must specify two, and in general only two, independent, consistent conditions.**

These conclusions will be generalized for other types of equations in later chapters. But in the next chapter we shall first reconsider the same type of equation from a different point of view.

CHAPTER III

VARIABLES, FUNCTIONS, AND GRAPHS

Algebra and Geometry

Algebra and geometry are usually treated in elementary mathematics as quite unrelated subjects. An occasional algebra problem may by chance, concern a particular geometric diagram. Or, an occasional geometry problem may involve a few incidental algebraic steps. But elementary algebra is usually thought of only as a way of solving essentially numerical problems by the device of letting unknown quantities be represented by letters in equations. And elementary geometry is thought of only as a way of demonstrating theorems concerning figures like triangles and circles by reasoning from axioms, postulates, and previously deduced theorems of the same kind.

There are historical reasons for this approach. Before the seventeenth century, both subjects were nearly always considered separately — algebra for several centuries, and geometry for over two thousand years.

As mentioned in Chapter I, however, a more advanced *approach* to mathematics may take a broader view of mathematical problems with resulting deeper insights into how these problems are to be understood and solved.

One such broader view is that which finds **a basic connection between algebra and geometry.** We shall see later in Section Three of this volume that geometric problems may be reduced — in analytic geometry — to algebraic problems of solving equations. In this present chapter we shall see that **algebraic equations may be clarified, and sometimes even solved, by geometric interpretations of the mathematical relationships which they express.**

What is this modern bridge which spans the historic gap between elementary algebra and geometry?

By way of anticipation, we may say that it consists of *four simple links.* Two of these links, from the *algebraic* side, are the concepts of *variables* and of *functions.* The other two links, from the *geometric* side, are the concepts of *coordinate systems* and of *graphs.*

We shall now explain each of these, using simple examples of linear equations in two unknowns. But as in the preceding chapter the ideas illustrated by these simple examples will apply in later chapters to much more complicated types of equations and problems.

Variables and Functions

Thus far we have referred to the quantities in equations as either *known* or *unknown.* For instance, in an equation like

$$y = \tfrac{1}{2}x,$$

we have called x and y *unknown quantities* as distinguished from the *known quantity,* $\frac{1}{2}$. This was because, unlike the value of $\frac{1}{2}$, the values of x and y were yet to be found.

Recall, however, what symbols like $\frac{1}{2}$, x, and y, represent in specific problems like that of Example 1A of the preceding chapter. There $\frac{1}{2}$ is the *fixed* rate, in feet per minute, at which liquid flows into a tank. But x represents all the *varying* numbers of minutes which the liquid may flow at this fixed rate. And y represents all the *varying* numbers of feet to which the liquid's level may rise in all these times.

From another point of view, therefore, we may think of so-called *known quantities,* like $\frac{1}{2}$, as **constants** — *quantities to which we may not assign different values during the course of a mathematical operation without changing the conditions of the problem.* And we may think of the so-called *unknown quantities* as **variables** — *quantities to which we may assign different values during the course of a mathematical operation.*

In the above example, for instance, $\frac{1}{2}$ is a

constant because we may not assign it any different value without changing the equation itself. But x and y are *variables* because we may assign any possible values to either in order to find the corresponding values of the other in the same equation.

More specifically, *if we first assign a definite value to x in order to find the corresponding value of y*, then x is called the **independent variable** and y is called **the dependent variable** and **a function of x.**

By calling x the *independent variable* we simply mean that the value of x is assigned *independently* of any consideration of y. By calling y the *dependent variable*, or a *function* of x, we simply mean that the value of y *depends upon* the previously selected value of x. Hence the use of the word *function* is also much the same in mathematics as in more common expressions like: "Health is a function of diet," which simply means that "how well we feel depends upon what we eat."

The brief way of stating in symbols the above mentioned functional relationship between y and x is:

$$y = f(x).$$

This is read: "y is a function, f, of x." Note that the symbol, "$f(x)$," does *not* mean that a quantity, f, is to be multiplied by another parenthetical quantity, x, as it would in elementary algebra. For this reason, a combination of letters and parentheses which is intended to indicate multiplication of factors, but which might be confused with a function symbol, is always avoided in more advanced mathematics.

Using the above symbol, we can now write the particular functional relationship between the variables in the previously discussed equation as —

$$y = f(x) = \tfrac{1}{2}x.$$

This is read: "y is the function, f, of x, equal to $\tfrac{1}{2}x$."

Then, for any particular values of the independent variable such as $x = 0, 1, 2$, etc., in the same equation, we may write:

$$f(0) = 0 \qquad [\text{since } \tfrac{1}{2}(0) = 0]$$
$$f(1) = \tfrac{1}{2} \qquad [\text{since } \tfrac{1}{2}(1) = \tfrac{1}{2}]$$
$$f(5) = 2\tfrac{1}{2}, \text{ etc.} \qquad [\text{since } \tfrac{1}{2}(5) = 2\tfrac{1}{2}]$$

These are read: "for $x = 0$, the function, f, of x, is 0; for $x = 1$, the function, f, of x, is $\tfrac{1}{2}$, etc."

When two different functions of the same independent variable appear in the same problem, they may be distinguished from each other by separate sub-script numbers, as in

$$y = f_1(x) = \tfrac{1}{2}x, \quad \text{and} \quad y = f_2(x) = -x + 9.$$

These are read: "y is the function, f-sub-one, of x, $= \tfrac{1}{2}x$; and y is the function, f-sub-two, of x, $= -x + 9$."

Two different function-letters may also be used for the same purpose, as in

$$y = f(x) = \tfrac{1}{2}x, \quad \text{and} \quad y = g(x) = -x + 9.$$

These are read: "y is the function, f, of x, $= \tfrac{1}{2}x$; and y is the function, g, of x, $= -x + 9$."

Tables of Values

When you are studying the mathematical behavior of a function it is usually convenient to arrange typical pairs of values for the dependent variable and the independent variable in a form called a table of values. This may be vertical in format, or it may be horizontal as in the following example for the function —

$$y = \tfrac{1}{2}x$$

$y = f(x) =$	0	$\tfrac{1}{2}$	1	$1\tfrac{1}{2}$	2	$2\tfrac{1}{2}$	3	$3\tfrac{1}{2}$	etc.
$x =$	0	1	2	3	4	5	6	7	etc.

Such a table gives us a kind of here-and-there accounting of how the dependent variable in an equation changes in value with changes in value of the independent variable. In the above case, for instance, the table of values tells us on its upper line how high the liquid-level, y, must be after each number of minutes, x, on its lower line — $1\tfrac{1}{2}$ feet after 3 minutes, 3 feet after 6 minutes, $3\tfrac{1}{2}$ feet after 7 minutes, etc.

Practice Exercise No. 6

A. According to the above table of values for the equation, $y = \tfrac{1}{2}x$, how high must the liquid level, y, be at the end of the following number of minutes?

(1) 0   (2) 1   (3) 2

B. According to the same, how many minutes, x, must have elapsed when the liquid level reaches the following numbers, y, of feet?

(1) 1   (2) 2   (3) $2\frac{1}{2}$

C. Compute similar tables of values for x and y in each of the following equations. Let x be the independent variable with values, 0, 1, 2, ..., 10.

(1) $y = 3x + 2$
(2) $y = -3x + 45$
(3) $y = \frac{1}{3}x + 7$
(4) $y = 2$
(5) $y = 2x - 6$
(6) $y = -2x + 7$

Explicit and Implicit Functions

When an equation is written with y on one side of the equality sign (usually the left side) and with all constants and terms involving x on the other side of the equality sign (usually the right side) then the equation is called an **explicit equation** and is said to express the dependent variable y as an **explicit function** of the independent variable x. That is the case, for instance, in the equation $y = \frac{1}{2}x$, where y is expressed as the explicit function, $f(x) = \frac{1}{2}x$.

However, the difference between independant and dependent variables is relative. An equation which expresses y as an explicit function of x may be re-written in an equivalent form which expresses x as an explicit function of y. Consider again, for instance, the equation,

$$y = f(x) = \tfrac{1}{2}x.$$

Interchanging sides and multiplying by $2 = 2$, we get the equivalent equation,

$$x = g(y) = 2y.$$

This states x as an explicit function of y, and is called the **inverse function** of $y = f(x)$.

Equations are often encountered, however, in *non-explicit form*. They are then said to be implicit equations. Their variables are called **co-variables**. And the equations are said to express their co-variables as **implicit functions** of each other. For instance, the equation,

$$f(x,y) = x - 2y + 8 = 0,$$

is in implicit form since neither of the co-variables, x or y, stands alone on one side of the equality sign. And the entire expression is to be read: "The function, f, of x and y, $= x - 2y + 8$, is zero."

In a case like this, however, we can derive two equivalent explicit equations from the above implicit equation — one expressing y as an explicit function of x, and the other expressing x as an explicit function of y:

$$y = f_1(x) = \tfrac{1}{2}x + 4, \quad \text{and} \quad x = f_2(y) = 2y - 8.$$

It usually shortens the work of computing a table of values if an implicit equation is first changed to explicit form. The following example in a vertical format, for instance, shows arithmetic details of computation in an added central column for the first explicit form of the above implicit equation:

x	$f(x) = \frac{1}{2}x + 4$	y
-10	$\frac{1}{2}(-10) + 4 = -5 + 4$	-1
.	.	.
.	.	.
-1	$\frac{1}{2}(-1) + 4 = -\frac{1}{2} + 4$	$3\frac{1}{2}$
0	$\frac{1}{2}(0) + 4 = 0 + 4$	4
1	$\frac{1}{2}(1) + 4 = \frac{1}{2} + 4$	$4\frac{1}{2}$
.	.	.
10	$\frac{1}{2}(10) + 4 = 5 + 4$	9

The dots in this table simply indicate the places where omissions have been made for the sake of brevity.

From this example we see again that *a table of values is a list of typical sets of corresponding values which the variables in an equation may have.* In other words, it is a kind of *arithmetic sampling of the mathematical behavior of the dependent variable in an equation as a function of the independent variable.* In this sense, *a table of values is an arithmetic illustration of the functional meaning of an equation.*

Practice Exercise No. 7

Re-write the following implicit equations in equivalent explicit forms expressing y as function of x, and x as a function of y. Then compute a table of values for each equation with x the independent variable as in the above illustration $x = 0, 1, \ldots, 10$.

(1) $3x - 6y + 5 = 0$
(2) $5x + 7y - 7 = 0$
(3) $x - 2y + \frac{5}{8} = 0$
(4) $2x + 5y - 6 = 0$

First Quadrant Graphs

We have seen that a table of values samples arithmetically the functional relationship between the variables in an equation. But on the basis of columns of numbers alone, it is often difficult to form a clear picture in your mind of the larger pattern of a function's mathematical behavior. Fortunately, a better picture of that behavior may be obtained, for most common functions, from a graph.

Graphs are usually explained in elementary texts as a visual means of presenting statistical data. A familiar graph of the growth of United States population from 1900 to 1950 is presented as follows —

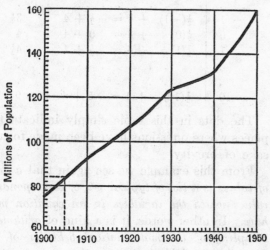

Growth of Population in the United States

Fig. 2

From the graph-line across such a diagram we can see at a glance what U. S. population figure corresponds to each census date noted along its bottom. Hence it literally enables us to "see" what the general trend of increase in this population has been over the years for which the graph has been constructed.

In terms of our previous definitions, however, the population of the United States is a *dependent variable*, or *function*, of the calendar year as an *independent variable*. And although it is somewhat different, the kind of functional relationship which we have been discussing here is, if anything, even better representable by graphs.

In a statistical situation like the census example, only a few points correspond to authentic values — one for the end of each ten-year period. But for the relationship between variables in an equation we can find all possible values of a dependent variable, y, corresponding to any possible value of an independent variable, x.

Consider again, for instance, the simple example above where the level, y feet, of the liquid in a tank is expressed as a function of x minutes by the equation, $y = f(x) = \frac{1}{2}x$.

Instead of census years, let values of x be represented along the bottom line of the graph, which we shall hereafter call **the x-axis**. Begin with $x = 0$ in the lower lefthand corner, and continue with $x = 1, 2, 3$, etc. at equal intervals to the right.

Instead of populations, let values of y be represented along the lefthand side of the graph, which we shall hereafter call **the y-axis**. Begin with $y = 0$ in the lower lefthand corner, and continue with $y = 1, 2, 3$, etc. at equal intervals above.

Now we can find points, on or above the horizontal x-axis and on or to the right of the vertical y-axis, corresponding to each pair of values for x and y in our previously computed table of values for the function, $y = f(x) = \frac{1}{2}x$. This is called **plotting** the values of the function.

As is shown in Figure 3, the point-plot for the pair of values, $x = 0, y = 0$ — designated

(0, 0) — is the lower lefthand corner of the diagram where the x-axis and the y-axis meet. The point-plot for the pair of values, $x = 1$, $y = \frac{1}{2}$ — designated $(1, \frac{1}{2})$ — is $\frac{1}{2}$ unit directly above the point on the x-axis where $x = 1$, and 1 unit directly to the right of the point on the y-axis where $y = \frac{1}{2}$. The point-plot for the pair of values, $x = 2$, $y = 1$ — designated $(2,1)$ — is one unit directly above the point on the x-axis where $x = 2$, and 2 units directly to the right of the point of the y-axis where $y = 1$. And so on for the others.

Additional points may, of course, be plotted for any other pairs of values for the variables x and y in in this first degree equation. But as you can see from the pattern which has already begun to emerge, all of these points will lie on the same straight line, designated L_1 — "L-sub-one" — in Figure 3.

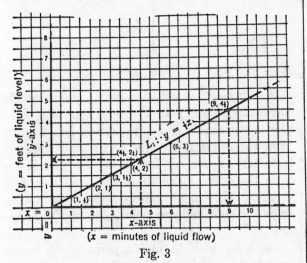

Fig. 3

That, by the way, is why a *first degree equation* is called a *linear equation*, and why a function expressed by a linear equation is called a *linear function*. We shall see later that functions expressed by equations of other degrees, or with different numbers of variables, have different types of graphs. But it is proven in analytic geometry that **the functional relationship between two variables in a first degree equation is such that its graph is always a straight line.**

Note, now, that this straight-line graph, L_1

in Figure 3, is a geometric picture of how the dependent variable, y, varies in value as a function of the independent variable, x, as x increases in value from 0 in the equation, $y = \frac{1}{2}x$. This line, L_1, in other words, is a kind of geometric image of all the possible solutions of the equation, $y = \frac{1}{2}x$, for zero or positive values of x.

Consequently, we can determine from a graph like that in Figure 3 — within the limits of accurate observation on its scale — the value of either variable corresponding to any selected value of the other variable. We can do this, moreover, without further computation or reference to a table of values.

As is indicated by the two pairs of broken arrow lines in Figure 3, for instance, we can tell at a glance that when $4\frac{1}{2}$ minutes have elapsed ($x = 4\frac{1}{2}$), the level, y, of the liquid must be $2\frac{1}{4}$ feet ($y = 2\frac{1}{4}$); or that when the liquid level is $4\frac{1}{2}$ feet ($y = 4\frac{1}{2}$), the time elapsed must be 9 minutes ($x = 9$).

Practice Exercise No. 8

Plot graphs of each of the functions for which you have computed a table of values in Practice Exercise No. 6. Tell by inspection on these graphs the values of y which correspond to $x = 4\frac{1}{2}$, $6\frac{1}{2}$, $8\frac{1}{2}$; also, the values of x which correspond to $y = 4\frac{1}{2}$, $6\frac{1}{2}$, $8\frac{1}{2}$; also, the values of x which correspond to $y = 1, 2, 3, 4\frac{1}{2}$.

Rectangular Coordinates

Like most statistical graphs, the graph in Figure 3 shows only positive values of x and y. No other values were needed for the example there interpreted because only positive values were meaningful for the variables, x-minutes of flow and y-feet of liquid level.

In other cases, however, we must also consider negative values of the variables. And for this purpose we require an extension of the above described graphing system.

As a convenient point of reference, let us call the point at which the horizontal x-axis and the vertical y-axis meet in Figure 3, **the origin.** This is the point where $x = 0$, $y = 0$ — designated either by the letter, O, or by the pair of values in parentheses (0, 0).

Next, let us extend both axes through this point, the x-axis to the left of the origin, and the y-axis below the origin, as in Figure 4.

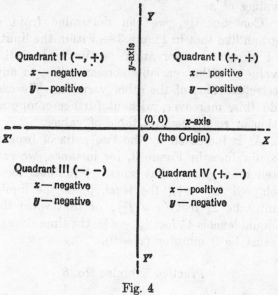

Fig. 4

Since we have already represented positive values of x by distances to the right of the origin, we can now represent negative values of x by distances to the left of the origin on the x-axis. And since we have already represented positive values of y by distances above the origin, we can represent negative values of y by distances below the origin on the y-axis.

Our extended pair of axes now divides the entire plane into four areas, called quadrants, conventionally numbered I, II, III, IV, in a counter-clockwise direction, as in Figure 4. In quadrant I, both x and y are positive. But in quadrant II, x is negative and y is positive. In quadrant III, both x and y are negative. And in quadrant IV, x is positive and y is negative.

As before, if we now wish to plot a point which corresponds to a pair of values like $x = 6$, $y = 4$, we can do so by measuring right 6 units from the origin along the x-axis, and

then up 4 units; or else by measuring up 4 units along the y-axis, and then right 6 units. Either procedure brings us to the same first-quadrant point, (6,4), labelled in Figure 5 as P_1 — to be read: "P-sub-one."

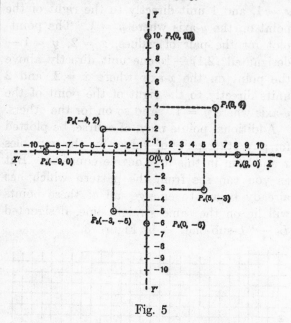

Fig. 5

If, however, we wish to plot a point which corresponds to a pair of values like $x = -4$, $y = 2$, we can do so by measuring left 4 units from the origin along the x-axis, and then up 2 units; or else by measuring up two units from the origin along the y-axis, and then left 4 units. Either procedure brings us to the same second-quadrant point, $P_2(-4,2)$ in Figure 5.

Or, if we wish to plot a point which corresponds to a pair of values like $x = -3$, $y = -5$, we can do so by measuring left 3 units from the origin along the x-axis, and then down 5 units; or else by measuring down 5 units from the origin along the y-axis, and then left 3 units. Either procedure brings us to the same third-quadrant point $P_3(-3, -5)$ in Figure 5.

And if we wish to plot a point which corresponds to a pair of values like $x = 5$, $y = -3$, we can do so by measuring right 5 units from the origin along the x-axis, and then down 3 units; or else by measuring down 3 units from the origin along the y-axis and then right 5 units. Either procedure brings us to the same fourth-quadrant point labelled P_4 $(5, -3)$ in Figure 5.

Special cases, of course, are points plotted for zero values of either variable. If only $y = 0$, then the point-plot is on the x-axis as in the cases of $P_5(8,0,)$ and $P_6(-9,0)$ in Figure 5. If only $x = 0$, then the point-plot is on the y-axis as in the cases of $P_7(0,10)$ and $P_8(0, -6)$ in Figure 5. But if both $x = 0$ and $y = 0$, then the point-plot is the origin itself.

In this graphing system the values of x and y which correspond to any point in the plane of the axes are called, respectively, the **x-coordinate** and the **y-coordinate** of that point. The *x-coordinate* of a point is also sometimes called its **abscissa** (plural: abscissae), and the *y-coordinate* is then called its **ordinate**.

As a whole, the graphing device is called a **system of rectangular coordinates**. This is because its axes are used *coordinately* (together) and meet *at rect-angles* (90°).

With this extended graphing system available, we can now plot a wider range of function values. In Figure 6, for instance, is shown a plot of the table of values computed above for the function $f(x,y) = x - 2y + 8 = 0$, or $y = f_1(x) = \frac{1}{2}x + 4$.

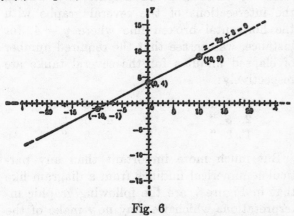

Fig. 6

As may be seen in Figure 6, the graph of this linear equation is also a straight line — broken at both ends in the diagram to indicate that it may be extended indefinitely in either direction depending upon the range of

values which we wish to consider for the variables.

Of course, for plotting a straight line graph like this it is necessary to compute the values of only two sets of coordinates to fix points which will guide the straight-edge of a ruler. A third set is usually enough to check the accuracy of the first two. For this reason you need hereafter compute the coordinates of only three points when graphing a linear function or equation. But we shall see in later chapters that longer tables of values are necessary for even the approximate plotting of other types of equations and functions.

Practice Exercise No. 9

Plot graphs of each of the functions for which you computed a table of values in Exercise No. 7. Determine by inspection of these graphs the values of y which correspond to $x = -7, -3, 5$; also, the values of x which correspond to $y = -6, -2, 4$.

Graphic Interpretations

Since the coordinates of the points on the graph of an equation represent all the possible solutions of that equation, **the coordinates of the points in which the graphs of a system of two or more equations coincide must represent all the possible simultaneous solutions of the system.**

Graphic methods for solving equations are generally not precise enough for practical use without further numerical check upon their accuracy. But such methods are extremely useful in helping the practical mathematician to visualize the underlying relationships between the variables in his equations, and thus to check or interpret the results of his algebraic operations.

In the preceding chapter, for instance, we considered a series of Examples — 1A, 2A, 3A, etc. — in which the level of liquid, y feet, in several tanks was expressed as a function of elapsed time, x minutes, by the linear equations in two variables —

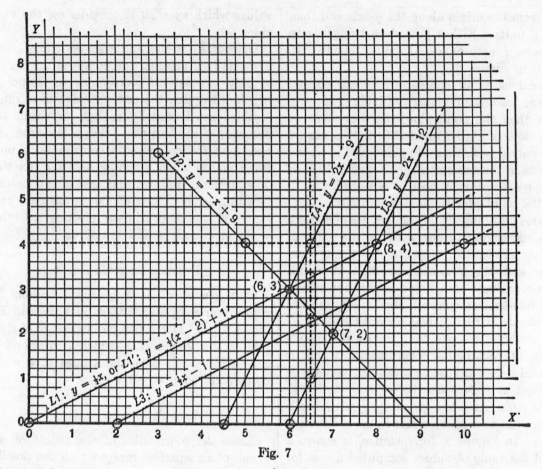

Fig. 7

A1. $y = \frac{1}{2}x,$

A2. $y = -x + 9,$

A1'. $y = \frac{1}{2}(x - 2) + 1,$

A3. $y = \frac{1}{2}x - 1,$

A4. $y = 2x - 9,$

A5. $y = 2x - 12.$

The graphs of these functions with respect to the same pair of coordinate axes are now shown in Figure 7 as the respective lines, L_1, L_2, L_1', L_3, etc.

In a composite graphic diagram like this we can see at a glance the relative heights of the liquid's level in each of the several tanks at any moment. By tracing the intersections of the several graphs with the broken vertical line where $x = 6\frac{1}{2}$, for instance, we can see that after $6\frac{1}{2}$ minutes the levels in the several tanks will be respectively:

1. $3\frac{1}{4}$ ft.   3. $2\frac{1}{4}$ ft.

2. $2\frac{1}{2}$ "   4. 4 "

1'. $3\frac{1}{4}$ "   5. 1 "

Or, we can see at a glance what intervals of time are required for the liquid in the several tanks to reach a given height. By tracing the intersections of the several graphs with the horizontal broken line where $y = 4$, for instance, we can see that the required number of elapsed minutes for the several tanks are respectively:

1. 8 min.   3. 10 min.

2. 5 "   4. $6\frac{1}{2}$ "

1'. 8 "   5. 8 "

But much more important than any particular numerical findings from a diagram like that in Figure 7 are the following graphic interpretations which we may now make of the main conclusions arrived at algebraically in the preceding chapter.

1. A determinate system of two first degree equations in two variables has one, and only one, solution because the two straight-

line graphs of these equations can intersect in only one point.

For instance, equations A1 and A2 have only the one solution, $x = 6$, $y = 3$ (page 20), because their graphs, L_1 and L_2 intersect only in the point (6,3).

2. A system of two inconsistent (incompatible) first degree equations in two variables is indeterminate with no possible solutions because the straight-line graphs of these equations are parallel and therefore cannot meet.

For instance, the inconsistent equations, A1 and A3, have the parallel graph-lines L_1 and L_3. Thus we see that the attempt to find a simultaneous solution for these equations in Example 2A of the preceding chapter was the algebraic equivalent of a geometric attempt to find a point of intersection for two parallel lines. No wonder then that this attempt resulted in a mathematical contradiction of the form $N = 0!$

3. A system of dependent (equivalent) first degree equations in two variables is indeterminate with an unlimited (infinite) number of solutions because the straight-line graphs of these equations coincide throughout and therefore in an unlimited (infinite) number of points.

For instance, the dependent equations A1 and A1′ have the common graph-line, L_1 or $L_{1'}$ in Figure 7. And the coordinates of each of the unlimited (infinite) number of points on this line are therefore solutions of both of these equations. For the same reason —

4. A defective system of one first degree equation in two variables is indeterminate because the coordinates of each of the unlimited (infinite) number of points on its straight-line graph corresponds to a different possible solution. And finally —

5. A redundant system of consistent, independent first degree equations in two variables is in general indeterminate with no possible solution because, in general, the straight-line graphs of each sub-pair of these equations will intersect in a different point.

For instance, the redundant system of equations, A1, A2, and A4, could be shown to have a unique solution, $x = 6$, $y = 3$, by exception in Example 4A of the preceding chapter because their graph-lines intersect by exception in the common point (6,3) in Figure 7. But the redundant system of equations, A1, A2, A5, was found to have no common solution because the graph-lines of these equations intersect more typically, by sub-pairs, in three distinct points — namely: (6,3), (7,2), and (8,4).

The above general principles of graphic interpretation may now be applied to any other systems of linear equations in two variables.

For instance, recall the set of equations which occurred in Examples 1, 2, 3, ... 7 of the preceding chapter:

$$(1)\ x - 2y = -8, \qquad (3)\ -3x + 6y = -6,$$
$$(2)\ x + y = 7, \qquad (4)\ 2x - y = -1,$$
$$(1')\ -\tfrac{1}{2}x + y = 4, \qquad (5)\ 2x - y = -10.$$

These are graphed in Figure 8 respectively by the straight lines, $L1$, $L2$, etc. And Examples 1, 2. ... 7 of the preceding chapter may be interpreted geometrically by these graphs as follows:

In Example 1 (page 18), we found #1 and #2 to have the unique solution, $x = 2$, $y = 5$, because these two independent, consistent equations have the straight-line graphs, $L1$ and $L2$, which intersect only in the point (2,5).

In Example 2 (page 20), we found #1 and #3 to have no simultaneous solution because these two inconsistent equations have the parallel graph-lines, $L1$ and $L2$, which cannot intersect in any point, etc., etc.

Practice Exercise No. 10

A. Complete the above geometric interpretations for Examples 3, 4, 5, 6, 7 of the preceding chapter.

B. Interpret graphically all the problems in Exercise No. 5 of the preceding chapter.

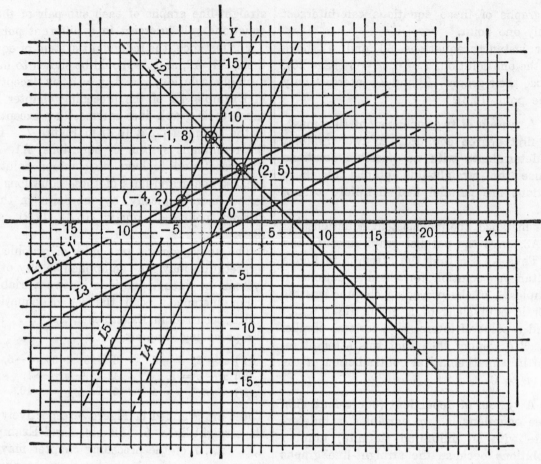

Fig. 8

Summary

In modern times a more advanced approach to mathematics has bridged the ancient gap between elementary algebra and geometry. The principal links in this bridge are the concepts of **variables, functions, coordinates,** and **graphs.**

A **table of values** is a *numerical sampling of the functional relationship between the variables in an equation.*

A **graph** is *a geometric picture of this functional relationship.* It consists of all those points whose coordinates are solutions of the equation which expresses the functional relationship algebraically.

Referred to a system of rectangular coordinates, **the graph of a first degree (linear) equation in two variables is a straight line of indefinite length.**

The **straight-line graphs** of *two linear equations intersect in a single point if the equations are independent and consistent, are parallel if the equations are inconsistent (incompatible), and coincide throughout if the equations are dependent (equivalent).*

These geometric facts, and their elaborations above, are graphic interpretations of the statements in the summary of the preceding chapter.

Note on Sequence of Study

This is one of the places mentioned in Chapter I where the reader may elect different sequences of topics for further study.

If you proceed to the next two chapters in

their given order, you will have a change of algebraic subject matter in equations of a different degree.

But if you would like first to follow through with your present study of linear equations — and that is not a bad idea at all — you may skip at once to Chapters VI and VII where the topic is rounded off with a consideration of more general cases and methods. Then you can return to Chapters IV and V later without any serious inconvenience.

Or, if you would like first to sample the beginnings of analytic geometry, you may also skip now to Chapters IX and X in Sec-tion Three. In that case, however, you will do well to turn back to Chapters VI and VII when you find it useful to apply determinants in handling linear equations (for instance, when you reach page 126). In any event, you will want to study Chapters IV and V on quadratic equations before attempting the topic of conic sections further on in analytic geometry.

The path you choose here is mainly a matter of personal preference. No time need be lost in the long run if you later change your mind after any of the mentioned choices. So why not give each a brief try at least, and see which sequence suits *you* best?

QUADRATIC EQUATIONS IN ONE VARIABLE

A Different Type of Problem

When a projectile is fired straight up from an elevation c feet above sea level at an initial velocity, b feet per second, its height, y, after x seconds (if we leave out of consideration the effect of air resistance) is given in feet by the formula,

$$y = -16x^2 + bx + c.$$

Suppose, then, that we are asked: "How many seconds will it take a projectile to reach a height of 15,600 feet above sea level if it is fired from an altitude of 2,000 feet with an initial velocity of 1,000 feet per second?"

To solve this problem we must first substitute in the above formula the values, $y = 15,600$, $b = 1,000$, and $c = 2,000$. That gives us the equation,

$$15,600 = -16x^2 + 1,000x + 2,000;$$

or, after terms are transposed and like-terms combined:

$$16x^2 - 1,000x + 13,600 = 0.$$

Now all we need do to find our answer is to solve this equation for x, the required number of seconds. The equation, however, is not one which can be solved by any methods which we have thus far discussed. Since it contains the *second-degree term*, $16x^2$, it is not linear. Hence the need for some new equation-solving technique, preceded by a few new —

Preliminary Definitions

A **second degree equation in one variable** is an equation which can be written in the typical form,

$$ax^2 + bx + c = 0,$$

where b and c are any constants and a is any constant except 0 (for, if $a = 0$, then $ax^2 = 0$, and the equation loses its second-degree term).

A **second degree function of one variable** is a function which can be written in the typical form,

$$y = f(x) = ax^2 + bx + c,$$

where b and c are again any constants, and a is any constant except 0.

Just as *first degree* equations and functions are also called *linear*, *second degree* equations and functions are also called **quadratic**, which means "squared."

Quadratic equations and functions are said to be **complete** if b and c are $\neq 0$ in the above definitions, and **pure** if $b = 0$. In the first case they are "complete" in the sense that they contain all the possible terms of a quadratic. In the second case they are "pure" in the sense that they contain only quadratic and constant terms.

For instance, $3x^2 = -5x + 2$ is a *complete quadratic equation* because it can be transposed to the form, $3x^2 + 5x - 2 = 0$, with $a = 3$, $b = 5$, $c = -2$. But $3x^2 - 2 = 0$ is an *incomplete, pure quadratic equation*, because $b = 0$ and the first degree, x-term, is therefore missing.

Solutions of equations are often called **roots**. This is because the solutions of equations of degrees higher than the first are found by processes involving, or equivalent to, the extraction of square roots, cube roots, etc.

Solution by Extracting Square Roots

The simplest kind of quadratic equation in one variable is that in the typical incomplete form,

$$x^2 = k,$$

where k is some positive number or 0.

As the student of elementary algebra already knows equations of this type, such as

$$x^2 = 4, \qquad x^2 = 9, \qquad x^2 = 10,$$

are satisfied, respectively, by the solutions,

$x = \sqrt{4} = 2$, $x = \sqrt{9} = 3$, $x = \sqrt{10} = 3.162 \ldots$

This is because —

$$2^2 = 4, \quad 3^2 = 9, \quad (3.162 \ldots)^2 = 10.$$

It is not always mentioned in elementary texts, however, that *every positive quantity has a second square root, equal in absolute value* — numerical magnitude without regard to a plus or minus sign — *to its positive square root, but negative in sign*. For instance,

$$(-2)^2 = 4, \quad (-3)^2 = 9, \quad (-3.162 \ldots)^2 = 10.$$

Hence the above equations are also satisfied, respectively, by the other possible roots,

$$x = -2, \quad x = -3, \quad x = -3.162 \ldots.$$

And the *complete* **solutions** of these quadratic equations are therefore the respective **pairs of roots,**

$$x = \pm 2, \quad x = \pm 3, \quad x = \pm 3.162 \ldots.$$

The symbol "$\pm$" here means: "plus *and* minus." Thus ± 2 is to be read: "plus 2 *and* minus 2."

Since the expression, $\sqrt{k}$, is ordinarily understood to mean only the positive square root of k, we must designate both square roots of this quantity by the combination of symbols, $\pm \sqrt{k}$, as in the summarizing solution formula —

A quadratic equation in the incomplete typical form $x^2 = k$, has the pair of roots, $x = \pm\sqrt{k}$.

Any equation of the form, $ax^2 + c = 0$, may of course be changed to the above typical form and solved in the same way.

EXAMPLE 1: Solve for x, the equation $9x^2 - 25 = 0$.

SOLUTION:

$$9x^2 = 25 \qquad \text{(Transposing } -25)$$
$$x^2 = \tfrac{25}{9} \qquad \text{(Dividing by } 9 = 9)$$
$$x = \pm\sqrt{\tfrac{25}{9}} = \pm\tfrac{5}{3}, \text{Ans.} \qquad \text{(Extracting square roots)}$$

Check:

$$9(\pm\tfrac{5}{3})^2 - 25 = 9(\tfrac{25}{9}) - 25$$
$$= 25 - 25 = 0 \;\checkmark \quad \text{(Substitution)}$$

Practice Exercise No. 11

Solve the following pure quadratic equations for both roots. Check all your answers as above.

(1) $4x^2 - 25 = 0$ (4) $2x^2 = 242$
(2) $3x^2 = 27$ (5) $5x^2 = 125$
(3) $3x^2 - 108 = 0$

Solution by Factoring

The equation, $ax^2 + bx + c = 0$, is relatively easy to solve when its lefthand side can be resolved by inspection into two linear factors. Each linear factor may then be separately equated to 0 because a product $= 0$ if any one of its factors $= 0$. Thus the problem is reduced to that of solving two separate linear equations.

For instance, suppose we are required to solve the quadratic equation,

$$x^2 - 2x = 0.$$

This may be rewritten in factored form as

$$x(x - 2) = 0.$$

Since the product on the left-hand side of this equation is 0 when either factor $= 0$, the condition of the equation is satisfied if we set

$$x = 0, \qquad \text{(Solution \#1)}$$

and if we set

$$x - 2 = 0. \qquad \text{(Equating the second factor to 0)}$$

By transposing -2 in the latter, we then get

$$x = 2. \qquad \text{(Solution \#2)}$$

And our *answer* is the pair of roots, $x = 0$ and $x = 2$, which can be written more briefly in the form $x = 0, 2$, understood to have the same meaning.

EXAMPLE 2: Solve for x,
$$3x^2 + 5x - 2 = 0.$$

SOLUTION: The left-hand side of this equation may be factored as follows:

$$(3x - 1)(x + 2) = 0 \qquad (MMS, \text{ Chapter XI})$$

Hence,

$$3x - 1 = 0 \qquad \text{(Factor \#1} = 0)$$
$$3x = 1 \qquad \text{(Transposing } -1)$$
$$x = \tfrac{1}{3}. \qquad \text{(Dividing by } 3 = 3, \text{ obtaining solution \#1)}$$

Also,

$$x + 2 = 0 \qquad \text{(Factor \#2 = 0)}$$
$$x = -2. \qquad \text{(Transposing 2, obtaining solution \#2)}$$

Answer: $x = \frac{1}{3}, -2$.

Check:

$$3(\tfrac{1}{3})^2 + 5(\tfrac{1}{3}) - 2 \qquad \text{(Substituting } x = \tfrac{1}{3})$$
$$= 3(\tfrac{1}{9}) + \tfrac{5}{3} - 2$$
$$= \tfrac{1}{3} + \tfrac{5}{3} - 2 = 2 - 2 = 0 \checkmark$$
$$3(-2)^2 + 5(-2) - 2 \qquad \text{(Substituting } x = -2)$$
$$= 3(4) - 10 - 2 = 12 - 12 = 0 \checkmark$$

Again we find *two roots* for our quadratic equation. In certain special cases, however, *these two roots may be equal*.

EXAMPLE 3: Solve for x,

$$x^2 - 6x + 9 = 0.$$

SOLUTION: In this case the two factors of the left-hand side are identical, since the expression is a perfect square:

$$(x - 3)(x - 3) = 0. \qquad \text{(Formula R30)}$$

Hence the same procedure as above gives us

$$x - 3 = 0 \text{ and } x - 3 = 0, \qquad \text{(Equating the factors to 0)}$$

or

$$x = 3 \text{ and } x = 3. \qquad \text{(Transposing } -3)$$

Answer: $x = 3, 3$.

Check:

$$3^2 - 6(3) + 9 = 9 - 18 + 9 = 0. \checkmark$$
$$\text{(Substituting } x = 3)$$

The facts that a quadratic equation in one variable has two roots, and that these two roots may be equal, are interpreted further below. We now ask, however —

Can a quadratic equation in one variable have more than two roots?

It is interesting to see what happens when we try to answer this question by the same algebraic device as that used for a pair of linear equations in Chapter II.

Considering only the root, $x = -2$, for the equation in Example 2 above, let us suppose any other possible root to differ from it by n. The other supposed root will then be $-2 + n$. And when we substitute this supposed solution into the original equation to check it, we get —

$$3x^2 + 5x - 2 \qquad \text{(The original expression)}$$
$$= 3(-2 + n)^2 + 5(-2 + n) - 2$$
$$\text{(Substituting } x = -2 + n)$$
$$= 3(4 - 4n + n^2) - 10 + 5n - 2$$
$$= 12 - 12n + 3n^2 - 12 + 5n$$
$$= 3n^2 - 7n = 0 \qquad \text{(Squaring and removing parentheses, etc.)}$$
$$n(3n - 7) = 0 \qquad \text{(Factoring)}$$
$$n = 0 \qquad \text{(Solution \#1)}$$
$$3n - 7 = 0 \qquad \text{(Setting factor \#2 = 0)}$$
$$n = \tfrac{7}{3} \qquad \text{(Solution \#2)}$$

Answer: $n = 0, \frac{7}{3}$.

Hence the supposed other roots must be

$$x = -2 + n = -2 + 0 = -2, \quad \text{(For } n = 0)$$

and

$$x = -2 + n = -2 + \tfrac{7}{3} = \tfrac{1}{3}. \quad \text{(For } n = \tfrac{7}{3})$$

But $x = \frac{1}{3}, -2$ was the solution we originally found in Example 2. The supposition of any solution other than either of these two roots therefore returns us algebraically to the same pair of roots.

Practice Exercise No. 12

Solve the following quadratic equations by the method of factoring. Check all solutions.

(1) $x^2 - 7x + 12 = 0$ (4) $x^2 = 6x - 8$
(2) $x^2 + 7x + 6 = 0$ (5) $x^2 + 3x = 10$
(3) $x^2 - 8x - 20 = 0$

Solution by Completing the Square

Factors of the quadratic expression in an equation of the form, $ax^2 + bx + c = 0$, are not always easy to recognize. Such an equation may nevertheless always be solved by a combination of the two preceding methods. This combined method, called **completing the square**, is based on prior knowledge of the fact that the square of a binomial, $x + k$, always has the form,

$$(x + k)^2 = x^2 + 2kx + k^2. \quad \text{(Formula R 30, page 15)}$$

Suppose, for instance, that we were not to recognize the factors of the previously considered equation,

$$3x^2 + 5x - 2 = 0. \qquad \text{(Example 2, page 39)}$$

The left-hand side of such an equation can always be rewritten in the form of the above perfect binomial-square by the following steps:

1. Transpose the constant term (in this case, -2):

$$3x^2 + 5x = 2.$$

2. Divide both sides by the coefficient of x^2 (in this case, 3):

$$x^2 + \tfrac{5}{3}x = \tfrac{2}{3}.$$

3. Take the resulting new coefficient of x (in this case, $\tfrac{5}{3}$), divide by 2 [getting here, $\tfrac{1}{2}(\tfrac{5}{3}) = \tfrac{5}{6}$], and add the square of the result [in this case, $(\tfrac{5}{6})^2 = \tfrac{25}{36}$] to both sides of the equation:

$$x^2 + \tfrac{5}{3}x + \tfrac{25}{36} = \tfrac{2}{3} + \tfrac{25}{36} = \tfrac{49}{36}.$$

4. Since the left-hand side is now in the form, $x^2 + 2kx + k^2$ (with k here $= \tfrac{5}{6}$), rewrite the entire equation with this side as a perfect square:

$$(x + \tfrac{5}{6})^2 = \tfrac{49}{36}.$$

5. Extract square roots, etc., as above:

$x + \tfrac{5}{6} = \pm\sqrt{\tfrac{49}{36}} = \pm\tfrac{7}{6}.$

$x = \tfrac{7}{6} - \tfrac{5}{6} = \tfrac{1}{3}$ (Transposing, etc., obtaining solution #1)

$x = -\tfrac{7}{6} - \tfrac{5}{6} = -2$ (Ditto, obtaining solution #2)

Answer: $x = \tfrac{1}{3}, -2.$

As may be seen by comparing this result with that of Example 2 above, the same solution is obtained by this method.

Practice Exercise No. 13

Rework the problems in Exercise No. 12 by completing the square.

Solutions by Formula

The method of completing the square is not often used in solving quadratic equations. This is because it is more convenient to solve such equations by means of a **general solution formula** which in turn is derived by applying the method of completing the square to the **equation in typical form,**

$$ax^2 + bx + c = 0.$$

The steps of this derivation are exactly the same as above.

Step 1:

$$ax^2 + bx = -c. \qquad \text{(Transposing } c\text{)}$$

Step 2:

$$x^2 + \frac{b}{a}x = -\frac{c}{a}. \qquad \text{(Dividing by } a = a\text{)}$$

Step 3:

$$x^2 + \frac{b}{a}x + \frac{b^2}{4a^2} = \frac{b^2}{4a^2} - \frac{c}{a} \qquad \left(\text{Adding } \frac{b^2}{4a^2} = \frac{b^2}{4a^2}\right)$$

$$= \frac{b^2 - 4ac}{4a^2} \qquad \text{(Adding fractions)}$$

Step 4:

$$\left(x + \frac{b}{2a}\right)^2 = \frac{b^2 - 4ac}{4a^2} \qquad \text{(Rewriting as a perfect square)}$$

Step 5:

$$x + \frac{b}{2a} = \pm\sqrt{\frac{b^2 - 4ac}{4a^2}} \qquad \text{(Extracting square roots)}$$

$$= \frac{\pm\sqrt{b^2 - 4ac}}{2a}$$

The **general solution of the typical quadratic equation, $ax^2 + bx + c = 0$,** is therefore given by the solution formula,

$$x = \frac{-b \pm \sqrt{b^2 - 4ac}}{2a}. \qquad \left(\text{Transposing } \frac{b}{2a}, \text{etc.}\right)$$

The quantity, $b^2 - 4ac$, in this formula is called the **discriminant** because it enables us to tell (discriminate) the mathematical nature of the roots of our equation, if necessary without actually computing their numerical values.

For instance, if the discriminant is positive, the roots are unequal. This condition may be stated algebraically as the case where $b^2 - 4ac > 0$, in which the symbol, $>$, means: "is greater than."

EXAMPLE 4: Solve by formula, $x^2 - 6x + 5 = 0$.

SOLUTION: Since here $a = 1$, $b = -6$, $c = 5$, we get by formula,

$$x = \frac{-(-6) \pm \sqrt{(-6)^2 - 4(1)(5)}}{2(1)} \qquad \text{(Substitution)}$$

$$= \frac{6 \pm \sqrt{36 - 20}}{2} = \frac{6 \pm 4}{2}$$ (Removing parentheses, etc.)

= 5 and 1, two unequal numbers, since the discriminant = 16 > 0.

When the discriminant "vanishes" — $b^2 - 4ac = 0$ — the roots are equal.

EXAMPLE 5: Solve by formula, $x^2 - 6x + 9 = 0$.

SOLUTION: Since $a = 1$, $b = -6$, $c = 9$, we get by formula,

$$x_1 = \frac{-(-6) \pm \sqrt{(-6)^2 - 4(1)(9)}}{2(1)}$$ (Substitution, etc., as above)

$$= \frac{6 \pm \sqrt{36 - 36}}{2} = \frac{6 \pm 0}{2}$$

= 3 and 3, two equal numbers, since the discriminant = 0.

Practice Exercise No. 14

A. Solve by formula the problems of Practice Exercise No. 12.

B. Determine the nature of the roots of each of the following equations by means of the discriminant. Verify your predictions by finding the roots by formula. Check all answers by substitution.

(1) $x^2 + 3x + 2 = 0$ (4) $5x^2 - 3x - 2 = 0$

(2) $x^2 + 3x - 2 = 0$ (5) $3x^2 - 7x + 4 = 0$

(3) $x^2 - 3x + 2 = 0$ (6) $2x^2 - 6x + 5 = 0$

"Imaginary" Roots

A complication arises when the solution of a quadratic equation contains the square roots of a negative quantity.

This may occur even in the simplest type of quadratic equation.

EXAMPLE 6: Solve, $x^2 + 4 = 0$.

SOLUTION:

$$x^2 = -4.$$ (Transposing 4)

$$x = \pm \sqrt{-4} = \pm 2\sqrt{-1}.$$ (Extracting square roots)

In general, however, the solution of a quadratic equation contains the square roots of a negative quantity if the discriminant is neg-

ative. This condition may be expressed algebraically as $b^2 - 4ac < 0$, in which the symbol, <, means: "is less than."

EXAMPLE 7: Solve by formula, $x^2 - 6x + 18 = 0$.

SOLUTION: Since here $a = 1$, $b = -6$, $c = 18$, we get by formula:

$$x = \frac{-(-6) \pm \sqrt{(-6)^2 - 4(1)(18)}}{2(1)}$$

$$= \frac{6 \pm \sqrt{36 - 72}}{2} = \frac{6 \pm \sqrt{-36}}{2}$$

$$= \frac{6 \pm 6\sqrt{-1}}{2} = 3 \pm 3\sqrt{-1}.$$ (Substitution, etc., as above)

These roots contain the square root of -1 because the discriminant = $-36 < 0$.

The difficulty now is that *all the numbers with which we are already familiar — positive, zero, or negative — have zero or positive squares.* For instance,

$$0^2 = 0, \quad (\pm 1)^2 = 1, \quad (\pm 9)^2 = 81, \quad \text{etc.}$$

What meaning, then, can we attach to an expression like $\sqrt{-1}$ when it appears in the solution of an equation?

To this question there are *two distinctly different answers.* Both should be understood very carefully from the very beginning of a study of quadratic and higher degree equations.

The first answer begins with the fact that it is possible to define a totally *new* "*number*," i, by the formula,

$$i = \sqrt{-1}, \text{ or } i^2 = -1.$$

This new number, i (pronounced "eye"), is called, by an unfortunate historical choice of names, the imaginary unit. Like the familiar arithmetical unit, 1, it may be multiplied by any ordinary number to form other imaginary numbers such as

$$2i = 2\sqrt{-1} = \sqrt{-4},$$
$$7i = 7\sqrt{-1} = \sqrt{-49},$$
$$\tfrac{1}{2}i = \tfrac{1}{2}\sqrt{-1} = \sqrt{-\tfrac{1}{4}}, \text{ etc.}$$

Hence the roots of the equation in Example 6 above may be written as: $x = \pm 2i$.

Once the imaginary unit has been thus defined, ordinary numbers not involving this unit are then distinguished as **real numbers**. Binomials involving a real number and an imaginary number are called **complex numbers**. And pairs of complex numbers which differ only in the sign of their imaginary terms are called **conjugate complex numbers**. For instance, the roots of the equation in Example 7 above are the *complex numbers*, $x = 3 + 3i$ and $x = 3 - 3i$, which together constitute the *pair of conjugate complex numbers*, $x = 3 \pm 3i$.

The fundamental operations of algebra may be defined for imaginary and complex numbers as for real numbers, provided the defining formula of the imaginary unit above is added to the definitions in our preceding Review Table of Elementary Formulas (page 14).

Imaginary roots then check exactly like real roots in an equation from which they derive.

For instance, substituting $x = \pm 2i$ in the equation of Example 6 above, we get —

$$x^2 + 4 = (\pm 2i)^2 + 4 \quad \text{(Removing parentheses)}$$
$$= 4i^2 + 4$$
$$= 4(-1) + 4 \quad \text{(Since } i^2 = -1$$
$$= -4 + 4 = 0. \checkmark \quad \text{by definition)}$$

Substituting $x = 3 + 3i$ in the equation of Example 7 above, we get —

$$x^2 - 6x + 18$$
$$= (3 + 3i)^2 - 6(3 + 3i) + 18 \quad \text{(Substitution)}$$
$$= 9 + 18i + 9i^2 - 18 - 18i + 18 \quad \text{(Removing parentheses)}$$
$$= 9 + 9i^2 = 9 + 9(-1) = 9 - 9 = 0. \checkmark \quad \text{(Since } i^2 = -1)$$

Or, substituting $x = 3 - 3i$ in the same equation, we get —

$$(3 - 3i)^2 - 6(3 - 3i) + 18 \quad \text{(Substitution)}$$
$$= 9 - 18i + 9i^2 - 18 + 18i + 18 \quad \text{(Removing parentheses)}$$
$$= 9 + 9i^2 = 0. \checkmark \quad \text{(As above)}$$

Moreover, when so-called imaginary numbers are investigated at length in a branch of higher mathematics called *the theory of functions of complex variables*, they are found to have many fascinating properties which can be applied to very practical matters like alternating electric currents. In such applications, these numbers are given an interpretation which is no more imaginary than that of real (ordinary) numbers in an income tax return.

For these and related reasons, the usual practice in courses which deal primarily with real variables is at least to identify complex roots when they occur. That is also our practice here.

As was mentioned above, however, there is a second answer to our question of what to make of $\sqrt{-1}$ when it occurs in the roots of quadratic equations. And that will be given in more detail after we have first considered other kinds of examples.

Practice Exercise No. 15

Solve the following quadratic equations by formula and check all roots as above.

(1) $2x^2 - 7x - 5 = 0$ (3) $7x^2 - 2x - 3 = 0$
(2) $3x^2 + 5x + 7 = 0$ (4) $4x^2 - x + 2 = 0$
(5) In a problem arising in alternating circuit theory, it was necessary to determine what x (in this case the resistance) should be to satisfy the equation:

$$.098x^2 - 3.6x - 4.2 = 0.$$

Equations Equivalent to Quadratics

Equations which do not at first appear to be quadratic may sometimes be changed to quadratic form and solved by the above methods.

1. *Certain fractional equations are quadratic when their fractions are simplified.*

EXAMPLE 8: Solve for x,

$$\frac{1}{x + 1} + \frac{1}{x - 1} = \frac{4}{3}.$$

SOLUTION: Simplifying fractions, this equation becomes —

$$3(x - 1) + 3(x + 1) = 4(x^2 - 1) \quad \text{(Formula R16, etc., page 15)}$$
$$3x - 3 + 3x + 3 = 4x^2 - 4 \quad \text{(Removing parentheses)}$$
$$-4x^2 + 6x + 4 = 0 \quad \text{(Transposing, etc.)}$$
$$2x^2 - 3x - 2 = 0 \quad \text{(Dividing by } -2 = -2)$$

Now we may use the quadratic formula with $a = 2$, $b = -3$, $c = -2$, as follows —

$$x = \frac{-(-3) \pm \sqrt{(-3)^2 - 4(2)(-2)}}{2(2)} \quad \text{(Substitution)}$$

$$= \frac{3 \pm \sqrt{9 + 16}}{4} = \frac{3 \pm 5}{4} \quad \begin{array}{l}\text{(Removing} \\ \text{parentheses,} \\ \text{etc.)}\end{array}$$

$$= 2 \text{ and } -\tfrac{1}{2}.$$

Check:

$$\frac{1}{2+1} + \frac{1}{2-1} = \frac{1}{3} + 1 = \frac{4}{3}. \checkmark$$

$$\frac{1}{-\tfrac{1}{2}+1} + \frac{1}{-\tfrac{1}{2}-1} = \frac{1}{\tfrac{1}{2}} + \frac{1}{-\tfrac{3}{2}} = 2 - \frac{2}{3} = \frac{4}{3}. \checkmark$$

$$\text{(Substitution)}$$

2. *Certain equations with the variable under a radical sign may be quadratic when cleared of radicals.*

EXAMPLE 9: Solve for x,

$$x - \sqrt{3x + 1} - 1 = 0.$$

SOLUTION: First we get the radical expression alone on one side of the equal sign:

$$x - 1 = \sqrt{3x + 1}. \quad \text{(Transposing } \sqrt{3x + 1})$$

Then we can clear the equation of radicals:

$$x^2 - 2x + 1 = 3x + 1 \quad \begin{array}{l}\text{(Squaring both sides,} \\ \text{by Axiom 5, page 16)}\end{array}$$

$$x^2 - 5x = 0. \quad \text{(Transposing, etc.)}$$

This quadratic equation may now be solved most conveniently by factoring to obtain the pair of solutions:

$$x = 0, 5. \quad \text{(Page 39)}$$

Of these two solutions for our intermediate quadratic equation, only $x = 5$ will check in the original radical equation if we are restricted to the positive square root of its radical expression. Specifically,

$$5 - \sqrt{3(5) + 1} - 1 = 5 - \sqrt{16} - 1$$

$$= 5 - 4 - 1 = 0, \text{ Check.} \quad \text{(Substituting } x = 5)$$

But,

$$0 - \sqrt{3(0) + 1} - 1 = 0 - \sqrt{1} - 1$$

$$= -1 - 1 = -2, \text{ No check!} \quad \text{(Substituting } x = 0)$$

However, if the conditions of our problem are such that the negative square root of the radical expression is acceptable, then the other solution, $x = 0$, will also check in the original equation. Specifically,

$$0 - (-\sqrt{1}) - 1 = -(-1) - 1 = 1 - 1 = 0.$$

$$\text{(Re-substituting } x = 0)$$

A complication of this sort must always be watched for when a radical sign is removed from an equation by squaring its sides as above. For, the operation of squaring a radical expression may produce a new equation which has solutions that cannot be admitted when we accept only the positive square root of the original radical expression.

3. *Certain incomplete equations of higher degrees may also be treated as quadratics.*

EXAMPLE 10: Solve for x,

$$3x^4 + 5x^2 - 2 = 0.$$

SOLUTION: Although this equation is of the fourth degree in x, it is incomplete in that it does not contain the possible x^3 or x terms of such an equation. By setting $z = x^2$, therefore, we may reduce it to the quadratic equation,

$$3z^2 + 5z - 2 = 0 \quad \text{(Substitution)}$$

However, we have already found the roots of this equation to be —

$$z = \tfrac{1}{3}, -2. \quad \text{(Example 2, above)}$$

Hence, when $x^2 = z = \tfrac{1}{3}$,

$$x = \pm\sqrt{\tfrac{1}{3}} = \pm\sqrt{\tfrac{1}{3}} = \pm\frac{\sqrt{3}}{3}$$

$$\text{(Extracting square roots)}$$

And when $x^2 = z = -2$,

$$x = \pm\sqrt{-2} = \pm\sqrt{2}\sqrt{-1} = \pm\sqrt{2}i. \quad \text{(Ditto)}$$

The original fourth degree equation therefore has the four roots, $x = \pm\sqrt{3}/3$ and $x = \pm\sqrt{2}i$. We shall see later that all fourth degree equations have four roots.

Check:

$$3\left(\pm\frac{\sqrt{3}}{3}\right)^4 + 5\left(\pm\frac{\sqrt{3}}{3}\right)^2 - 2 = \frac{1}{3} + \frac{5}{3} - 2 = 0.$$

$$3(\pm\sqrt{2}i)^4 + 5(\pm\sqrt{2}i)^2 - 2 = 12 - 10 - 2 = 0.$$

$$\text{(Substitution)}$$

Note in the above check that since $i^2 = -1$ (page 42), it follows that $i^4 = i^2(i^2) = -1(-1) = 1$. Hence $12i^4 = 12$, etc.

Practice Exercise No. 16

Solve the following equations equivalent to quadratics. Check all your answers.

(1) $\dfrac{1}{x-1} + \dfrac{2}{x-3} = 4$

(2) $x^4 - 5x^2 + 3 = 0$

(3) $x^6 - 26x^3 + 25 = 0$

(4) $\dfrac{1}{x-2} + \dfrac{2}{x-3} = \dfrac{3}{(x-2)(x+1)}$

Graphic Interpretation of Quadratic Roots

Why does a quadratic equation in one variable, unlike a linear equation in one variable, always have two roots?

Why are these roots sometimes unequal, sometimes equal, and sometimes imaginary or complex?

For answers to these and many similar questions, let us first analyze quadratic functions graphically, as we did linear functions in the preceding chapter.

We begin with the simplest case of the incomplete quadratic function,

$$y = f(x) = ax^2.$$

Since $(-x)^2 = x^2$ (page 39), a table of typical values for this function when $a = 1$ may be abbreviated as follows:

x	$f(x) = x^2$	y
0	0^2	0
$\pm\frac{1}{2}$	$(\pm\frac{1}{2})^2$	$\frac{1}{4}$
± 1	$(\pm 1)^2$	1
± 2	$(\pm 2)^2$	4
± 3	$(\pm 3)^2$	9
± 4	$(\pm 4)^2$	16
± 5	$(\pm 5)^2$	25
etc.	. . .	. . .

When plotted and connected by a continuous curve, these pairs of values for x and y produce the curved graph of $y = x^2$ in the upper half of Figure 9.

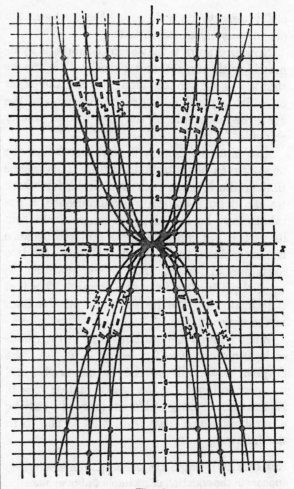

Fig. 9

If the coefficient, a, of x^2 has any other value in the function, $y = f(x) = ax^2$, then all the numbers in the second and third columns of the above table of values are simply multiplied by this other value of a. And when plotted in the same manner, the resulting pairs of values for x and y produce the curved-line graphs of $y = 2x^2$, $y = \frac{1}{2}x^2$, $y = -x^2$, etc., in Figure 9.

All of these graphs are called **parabolas**. The exact definition of such a curve is given later in analytic geometry. But for the present these basic facts about such a curve should be noted.

An added straight line down its center — in these cases the y-axis itself — is called the parabola's **axis**. The point at which a parabola intersects its axis — in these cases, the origin — is called its **vertex**.

Note in Figure 9 that each parabola is **symmetric** with respect to its axis. This means that if you were to fold along its axis, the paper on which a parabola is drawn, the parts of the curve on each side of the axis would exactly coincide.

Also note in Figure 9 that parabolas are **open curves of unlimited length**. This means that unlike a figure such as a circle, for which we can draw a complete circumference, a parabola may be traced as far as we wish from its vertex, but can never be completed because it never closes in on itself. Rather, its two parts spread further and further apart as we trace them further from the vertex. This fact is indicated by the broken ends on the lines in the diagram.

When b and c are not both zero in the quadratic function, $y = f(x) = ax^2 + bx + c$, the graph's vertex will not lie at the origin, and its axis will not be the y-axis unless $b = 0$. Nevertheless, the graph's contour will still be a parabola which may be plotted as shown in the following —

EXAMPLE 4G: Graph the quadratic function which appears in the equation of Example 4 above; namely,

$$y = f(x) = x^2 - 6x + 5.$$

SOLUTION: Beginning as a matter of routine with the three simplest substitutions possible, we find:

$f(-1) = (-1)^2 - 6(-1) + 5 = 1 + 6 + 5 = 12.$
(Substituting $x = -1$)

$f(0) = (0)^2 - 6(0) + 5 = 0 - 0 + 5 = 5.$
(Substituting $x = 0$)

$f(1) = (1)^2 - 6(1) + 5 = 1 - 6 + 5 = 0.$
(Substituting $x = 1$)

Hence three of our required points must be:

$$(-1,12), \quad (0,5), \quad (1,0) \quad \text{(Figure 10)}$$

Since these last two points are closer together, we know from the general shape of a parabola in Figure 9 that the required graph's vertex must lie to the right rather than to the left. We therefore continue with additional substitutions in this direction, $x = 2, 3, 4$, finding

$f(2) = -3, \quad f(3) = -4, \quad f(4) = -3.$

Hence three more of our required points must be:

$$(2,-3), \quad (3,-4), \quad (4,-3). \quad \text{(Figure 10)}$$

From the fact that the point $(3,-4)$ lies lower than the others, and from our familiarity with the general shape of a parabola, we now know that our plot must have passed the vertex of the required graph. But since a curve drawn through the six points thus far plotted is still an incomplete looking, hook-shaped figure which has crossed the x-axis only once, we continue with additional substitutions in the same direction, $x = 5, 6, 7$, finding the additional points,

$$(5,0), \quad (6,5), \quad (7,12).$$

As may be seen in Figure 10, a curved line drawn through the nine points now plotted is a reasonably

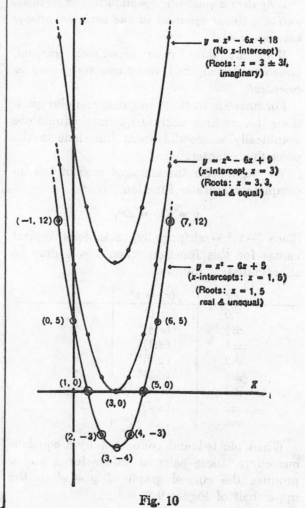

Fig. 10

adequate, symmetrical segment of a parabola. The parabola graph may be continued, however, for as many additional pairs of points on either side of the vertex as we desire. This is indicated by the "open ends" on the following table of values which summarizes the above computations:

$$y = f(x) = x^2 - 6x + 5$$

y	12	5	0	−3	−4	−3	0	5	12
x	−1	0	1	2	3	4	5	6	7

In Figure 10, similar parabola graphs are also shown for the quadratic functions, $y = f_1(x) = x^2 - 6x + 9$, and $y = f_2(x) = x^2 - 6x + 18$.

Observe now that the parabola graph of the quadratic function, $y = f(x) = x^2 - 6x + 5$ in this Figure, intersects the x-axis in the *two distinct points* where $x = 1$ and $x = 5$; and that the corresponding quadratic equation, $x^2 - 6x + 5 = 0$ in Example 4 above (page 41) has the pair of *unequal, real roots*, $x = 1, 5$.

Observe also that the parabola graph of the quadratic function, $y = f_1(x) = x^2 - 6x + 9$, intersects the x-axis only in the one point where $x = 3$; and that the corresponding quadratic equation, $x^2 - 6x + 9 = 0$ in Example 5 above (page 42), has the *two equal roots*, $x = 3, 3$.

Observe, finally, that the parabola graph of the quadratic function, $y = f_2(x) = x^2 - 6x + 18$, *does not intersect the x-axis at all;* and that the corresponding quadratic equation, $x^2 - 6x + 18 = 0$ in Example 7 above, has the two *complex conjugate roots*, $x = 3 \pm 3i$.

These cases are *not* coincidences. *An equation in the form, $ax^2 + bx + c = 0$, is simply an algebraic demand for those values of x which make $y = f(x) = 0$ in the corresponding quadratic function, $y = f(x) = ax^2 + bx + c$.* We see from the pattern of graphs and roots in Figure 10, therefore, that —

A quadratic equation of the form, $ax^2 + bx + c = 0$, may have —

(1) **two unequal, real roots,**

(2) **two equal, real roots, or**

(3) **two conjugate complex roots,**

depending upon whether the graph of the corresponding quadratic function, $y = f(x) = ax^2 + bx + c$, intersects the y-axis in —

(1) **two distinct points,**

(2) **one point, or**

(3) **no points.**

This means that the lower two graphs in Figure 10 are actually **graphic solutions** of the corresponding quadratic equations.

It is also possible, by a complicated set of rules, to tell from the position of the vertex of the upper graph what the complex solutions of its corresponding equation must be. For our present purposes, however, it is enough to observe that *when the roots of a quadratic equation involve the imaginary unit, i, this is because the parabola graph of the corresponding quadratic function does not intersect the x-axis, and therefore no real value of x can possibly satisfy the mathematical condition expressed by the equation.*

Practice Exercise No. 17

Interpret graphically your solutions of the problems in Practice Exercises Nos. 11(1), 14(3)(5)(6), and 15(2)(3), above.

Physical Interpretation of Quadratic Roots

When a quadratic equation expresses a condition of a physical problem, what physical interpretation are we to place upon the fact that the equation may have two different solutions?

In general, the answer to this question depends upon the specific kind of physical problem with which we are dealing.

In the simplest type of case, *the problem may be one which actually admits of two different answers* even though we had only one possibility in mind when posing the question. For instance —

EXAMPLE 11: Now finish solving the problem with which this chapter started (page 38): "How many seconds will it take a projectile to reach a height of 15,600 feet above sea level if it is fired from an altitude of 2,000 feet, etc.?"

SOLUTION: The equation at which we had previously arrived in beginning the solution of this problem was:

$$16x^2 - 1{,}000x + 13{,}600 = 0. \qquad \text{(Page 38)}$$

To solve this equation for x, the required number of seconds, we may now use the quadratic formula with $a = 16$, $b = -1{,}000$, $c = 13{,}600$, as follows:

$$x = \frac{-(-1{,}000) \pm \sqrt{(-1{,}000)^2 - 4(16)(13{,}600)}}{2(16)}$$

$$= \frac{1{,}000 \pm \sqrt{129{,}600}}{32} = \frac{1{,}000 \pm 360}{32}$$

$$= \frac{640}{32} \text{ and } \frac{1360}{32} = 20 \text{ and } 42.5$$

Answer: The projectile will be at a height of 15,600 feet at the end of 20 seconds *and* at the end of 42.5 seconds.

The *two different answers* obtained here are explained as follows —

Perhaps in asking the above question we may have had in mind only the *ascent* of the projectile. The quadratic formula given in connection with the problem, however, states the height, y, of the projectile *at any time*, including the period of its *descent*. The solution, $x = 20$, therefore tells us the number of seconds it will take the projectile to rise to a height of 15,600 feet on its way up. And the other solution, $x = 42.5$, tells us the number of seconds it will take the projectile to reach its greatest elevation and then fall back down again to a height of 15,600 feet.

In the case of Example 11, therefore, both solutions of our quadratic equation are physically possible. Only correct physical interpretation is needed to explain the fact that the quadratic equation which expresses the condition of the problem has two different solutions.

In some problems, on the other hand, *a pair of equal solutions may appear from our work where two different solutions are anticipated.*

EXAMPLE 12: How many seconds will it take the projectile in Example 11 to reach a height of 17,625 feet? And how many seconds will it take the same projectile to fall back down to this same height?

SOLUTION: Since here $y = 17{,}625$, our quadratic equations become

$$17{,}625 = -16x^2 + 1{,}000x + 2{,}000, \text{ (Substitution)}$$

or,

$$16x^2 - 1{,}000x + 15{,}625 = 0. \qquad \text{(Transposing, etc.)}$$

Hence, as before, —

$$x_1 = \frac{-(-1{,}000) \pm \sqrt{(-1{,}000)^2 - 4(16)(15{,}625)}}{2(16)}$$

$$= \frac{1{,}000 \pm \sqrt{1{,}000{,}000 - 1{,}000{,}000}}{32}$$

$$= \frac{1{,}000 \pm 0}{32} = 31.25 \text{ and } 31.25$$

Answer: The projectile will be at a height of 17,625 feet at the end of 31.25 seconds and at the end of 31.25 seconds.

Explanation: Although two different solutions may have been expected by the form of the question in Example 12, the fact that the two roots of the resulting quadratic equation are equal tells us that the projectile reaches the height of 17,625 feet only once. This means that, by coincidence, it must be at its maximum altitude at the end of this number of seconds.

In other words, *a mathematical solution will be affected only by the mathematical facts expressed in its equations. It will not be influenced by the grammatical or rhetorical form of a question any more than by the tone of voice in which the question is asked.*

Observe, however, that *in some cases one solution of a quadratic equation may have to be rejected as impossible under the physical conditions of a given problem, even though it is numerically possible under the algebraic conditions of the quadratic equation by means of which the problem is solved.*

EXAMPLE 13: After one side of a square field has been reduced by 50 feet, and the adjacent side has been reduced by 80 feet, the remaining rectangular area is 400 square feet. How large was each side of the original square field?

SOLUTION: Let $x =$ the side of the square field originally. Then the reduced sides are $x - 50$, $x - 80$, and

$$(x - 50)(x - 80) = 400 \text{ (The given condition)}$$
$$x^2 - 130x + 4{,}000 = 400 \text{ (Removing parentheses)}$$
$$x^2 - 130x + 3{,}600 = 0 \text{ (Transposing 400)}$$

Hence, by formula —

$$x = \frac{-(-130) \pm \sqrt{(-130)^2 - 4(1)(3,600)}}{2(1)}$$

$$= \frac{130 \pm \sqrt{2,500}}{2} = \frac{130 \pm 50}{2}$$

$$= 90 \text{ and } 40.$$

Check:

$$(90 - 50)(90 - 80) = (40)(10) = 400 \checkmark$$
(Substituting $x = 90$)
$$(40 - 50)(40 - 80) = -10(-40) = 400 \checkmark$$
(Substituting $x = 40$)

In Example 13, both of the solutions, $x = 90, 40$, are equally plausible in that they are both positive, and both check numerically in the original equation.

However, a field which is 90 feet square can have its sides reduced by 50 feet and by 80 feet, and there is still an actual rectangle left. But a field which is 40 feet square cannot have its sides reduced by these amounts with an actual rectangle remaining. Although numerically sound, therefore, the second root in this case must be rejected for the purposes of the given problem as *physically inappropriate*.

Finally, when a quadratic problem requires a real answer, the emergence of imaginary roots is evidence that the conditions of the problem are physically impossible.

EXAMPLE 14: In how many seconds will the projectile in Example 11 reach a height of 20,125 feet?

ATTEMPTED SOLUTION: Substituting $y = 20,125$, we this time get the quadratic equation,

$$20,125 = -16x^2 + 1,000x + 2,000,$$

or,

$$16x^2 - 1,000x + 18,125 = 0. \quad \text{(Transposing, etc.)}$$

Hence, by substitution again in the quadratic formula —

$$x = \frac{-(-1,000) \pm \sqrt{(-1,000)^2 - 4(16)(18,125)}}{2(16)}$$

$$= \frac{1,000 \pm \sqrt{1,000,000 - 1,160,000}}{32}$$

$$= \frac{1,000 \pm 400i}{32} = 31.25 \pm 12.5i.$$

In view of the geometric interpretation of complex roots for quadratic equations we can only conclude from the above result that it is physically impossible for a projectile fired from the given altitude at the given initial velocity to rise to a height of 20,125 feet. This conclusion is indirectly confirmed by our incidental finding in Example 12 above that the highest elevation the projectile reaches is actually only 17,625 feet (page 48).

Practice Exercise No. 18

Solve the following problems involving quadratic equations. Interpret your results physically as in the above illustrative examples.

1. For the diameter x of the shortest rectangular beam to be cut out of a log, one has to solve the following equation:

$$5.2x^2 - 4.3x - 6.7 = 0$$

2. For two resistances R in a parallel circuit to have a total resistance of 5 ohms, and if one is 4 ohms greater than the other, the equation to be solved is:

$$\frac{1}{R} + \frac{1}{R + 4} = \frac{1}{5}$$

3. If a bomber 2000 ft. up fires upward some shells with an initial speed of 1000 ft/sec., the height h of the shells off the ground at any time t is given by the formula $h = -16t^2 + 1000t + 2000$.

(a) When are the shells at the height of the bomber?

(b) When do the shells hit the ground?

Summary

Whereas a linear equation in one variable has only one root, **a quadratic equation in one variable has two roots.**

These roots may be found by the **methods of extracting square roots, of factoring, or of completing the square.**

When the latter method is applied to the **general quadratic equation in one variable, $ax^2 + bx + c = 0$, we derive the general solution formula,**

$$x = \frac{-b \pm \sqrt{b^2 - 4ac}}{2a}.$$

Depending upon whether the discriminant, $b^2 - 4ac$, is > 0, 0, or < 0, the two roots yielded by this formula are real and unequal, real and equal, or conjugate imaginaries.

The graph of a quadratic function, $y = f(x) = ax^2 + ax + c$, is a symmetric curve called a parabola. This parabola graph intersects the x-axis in 2, 1, or no points, depending, respectively, upon the three cases of roots and discriminant values mentioned above.

When physical problems lead to a quadratic equation of the above form, two unequal real roots may in some cases relate to different aspects of the same physical problem. In some cases, however, one root (usually a negative root) may have to be regarded as physically inapplicable to the physical situation, even though it checks numerically.

Complex numbers are of great interest and practical value in a branch of *higher* mathematics called *the theory of functions of complex variables*. Their main significance here, however, is that complex roots for any equation deriving from a physical problem which requires a real answer are a sign that the stated conditions of the problem are physically impossible.

CHAPTER V

SYSTEMS INVOLVING QUADRATICS

Preliminary Definitions

When discussing systems of equations in Chapters II and III we were limited to linear examples. Now that quadratic equations have been introduced in Chapter IV, we may extend the definitions and principles of those earlier chapters to systems of equations involving quadratics.

The most general type of quadratic equation in two variables is one which can be written in the typical form,

$$ax^2 + bxy + cy^2 + dx + ey + f = 0,$$

with x and y as variables and the other letters representing constants.

Here f is called the **constant term** since it contains neither variable. The quantities, dx and ey, are called **linear terms** for reasons already explained (page 18). And the quantities, ax^2, bxy, and cy^2, are called **quadratic terms** since, in each, the sum of the exponents of the variables is 2 (Recall that xy is defined as meaning x^1y^1 by formula R22, page 15).

Obviously, the constants, a, b, c, cannot all $= 0$ in the above typical form, since then all the quadratic terms would be missing and the equation would no longer be of the *second degree* (Chapter IV).

Note on Sequence of Topics

As in the linear case, a **system of quadratic equations** may be *determinate* (page 19), or it may be *indeterminate* because it is *defective* (page 23), *redundant* (page 23), *dependent* (page 22), or *inconsistent* (page 20).

Certain types of systems involving quadratics may also be solved and interpreted by much the same devices as those already explained in the three preceding chapters.

In other respects, however, the topic of systems of equations involving quadratics is more of a "mathematical crossroads" than a straight stretch on the "highway" of intermediate algebra.

Attempts to solve many systems of this type lead to equations of the fourth degree or higher. And the latter cannot be discussed very satisfactorily until the techniques of differential calculus are closer to hand than they are at this point. On the present level, therefore, the discussion of such systems is usually limited to the display of a kind of "bag of tricks" which apply only to special cases.

Moreover, when we attempt to interpret geometrically the more complicated patterns of roots which emerge from systems involving quadratics, it is easy to get side-tracked into topics of analytic geometry which have yet to be introduced in adequate detail.

The present "crossroads" situation is therefore this: When you begin to study systems of equations involving quadratics — a standard topic of high school Intermediate Algebra — you have reached a point where several of the traditionally separated branches of mathematics are so interrelated that a discussion of any one leads to basic topics of the others (Recall the introductory note on Recommended Sequence of Study, page 12 above).

Notwithstanding these complications, however, a limited consideration of systems of equations involving quadratics is both possible and desirable at this stage of your study.

Some of the devices for treating these systems without recourse to higher degree equations, are *genuine short-cuts* to be preferred even later when other methods are fully available. And they may also serve usefully now as further exercises in applying the general principles of the three preceding chapters.

Later in analytic geometry, moreover, the focus of attention on the graphic representation of quadratic equations is somewhat different from our present one. It is true that more satisfactory demonstrations are possible there. But then the central concern is to derive general methods for studying the properties of certain figures which are of primarily *geometric* interest. The use of equations for this purpose, although basic to the technique, is often only incidental to the *geometric ends* for which that technique is used.

In the present study of algebra, on the other hand, we are more concerned with graphic representation as a device to depict the relationship between the variables in our equations (Chapter III, page 30), and thus to check, and/or interpret, the patterns of roots which emerge, or sometimes fail to emerge, from our formal algebraic operations.

While working through this chapter, therefore, you should keep these two considerations in mind: (1) *Treatment of the topic is necessarily incomplete at several points where a more complete treatment would jump us ahead too soon into more advanced topics.* But (2) *this partial treatment usefully serves certain interim needs and is an integral part of your general preparation for studying those other topics with a different emphasis at the appropriate time.*

Defective Systems

Whereas a single quadratic equation in one variable is determinate with a unique pair of roots (Chapter IV, page 20), **a single quadratic equation in two variables is indeterminate with an unlimited (infinite) number of roots.**

As in the linear case (Chapter II, page 23), a defective system of one quadratic equation in two variables can best be treated by first being rewritten, if necessary, in explicit form (Chapter III, page 29). Then substitutions may be made for the independent variable to find corresponding values of the dependent variable. And the resulting pattern of solutions can be summarized in a table of values and/or plotted (when real) in as much detail as desired.

In the simplest instances, a quadratic equation may be solved for one variable in terms of another by *the method of extracting square roots* (Chapter IV, page 38).

EXAMPLE 1: Solve for y in terms of x, and graph the resulting solutions:

$$4y^2 - x^2 = 0.$$

SOLUTION:

$4y^2 = x^2$	(Transposing $-x^2$)
$y^2 = \frac{1}{4}x^2$	(Dividing by $4 = 4$)
$y = \pm\frac{1}{2}x$	(Extracting square roots)

Check:

$4(\pm\frac{1}{2}x)^2 - x^2 = x^2 - x^2 = 0.$ ✓ (Substitution)

From this algebraic result we see that for each real value of the selected independent variable, x, the dependent variable has two real values, $y = \pm\frac{1}{2}x$, which in a special case may be equal. For instance, when $x = 4$, then $y = \pm\frac{1}{2}(4) = \pm2$; when $x = -4$, then $y = \pm\frac{1}{2}(-4) = \mp2$; but when $x = 0$, then $y = \pm\frac{1}{2}(0) = \pm0 = 0$.

Since these pairs of values are given by two separate linear equations, $y = \frac{1}{2}x$ and $y = -\frac{1}{2}x$, we need not sample the solutions further, but may sketch the graph directly (Chapter III, page 23) as the pair of straight lines, L_1 and L_2, shown intersecting at the origin in Figure 11.

The same result could have been obtained in Example 1 by *the method of factoring*, as follows:

$(2y - x)(2y + x) = 0$	(Factoring $4y^2 - x^2$ by formula R29)

Then – –

$2y - x = 0$	(Equating factor #1 to 0)
$y = \frac{1}{2}x$	(Transposing, etc.)

And – –

$2y + x = 0$	(Equating factor #2 to 0)
$y = -\frac{1}{2}x$, etc.	(Transposing, etc.)

This method may be available when the method of extracting square roots is not.

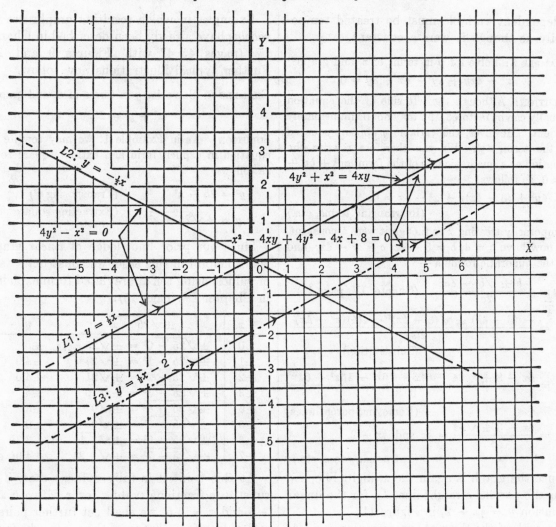

Fig. 11

EXAMPLE 2: Solve for y in terms of x, and graph:
$$4y^2 + x^2 = 4xy.$$

SOLUTION: We cannot transpose this equation directly into explicit form. Hence we rewrite it as a quadratic expression equated to 0:

$$4y^2 - 4xy + x^2 = 0. \qquad \text{(Transposing } 4xy)$$

Then, by factoring:

$$(2y - x)(2y - x) = 0 \qquad \text{(Formula R30)}$$
$$2y - x = 0 \qquad \text{(Equating either}$$
$$\text{factor to 0)}$$
$$y = \tfrac{1}{2}x \text{ and } \tfrac{1}{2}x \qquad \text{(Transposing } x, \text{ etc.)}$$

Check:

$$4(\tfrac{1}{2}x)^2 - 4x(\tfrac{1}{2}x) + x^2 = \qquad \text{(Substitution)}$$
$$x^2 - 2x^2 + x^2 = 0. \checkmark \qquad \text{(Removing parentheses, etc.)}$$

In this case we find that, by coincidence, each pair of solutions for our quadratic equation is equal to the corresponding solution of the linear equation, $y = \tfrac{1}{2}x$. Hence the quadratic equation is equivalent to a pair of identical linear equations, $y = \tfrac{1}{2}x$, $y = \tfrac{1}{2}x$, and the relationship between its variables may be graphed by the single straight line, L_1, in Figure 11.

When simpler methods fail or are not seen to apply, *a quadratic equation in two variables may always be solved for either variable, provided it appears squared, by means of the quadratic formula.* In this procedure, the selected independent variable must be treated temporarily as though it were a constant.

EXAMPLE 3: Solve for y in terms of x, and graph:

$$x^2 - 4xy + 4y^2 - 4x + 8y = 0.$$

SOLUTION: Although the left side of the equation actually can be factored, this may not be immediately evident. However, the variable y appears squared in the term, $4y^2$. We can therefore rewrite the equation as a quadratic in y of the form, $ay^2 + by + c = 0$, as follows:

$$4y^2 + (8 - 4x)y + x^2 - 4x = 0.$$

(Rearranging terms, etc.)

Temporarily treating x as though it were a constant, we now have $a = 4$, $b = 8 - 4x$, $c = x^2 - 4x$; and by the quadratic formula —

$$y = \frac{-b \pm \sqrt{b^2 - 4ac}}{2a} \quad \text{(Chapter IV, page 41)}$$

$$= \frac{-(8 - 4x) \pm \sqrt{(8 - 4x)^2 - 4(4)(x^2 - 4x)}}{2(4)}$$

(Substitution)

$$= \frac{4x - 8 \pm \sqrt{64 - 64x + 16x^2 - 16x^2 + 64x}}{8}$$

(Removing parentheses)

$$= \frac{4x - 8 \pm \sqrt{64}}{8} = \tfrac{1}{2}x \text{ and } \tfrac{1}{2}x - 2.$$

(Extracting roots, etc.)

Hence the graph is again the straight line, L_1, plus the parallel straight line, L_3, for the linear equation $y = \tfrac{1}{2}x - 2$, in Figure 11.

In each of the three above examples, a single quadratic equation in two variables has been found to be the equivalent of a pair of linear equations. Hence the dependent variable has *a pair of real values* — in certain cases, equal real values — for each real value assigned to the independent variable. And the resulting graphs are pairs of straight lines — coinciding in Example 2, but intersecting in Example 1 and parallel in Example 3.

In some cases, however, *one variable in a quadratic equation may have no real value, or no definite value at all, corresponding to some, one, or all, real values of the other variable.* Then the algebraic relationship between the variables, and its graphic depiction, are very different from those of linear equations.

Examples in which resulting graphs are *parabolas* have already been described in Chapter IV (pages 45–47 with Figures 9 and 10). Further typical illustrations now follow.

EXAMPLE 4: Solve for y in terms of x, and graph:

$$x^2 + y^2 = 25.$$

SOLUTION: As in Example 1, we may rewrite the equation in explicit form and extract square roots, as follows:

$$y^2 = 25 - x^2 \quad \text{(Transposing } x^2\text{)}$$

$$y = \pm\sqrt{25 - x^2} \quad \text{(Extracting square roots)}$$

Since our prior knowledge of *linear* graphs is not helpful in this case, we construct a table of values, using the abbreviated form explained in Chapter IV (page 45):

x	$\pm\sqrt{25 - x^2}$	y
0	$\pm\sqrt{25 - 0^2} = \pm\sqrt{25}$	± 5
± 1	$\pm\sqrt{25 - 1^2} = \pm\sqrt{24}$	± 4.9
± 2	$\pm\sqrt{25 - 2^2} = \pm\sqrt{21}$	± 4.6
± 3	$\pm\sqrt{25 - 3^2} = \pm\sqrt{16}$	± 4
± 4	$\pm\sqrt{25 - 4^2} = \pm\sqrt{9}$	± 3
± 5	$\pm\sqrt{25 - 5^2} = \pm\sqrt{0}$	± 0
± 6	$\pm\sqrt{25 - 6^2} = \pm\sqrt{-11}$	$\pm 3.3i$
± 7	$\pm\sqrt{25 - 7^2} = \pm\sqrt{-24}$	$\pm 4.9i$, etc.

At this point it is fully evident that, if we substitute fractional values for x between $x = -5$ and $x = +5$, we shall get further pairs of real values for y between those already found by substituting integral values for x within these limits. But no matter what other values we substitute for x, less than $x = -5$ or greater than $x = +5$, we shall always get a pair of imaginary values for y (Chapter IV, page 42).

We can therefore plot real solutions for our equation only from $x = -5$ to $x = +5$, as in Figure 12. And when we draw a smooth curve through these points, the graph of the equation is the rounded contour at the center of Figure 12, which later in analytic geometry we shall prove to be a circle.

In the equation of Example 4, the selected dependent variable, y, is again a **two-valued** function of the corresponding independent va-

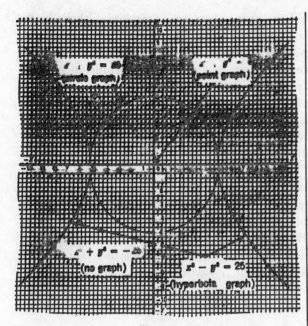

Fig. 12

riable, x. But since it has only imaginary values for other values of x, it is said to be a **real function** of x only for values of x from $x = -5$ to $x = +5$.

Suppose now that the equation of Example 4 had been of the slightly different form,

$$x^2 + y^2 = 0.$$

Replacing 25 by 0 in the above table, we can see at once that y is a real function of x — in this case with the pair of equal values, $y = \pm 0$ — only for the single real value of the independent variable, $x = 0$. The resulting graph is then only a *single point* located, in this instance, at the origin. For obvious reasons, such a graph is sometimes called a "degenerate point-circe."

Or suppose that the equation in Example 4 had been of the still different form,

$$x^2 + y^2 = -25.$$

Replacing 25 by -25 in the above table, we can see at once that y has no real value for any real value of x. In such a case the equation has *no real solution*, and therefore *no graph* with respect to a pair of real coordinate axes.

Suppose, finally, that the equation in Example 4 had been of the form,

$$x^2 - y^2 = 25.$$

Transposed to explicit form, this becomes

$$y = \pm \sqrt{x^2 - 25}.$$

And our table of values, computed as above but further abbreviated in format, is —

$x =$	0	$\ldots$	± 4	± 5	± 6	± 7	± 8	± 9	± 10	$\ldots$
$y =$	$\pm 5i$	$\ldots$	$\pm 3i$	± 0	± 3.3	± 4.9	± 6.2	± 7.5	± 8.6	$\ldots$

At this point it is fully evident that the relationship between the variables in our equation has a pattern directly opposite to that in Example 4. If we substitute other values for x between $x = -5$ and $x = +5$, we shall always get a pair of imaginary values for y. But when we substitute larger plus or minus values for x, we get larger and larger plus-minus values for y.

With respect to a pair of real coordinates, therefore, we can plot typical (real) solutions for this equation only from $x = -5$ to the left, and from $x = +5$ to the right, as in Figure 12. And when we draw smooth curves through these points, the graph of the equation is the pair of open contours (Chapter IV, page 46). at the sides of Figure 12, which later in analytic geometry we shall prove to be the two branches of a hyperbola.

A final peculiarity of the relationship of variables in a quadratic equation, as contrasted to that in a linear equation, is illustrated by —

EXAMPLE 5: Solve for y in terms of x, and graph:

$$xy = 1.$$

SOLUTION: This is the case in which neither variable appears squared (page 53), and to which the quadratic formula therefore cannot be applied. However, the equation is easily rewritten in the explicit form,

$$y = \frac{1}{x}. \qquad \text{(Dividing by } x = x\text{)}$$

It is a simple routine to compute a table of values for this function when x has integral real values other than $x = 0$. In abbreviated form:

$x =$	-10	-5	-2	-1	$\ldots$	1	2	5	10
$y =$	-0.1	-0.2	-0.5	-1	$\ldots$	1	0.5	0.2	0.1

From the arithmetic pattern of this table it is clear that, as x assumes larger and larger plus or minus values, the function $y = f(x) = 1/x$ assumes smaller and smaller values of the same sign. By taking a value of x sufficiently large, we can find a value of y as close to zero as we wish without the function ever actually becoming zero.

The graph of $y = f(x) = 1/x$, for values of x from $x = -1$ to the left, and from $x = +1$ to the right, is therefore indicated by the pair of heavily drawn curve-segments which lie close to the x-axis on either side of Figure 13.

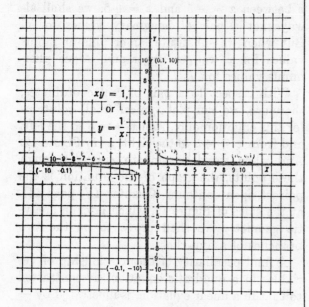

Fig. 13

But when $x = 0$, the function $y = f(0) = 1/0$ has no definite value. For, in the original equation — $xy = 1$ — there is no definite numerical value which we can assign to y so that $xy = 0(y) = 1$. This is a point to which we shall return in more detail later (Chapter VII, page 88). But in the present connection, the fact that we can assign no definite value to the function when $x = 0$, suggests that we should explore its mathematical behavior further when x has other values between -1 and $+1$, as follows:

$x =$	-1	$-.5$	$-.2$	$-.1$	0	$.1$	$.2$	$.5$	1
$y =$	-1	-2	-5	-10	?	10	5	2	1

When these values are compared with those on the previous table, we see that we can find a plus or minus value for y as large as we wish provided we take a value of x sufficiently close to zero and having the same sign. The rest of the graph for the function, $y = f(x) = 1/x$, is therefore indicated by the dotted-line curve-segments which lie close to the y-axis in Figure 13.

That the graph in Figure 13 is also a hyperbola, and better language for describing its relationship to the coordinate axes, will be explained in analytic geometry (page 159). Here, however, we are mainly concerned to note from Example 5 that either variable in a quadratic equation may also be a single-valued real function of the other for all values except one.

Practice Exercise No. 19

Solve the following for y in terms of x, and plot the resulting patterns of solutions:

(1) $x^2 + y^2 = 9$ (5) $x - 2 = y^2 + 4y + 4$
(2) $y^2 = x - 1$ (6) $(x - 1)^2 + (y - 2)^2 = 25$
(3) $x^2 - 2y^2 = 1$ (7) $3x^2 + 4y^2 = 12$
(4) $xy = 5$ (8) $x^2 - y^2 = 0$

Systems with a Linear Equation

A simultaneous system (Chapter II, page 18) may contain equations of *different degrees*. The simplest instance is **a system of one linear equation and one quadratic equation in two variables.**

To solve such a system, first substitute into the quadratic equation the value found from the linear equation for one variable in terms of the other. This substitution will produce either —

(A) A quadratic equation in one variable.
(B) A linear equation in one variable. Or,
(C) A result of the form, $0 = 0$ or $N = 0$!

In case (A), the system is determinate with a unique pair of roots. These may be found by solving the resulting quadratic equation for the two values of its variable, and then substituting these back into the linear equation to find the corresponding values of the other variable.

In this case the roots are real and unequal, real and equal, or conjugate imaginaries, depending upon whether the straight-line graph of the linear equation intersects the conic-section graph of the quadratic equation in two, one, or no points, respectively.

EXAMPLE 6: Solve, and interpret graphically, the system:

$$x^2 + y^2 = 25,$$ (Equation #6A)
$$x - y - 1 = 0.$$ (Equation #6B)

SOLUTION: From the linear equation,
$$x = y + 1.$$ (Transposing in #6B)

Hence,

$(y + 1)^2 + y^2 = 25.$	(Substitution in #6A)
$y^2 + 2y + 1 + y^2 = 25$	(Removing parentheses)
$2y^2 + 2y - 24 = 0$	(Transposing 25, etc.)
$y^2 + y - 12 = 0$	(Dividing by 2 = 2)
$(y + 4)(y - 3) = 0$	(Factoring)
$y + 4 = 0$	(Equating factor #1 to 0)
$y = -4.$	(Transposing 4)
$y = 3.$	(Ditto with factor #2)

When $y = -4$,

$$x = y + 1 = -4 + 1 = -3.$$ (Substitution)

When $y = 3$,

$$x = y + 1 = 3 + 1 = 4.$$ (Substitution)

Our *two, real, unequal roots* are therefore $x = -3$, $y = -4$, and $x = 4$, $y = 3$. This may also be written: $x,y = -3,-4$; $4,3$.

Check:

$$(-3)^2 + (-4)^2 = 9 + 16 = 25. \checkmark$$
$$-3 - (-4) - 1 = -3 + 4 - 1 = 0. \checkmark$$
(Substituting $x,y = -3,-4$)
$$4^2 + 3^2 = 16 + 9 = 25. \checkmark$$
$$4 - 3 - 1 = 0. \checkmark$$ (Substituting $x,y = 4,3$)

A further *geometric check*, and *interpretation*, of the above work is given in Figure 14. There the straight-line graph of the linear equation, $x - y - 1 = 0$, intersects the graph of $x^2 + y^2 = 25$ at the *two points*, $(-3, -4)$ and $(4, 3)$, whose coordinates are the values of the variables in the two roots of the above simultaneous solution.

When we apply the method of Example 6 to the system of equations,

$$x^2 + y^2 = 25 \quad \text{and} \quad 3x - 4y + 25 = 0,$$

we obtain the pair of *real, equal roots*,

$$x,y, = -3,4; \; -3,4.$$

A geometric check and interpretation of this result is given in Figure 14 by the fact that the straight-line graph of the linear equation, $3x - 4y + 25 = 0$, intersects the graph of $x^2 + y^2 = 25$ only at the *one point* $(-3,4)$.

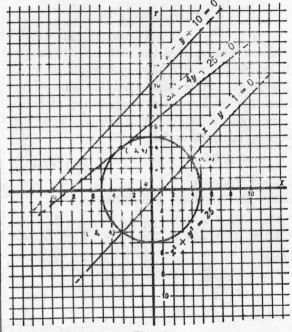

Fig. 14

Also note in Figure 14 that the straight-line graph of the linear equation, $x - y + 10 = 0$, does not intersect at all with the graph of the quadratic equation, $x^2 + y^2 = 25$. From this we may anticipate that a simultaneous treatment of the two equations by the above method will produce only *imaginary solutions*.

EXAMPLE 7: Solve simultaneously, and check:

$$x^2 + y^2 = 25,$$ (Equation #7A)
$$x - y + 10 = 0.$$ (Equation #7B)

SOLUTION: From the linear equation,
$$y = x + 10.$$ (Transposing $-y$, etc.)

Hence,

$$x^2 + (x + 10)^2 = 25 \qquad \text{(Substitution in #7A)}$$

$$x^2 + x^2 + 20x + 100 = 25 \qquad \text{(Removing parentheses)}$$

$$2x^2 + 20x + 75 = 0 \qquad \text{(Transposing 25)}$$

$$x = \frac{-20 \pm \sqrt{(20)^2 - 4(2)(75)}}{2(2)}$$

$$\text{(The quadratic formula)}$$

$$= \frac{-20 \pm \sqrt{400 - 600}}{4} = \frac{-20 \pm \sqrt{-200}}{4}$$

$$\text{(Removing parentheses)}$$

$$= \frac{-20 \pm 10\sqrt{2}i}{4} = -5 \pm 2.5\sqrt{2}i.$$

$$\text{(Extracting square roots)}$$

With recognition that the roots are imaginary, the work for many purposes would stop at this point. Since higher applications sometimes require the exact computation of imaginary roots (Chapter IV, page 43), however, we *may* continue as follows:

$$y = -5 \pm 2.5\sqrt{2}i + 10 \qquad \text{(Substitution)}$$

$$= 5 \pm 2.5\sqrt{2}i.$$

Hence the imaginary solutions are:

$$\begin{cases} x = -5 + 2.5\sqrt{2}i \\ y = 5 + 2.5\sqrt{2}i \end{cases} \text{ and } \begin{cases} x = -5 - 2.5\sqrt{2}i \\ y = 5 - 2.5\sqrt{2}i. \end{cases}$$

Check:

$$(-5 + 2.5\sqrt{2}i)^2 + (5 + 2.5\sqrt{2}i)^2$$

$$= 25 - 25\sqrt{2}i - 12.5 + 25 + 25\sqrt{2}i - 12.5$$

$$= 25 + 25 - 25 = 25. \checkmark$$

$$\text{(Substitution in #7A)}$$

$$-5 + 2.5\sqrt{2}i - (5 + 2.5\sqrt{2}i) + 10$$

$$= -5 + 2.5\sqrt{2}i - 5 - 2.5\sqrt{2}i + 10$$

$$= -10 + 10 = 0. \checkmark$$

$$\text{(Substitution in #7B)}$$

$$(-5 - 2.5\sqrt{2}i)^2 + (5 - 2.5\sqrt{2}i)^2$$

$$= 25 + 25\sqrt{2}i - 12.5 + 25 - 25\sqrt{2}i - 12.5$$

$$= 25 + 25 - 25 = 25. \checkmark$$

$$\text{(Substitution in #7A)}$$

$$-5 - 2.5\sqrt{2}i - (5 - 2.5\sqrt{2}i) + 10$$

$$= -5 - 2.5\sqrt{2}i - 5 + 2.5\sqrt{2}i + 10$$

$$= -10 + 10 = 0. \checkmark$$

$$\text{(Substitution in #7B)}$$

In case (B), a system of one linear and one quadratic equation is determinate with one pair of real values for its variables as the solution.

The relationship between the variables in the quadratic equation is then such that it is depictable by straight-line graphs which intersect the straight-line graph of the linear equation in only one point.

EXAMPLE 8: Solve, and interpret graphically, the system:

$$4y^2 - x^2 = 0, \qquad \text{(Equation #8A)}$$

$$y - \tfrac{1}{2}x - 4 = 0 \qquad \text{(Equation #8B)}$$

SOLUTION: From the linear equation,

$$y = \tfrac{1}{2}x + 4. \qquad \text{(Transposing in #8B)}$$

Hence,

$$4(\tfrac{1}{2}x + 4)^2 - x^2 = 0 \qquad \text{(Substitution in #8A)}$$

$$4(\tfrac{1}{4}x^2 + 4x + 16) - x^2 = 0 \qquad \text{(Expanding the binomial)}$$

$$x^2 + 16x + 64 - x^2 = 0 \qquad \text{(Removing parentheses)}$$

$$16x = -64 \qquad \text{(Transposing, etc.)}$$

$$x = -4 \qquad \text{(Dividing by } -4 = -4)$$

Since, in this type of case, our substitution into the quadratic equation produces a linear equation with only one solution for the first variable, we have only one further substitution to make:

$$y = \tfrac{1}{2}x + 4 = \tfrac{1}{2}(-4) + 4$$

$$\text{(Substituting } x = -4)$$

$$= -2 + 4 = 2.$$

Hence our *unique solution* for the system is

$$x, y = -4, 2.$$

Check:

$$4(2)^2 - (-4)^2 = 16 - 16 = 0. \checkmark$$

$$\text{(Substitution)}$$

$$2 - \tfrac{1}{2}(-4) - 4 = 4 - 4 = 0. \checkmark$$

As we saw in Example 1 (page 52), all the possible solutions of the equation, $4y^2 - x^2 = 0$, are depicted by the pair of intersecting straight lines, $y = \tfrac{1}{2}x$ and $y = -\tfrac{1}{2}x$, duplicated here in Figure 15. And you can see in this Figure that the straight-line graph of the linear equation, $y - \tfrac{1}{2}x - 4 = 0$, intersects the

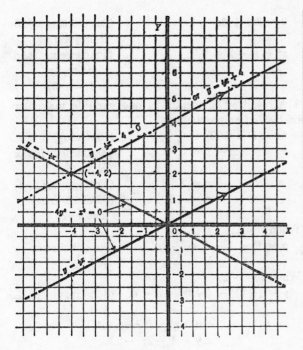

Fig. 15

latter of these in the point $(-4,2)$ with coordinates that are the values of the variables in the above solution. But the linear equation, $y - \frac{1}{2}x - 4 = 0$ — which can be transposed as $y = \frac{1}{2}x + 4$ — is *inconsistent* with the other equation, $y = \frac{1}{2}x$. Hence the corresponding graphs must be parallel (Chapter III, page 35); and $(-4,2)$ is the only point in which the two complete graphs can meet. This is a typical graphic interpretation of Case B.

In case (C), the system of one linear and one quadratic equation is indeterminate. It is *dependent* (Chapter II, page 22) with an unlimited (infinite) number of possible simultaneous solutions when the typical result, $0 = 0$, emerges from the above described substitution. Or it is *inconsistent* (Chapter II, page 20) with no possible simultaneous solutions when the typical result, $N = 0$, emerges from this substitution.

In such instances, the relationship between the variables in the quadratic equation is always such that it is depictable by straight lines which are identical to, or parallel with, the straight-line graph of the linear equation.

Suppose, for instance, that we were to try to solve as a simultaneous system:

$$4y^2 - x^2 = 0, \qquad \text{(Equation A)}$$
$$y = \frac{1}{2}x. \qquad \text{(Equation B)}$$

By the above method, we should then get:

$$4(\tfrac{1}{2}x)^2 - x^2 = 0, \qquad \text{(Substitution from B in A)}$$
$$x^2 - x^2 = 0, \qquad \text{(Removing parentheses)}$$
$$0 = 0! \qquad \text{(Subtraction)}$$

This *trivial result* is to be expected (Chapter II, page 22), since we saw in Example 1 that one set of solutions for the above quadratic equation is given by the linear equation, $y = \frac{1}{2}x$ (page 52). However, the system still has an unlimited (infinite) number of possible solutions, represented in Figures 11 and 15 by the graph of $y = \frac{1}{2}x$.

Or, suppose that we were to try to solve as a simultaneous system:

$$4y^2 + x^2 = 4xy \qquad \text{(Equation C)}$$
$$y = \frac{1}{2}x + 4 \qquad \text{(Equation D)}$$

By the same method, we should then get:

$$4(\tfrac{1}{2}x + 4)^2 + x^2 = 4x(\tfrac{1}{2}x + 4) \;\text{(Substitution from D in C)}$$
$$x^2 + 16x + 64 + x^2 = 2x^2 + 16x \;\text{(Removing parentheses)}$$
$$64 = 0! \qquad \text{(Transposing, etc.)}$$

This *self-contradictory result* is to be expected (Chapter II, page 20), since we saw in Example 2 above (page 53) that both sets of possible solutions for the above quadratic equation are identical with those of the linear equation, $y = \frac{1}{2}x$, which is inconsistent with the linear equation, $y = \frac{1}{2}x + 4$, of the system. Figure 15 also illustrates this situation geometrically by the fact that its graph-lines, $y = \frac{1}{2}x$ and $y = \frac{1}{2}x + 4$, are parallel and cannot meet in a point whose coordinates would be the values of the variables in a solution for the above simultaneous system of equations.

Practice Exercise No. 20

Solve the following systems of equations simultaneously, and interpret (3) graphically:

(1) $x + y = 2$ (2) $x^2 + y^2 = 9$
      $y^2 = x - 1$ $y = 3x + 8$

(3) $3x^2 + 4y^2 = 12$
$x - y = 2$

(5) $(x - 1)^2 + (y - 2)^2 = 9$
$x - 1 = y$

(4) $y^2 = x - 2$
$y = 2x + 1$

Systems Quadratic in Only One Variable

The general observations which we have made, in considering systems with one linear equation at some length above, will now enable us to cover other kinds of systems involving quadratics more briefly.

A system is said to be quadratic in only one variable if all equations of the system contain quadratic terms only in the same variable (definition, page 38). For instance, the following system is *quadratic only in x* because it contains x^2-terms but no y^2-terms and no xy-terms:

$y = x^2 - x - 2$,   (Equation #9A)
$y = -2x^2 - 4x + 4$.   (Equation #9B)

To solve such a system, first eliminate the variable in which the system is not quadratic. That may be done by comparison, as for linear equations (Chapter II, page 18). And, as in the case of a system with one linear equation, this first operation will produce either —

(A) A quadratic equation in one variable.
(B) A linear equation in one variable. Or,
(C) A result of the form, $0 = 0$ or $N = 0$!

This method of completing the solution, and the results to be anticipated in each case, are also exactly the same as for systems with a linear equation (pages 56–59 above). You will find by plotting, however, that *the graphs of a system quadratic only in one variable are always parabolas with their axes along, or parallel to, the coordinate axis of the variable in which the system is* not *quadratic*. The several kinds of results which emerge must therefore be interpreted geometrically in terms of the various possibilities of intersection, or non-intersection, between parabola-shaped graphs.

EXAMPLE 9: Solve, and interpret graphically, the above system.

SOLUTION: First we eliminate y by comparison:

$0 = 3x^2 + 3x - 6$ (Subtracting #9B from #9A)

$x^2 + x - 2 = 0$ (Transposing and dividing by $-3 = -3$)

The resulting quadratic in one variable may now be treated as previously:

$(x - 1)(x + 2) = 0$,   (Factoring)
$x - 1 = 0$,   (Equating factor #1 to 0)
$x = 1$.   (Transposing -1)
$x = -2$.   (Similarly from factor #2)

When $x = 1$,
$y = 1^2 - 1 - 2 = -2$.   (Substitution in #9A)

When $x = -2$,
$y = (-2)^2 - (-2) - 2 = 4$.   (Substitution in #9A)

Hence the *two, real, unequal roots* are
$$x, y = 1, -2; \quad -2, 4.$$

Check:
$-2 = 1^2 - 1 - 2 = -2$. ✓
$-2 = -2(1) - 4(1) + 4 = -2$. ✓
(Substituting $x, y = 1, -2$)
$4 = (-2)^2 - (-2) - 2 = 4$. ✓
$4 = -2(-2)^2 - 4(-2) + 4 = 4$. ✓
(Substituting $x, y = -2, 4$)

The above equations are plotted and graphed in Figure 16 by the method of Chapter IV (Example 4G, page 46). As can be seen in the figure, the *parabola-graphs* intersect only in the *two points* $(-2, 4)$ and $(1, -2)$, whose co-ordinates are the values of the variables in the two roots of the above simultaneous solution.

By comparing the steps, results, and diagrams, of Examples 6 (page 57) and 9, you may note the similarity between the corresponding cases in the two types of systems notwithstanding the difference in the form of their graphic depiction.

Suppose, moreover, that the system in Example 9 had consisted of the equations,

$y = x^2 - x - 2$,   (Equation #9A)
$y = x^2 - x + 8$.   (Equation #9C)

By the same method, we should then get:

$0 = -10!$ (Subtracting #9C from #9A)

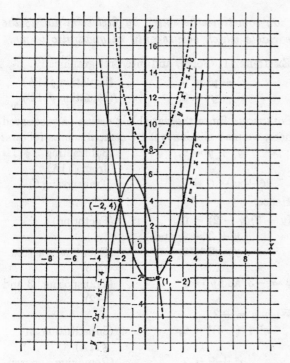

Fig. 16

This self-contradictory result is to be expected since the equations are inconsistent (page 20) in that they differ only in their constant terms, -2 and 8, respectively. Note also in Figure 16, that the (dotted-line) graph of $y = x^2 - x + 8$ therefore has the same shape, axis, and direction as that of $y = x^2 - 2x - 2$, but lies so that its ordinates are greater by the same amount — 10 units — at every point. In this sense, the graphs are "parallel parabolas" corresponding to the parallel-line graphs of inconsistent equations previously considered (Chapter III, page 35, and Chapter V, page 54 above).

Other correspondences between systems quadratic in one variable and systems with one linear equation may be recognized in the examples of —

Practice Exercise No. 21

Solve simultaneously.

(1) $3x^2 + 4y^2 = 12$
    $x^2 + y^2 = 25$

(2) $x^2 - 2x + y^2 = 21$
    $y^2 = x - 1$

(3) $x^2 + y^2 = 9$
    $(x - 1)^2 + y^2 = 9$

(4) $y^2 = x$
    $x^2 = y$

(5) $y = x^2$
    $y = -x^2 + 1$

(6) $xy = 2$
    $x^2 + y^2 = 9$

Other Systems Treatable by Comparison

Systems quadratic in one variable may be solved without recourse to higher-degree equations, as above, only because *we can first eliminate one variable from the system by the method of comparison. The same method may also be extended to many other types of systems involving quadratics.*

Systems Linear in Quadratic Terms

A simple instance is *a system of quadratics which contains only constant terms and terms in which the variables are squared.* Such a system is sometimes said to be linear in quadratic terms. This is because it may first be treated as a linear system in the squares of the variables, rather than as a system in the original variables themselves.

EXAMPLE 10: Solve, and interpret graphically:

$x^2 + y^2 = 13$,   (Equation #10A)
$2x^2 + 3y^2 = 30$.   (Equation #10B)

SOLUTION: Regarding the system as though it were a pair of linear equations in x^2 and y^2, we may first solve it for either x^2 or y^2 by the method of comparison. For instance:

$3x^2 + 3y^2 = 39$,   (Multiplying #10A by
                                                3 = 3)
$x^2 = 9$.   (Subtracting #10B from
                                                the above)

But now, treating x^2 as a quadratic term:

$x = \pm 3$.   (Extracting square
                                                roots)

And,

$(\pm 3)^2 + y^2 = 13$,   (Substitution in #10A)
$y^2 = 13 - 9 = 4$,   (Removing parentheses, etc.)
$y = \pm 2$.   (Extracting square roots)

The *four, unequal, real roots* of the system are therefore $x;y = 3,2; \; -3,2; \; 3,-2; \; -3,-2$.

Check:

$(\pm 3)^2 + (\pm 2)^2 = 9 + 4 = 13. \checkmark$ (Substitution in #10A)

$2(\pm3)^2 + 3(\pm2)^2 = 18 + 12 = 30. \checkmark$ (Substitution in #10B)

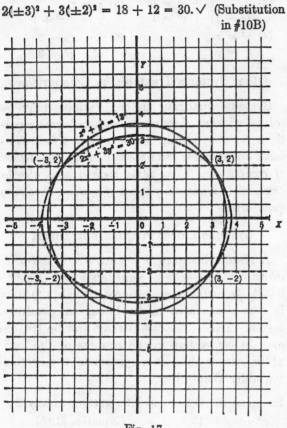

Fig. 17

Graphs of the equations, as plotted by the method of Example 4 (page 54), are shown in Figure 17. Later in analytic geometry we shall see that these curves are a circle and an ellipse respectively. But the main interest of the diagram now is the fact that the graphs intersect in the four points, (3,2,) (−3,2), etc., whose coordinates are the values of the variables in the roots of the above simultaneous system.

The diagram of Figure 17 also illustrates the general fact that a system of two quadratic equations in two variables may have four, unequal, real roots, when the relationships between the variables in its equations are such that the conic section graphs intersect in four distinct points.

Systems with Only One Linear Term

When only one linear term occurs in a pair of quadratic equations in two variables, the system may sometimes be solved by first eliminating the variable which occurs only in quadratic terms.

EXAMPLE 11: Solve, and interpret graphically:

$$x^2 + y^2 = 9, \qquad \text{(Equation #11A)}$$
$$x^2 - y = 3. \qquad \text{(Equation #11B)}$$

SOLUTION: Since the variable x occurs only squared here, we may eliminate it by comparison as follows:

$$y^2 + y = 6, \qquad \text{(Subtracting #11B from #11A)}$$
$$y^2 + y - 6 = 0. \qquad \text{(Transposing 6)}$$

Hence,

$$(y + 3)(y - 2) = 0, \qquad \text{(Factoring)}$$
$$y + 3 = 0, \text{ and}$$
$$y - 2 = 0. \qquad \text{(Equating factors to 0)}$$
$$y = -3, \text{ and } y = 2. \qquad \text{(Transposing 3 and } -2)$$

When $y = -3$,

$$x^2 + (-3)^2 = 9, \qquad \text{(Substitution in #11A)}$$
$$x^2 + 9 = 9, \qquad \text{(Removing parentheses)}$$
$$x = 0. \qquad \text{(Transposing 9)}$$

When $y = 2$,

$$x^2 + 2^2 = 9, \qquad \text{(Substitution in #11A)}$$
$$x^2 = 9 - 4 = 5, \qquad \text{(Transposing, etc.)}$$
$$x = \pm\sqrt{5}. \qquad \text{(Extracting square roots)}$$

The *four real roots* — one pair equal, and one pair unequal — are therefore:

$$x,y = 0,-3; \ 0,-3; \ \sqrt{5},2; \ -\sqrt{5},2.$$

Check:

$$0^2 + (-3)^2 = 9. \checkmark$$
$$0^2 - (-3) = 3. \checkmark \qquad \text{(Substituting } x,y = 0,-3)$$
$$(\pm\sqrt{5})^2 + 2^2 = 5 + 4 = 9. \checkmark$$
$$(\pm\sqrt{5})^2 - 2 = 5 - 2 = 3. \checkmark$$
$$\text{(Substituting } x,y = \pm\sqrt{5},2)$$

And in Figure 18 the graphs intersect in the points whose coordinates are the values of the variables in these roots.

Note in Figure 18 that the point $(0, -3)$ corresponds to the system's pair of equal roots. In this detail, Figure 18 illustrates the general fact that a system of equations involving

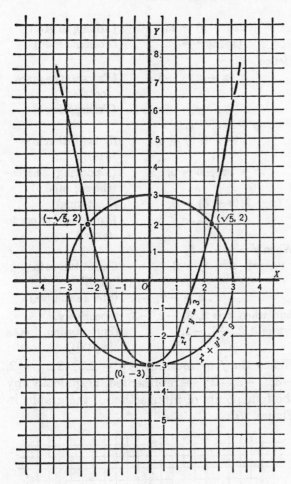

Fig. 18

quadratics has a pair of real, equal roots wherever the relationships between the variables in its equations are such that the graphs "touch in a single point without crossing."

Whenever it is possible to obtain a linear equation by the comparison of two quadratic equations, the system has only two possible solutions, if it is determinate. Then the graphs are of a type which can intersect at most in two distinct points (unless they coincide throughout, in which case the system is dependent and therefore indeterminate).

EXAMPLE 12: Solve, and interpret graphically, the system:

$$x^2 + y^2 = 25, \qquad \text{(Equation #12A)}$$
$$x^2 + y^2 - 24x = -119. \qquad \text{(Equation #12B)}$$

SOLUTION: By comparison we obtain the linear equation:

$$24x = 144, \qquad \text{(Subtracting #12B from #12A)}$$
$$x = 6. \qquad \text{(Dividing by } 24 = 24)$$

Hence,
$$6^2 + y^2 = 25, \qquad \text{(Substitution in #12A)}$$
$$y^2 = 25 - 36 = -11, \qquad \text{(Transposing, etc.)}$$
$$y = \pm\sqrt{-11} \qquad \text{(Extracting square roots, etc.)}$$
$$= \pm\sqrt{11}\,i = \pm 3.3i, \text{ approximately.}$$

And the *two imaginary roots* are: $x, y = 6, \pm\sqrt{11}i$.

Check:

$$6^2 + (\pm\sqrt{11}i)^2 = 36 - 11 = 25 \checkmark \quad \text{(Substitution in #12A)}$$
$$6^2 + (\pm\sqrt{11}i)^2 - 24(6) = 36 - 11 - 144 = -119. \checkmark \quad \text{(Substitution in #12B)}$$

The reason why the roots are imaginary in this case is shown in Figure 19 where you can see that y is a real function of x in equation 12A only for values of x from $x = -5$ to $x = +5$, and in equation 12B only for values of x from $x = 7$ to $x = 17$.

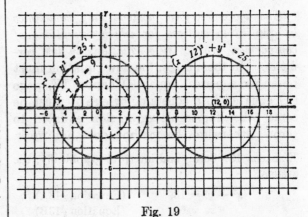

Fig. 19

In the above-mentioned detail, Figure 19 illustrates the general fact that a system of equations has a pair of imaginary roots whenever the relationships between its variables are such that the graphs fail to intersect in a possible point of intersection.

Suppose, however, that the equations in Example 12 had been,

$$x^2 + y^2 = 25, \qquad \text{(Equation \#12A)}$$
$$x^2 + y^2 = 9. \qquad \text{(Equation \#12C)}$$

The same method would then give us,

$$0 = 16! \qquad \text{(Subtracting \#12C from \#12A)}$$

This self-contradictory result is to be expected, since the equations differ only in their constant terms and are therefore inconsistent (page 20). Note, moreover, that the graphs of these equations in Figure 19 are concentric circles of different radii (*MMS*, Chapter XIV). This again is the relationship, between conic sections, equivalent to parallelism between straight lines (compare with "parallel parabolas," page 61 above).

Practice Exercise No. 22

Solve simultaneously.

(1) $4x^2 + 25y^2 = 100$ (3) $\qquad xy = 4$
      $x^2 + y^2 = 9$ $x^2 + y^2 = 2$

(2) $9x^2 - 16y^2 = 144$ (4) $\qquad y = 4x^2$
      $4x^2 + 25y^2 = 100$ $4x^2 + 25y^2 = 100$

Systems of Other Special Types

The above-described methods may sometimes be applied to systems for which operations of comparison produce new equations in two variables. In these cases, the new equation must be either a perfect square (Chapter IV, page 41) equated to a constant, or else it must be some other factorable expression equated to zero. Otherwise, the method does not work.

EXAMPLE 13: Solve, and interpret graphically, the system:

$$x^2 + y^2 = 13, \qquad \text{(Equation \#13A)}$$
$$xy = 6. \qquad \text{(Equation \#13B)}$$

SOLUTION: Foreseeing that we can combine the terms on the left sides of these equations to form a perfect square, we proceed by comparison as follows:

$$2xy = 12, \qquad \text{(Multiplying \#13B by } 2 = 2)$$

$$x^2 + 2xy + y^2 = 25, \qquad \text{(Adding to \#13A)}$$
$$(x + y)^2 = 25. \qquad \text{(Formula R 30)}$$

Hence,

$$x + y = \pm 5, \qquad \text{(Extracting square roots)}$$
$$y = -x \pm 5. \qquad \text{(Transposing } x)$$

When $y = -x + 5$,
$$x(-x + 5) = 6, \qquad \text{(Substitution in \#13B)}$$
$$x^2 - 5x + 6 = 0, \qquad \text{(Removing parentheses, etc.)}$$
$$(x - 2)(x - 3) = 0, \qquad \text{(Factoring)}$$
$$x = 2, 3. \qquad \text{(Equating factors to 0, etc.)}$$

When $y = -x - 5$,
$$x(-x - 5) = 6, \qquad \text{(Substitution in \#13B)}$$
$$x = -2, -3. \qquad \text{(Steps as before)}$$

And when $x = 2, 3, -2, -3$, respectively:
$$y = 3, 2, -3, -2. \qquad \text{(Substitution, etc., in \#13B)}$$

Hence the *four roots* are $x, y = 2, 3$; $3, 2$; $-2, -3$; $-3, -2$; — represented graphically by the four points of intersection of the graphs in Figure 20.

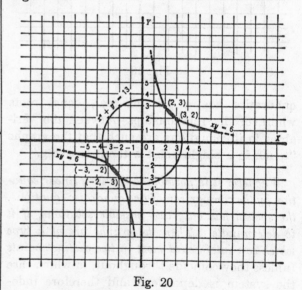

Fig. 20

Practice Exercise No. 23

Solve simultaneously.

(1) Do Example 13 by subtraction.

(2) $x^2 - xy + y^2 = 1$
           $xy = 1$

(3) $x^2 + xy = 2$
$xy + y^2 = 7$

(4) $x^2 + 4y^2 = 7$
$x + 2y = 3$

(5) $x^2 + 3y^2 = 31$
$7x^2 - 2y^2 = 10$

(6) $x^2 + y^2 - 3x + 2y - 39 = 0$
$3x^2 - 17xy + 10y^2 = 0$

(7) $x^3 - y^3 = 63$
$x - y = 3$

Hint: First divide each side of equation #1 by the corresponding side of equation #2.

Systems Treatable by Substitution

Some systems of quadratic equations may also be solved without recourse to higher degree equations if we first make certain special substitutions for the variables.

A simple case in two variables is a system of two **pure quadratic equations** — defined as: *containing only quadratic and constant terms* (page 38). Such a system may always be solved by a substitution of the form, $y = kx$, for one of the variables.

For instance, the equations of Example 13 are pure quadratics in which this substitution gives us the new pair of equations in k and x:

$x^2 + (kx)^2 = x^2 + k^2x^2$ (Substituting $y = kx$)

$= (k^2 + 1)x^2 = 13$ (New equation #13A′)

and

$x(kx) = kx^2 = 6.$ (New equation #13B′)

From these we may eliminate x to solve for k, as follows:

$\dfrac{k^2 + 1}{k} = \dfrac{13}{6},$ (Dividing #13A′ by 13B′)

$6k^2 + 6 = 13k,$ (Multiplying by $6k - 6k$)

$6k^2 - 13k + 6 = 0,$ (Transposing $13k$)

$k = \dfrac{-(-13) \pm \sqrt{(-13)^2 - 4(6)(6)}}{2(6)},$ (The quadratic formula)

$= \dfrac{13 \pm 5}{12} = \dfrac{3}{2}, \dfrac{2}{3}.$ (Removing parentheses, etc.)

But when $k = \dfrac{3}{2}$,

  (Substitution in $y - kx$)

$y = \dfrac{3}{2}x,$

$x\left(\dfrac{3}{2}x\right) = \dfrac{3x^2}{2} = 6,$ (Substitution in #13B)

$x^2 = (\tfrac{2}{3})6 = 4,$ (Multiplying by $\tfrac{2}{3} - \tfrac{2}{3}$)

$x = \pm 2.$ (Extracting square roots)

Similarly, when $k = \tfrac{2}{3}$,

$x = \pm 3.$ (Steps as above)

And from this point the same solution follows as before in Example 13.

Direct substitution from one quadratic equation into another may also, in special cases, lead to a fourth degree equation which is solvable by quadratic means (Chapter IV, page 44).

In the system of Example 13 again, for instance, we may find y in terms of x from equation 13B, as follows:

$y = \dfrac{6}{x}.$ (Dividing by $x = x$)

Hence,

$x^2 + \left(\dfrac{6}{x}\right)^2 = 13,$ (Substitution in #13A)

$x^2 + \dfrac{36}{x^2} = 13,$ (Removing parentheses)

$x^4 + 36 = 13x^2,$ (Multiplying by $x^2 = x^2$)

$x^4 - 13x^2 + 36 = 0.$ (Transposing $13x^2$)

But this fourth-degree equation may be treated as a quadratic in x^2 (reference above) as follows:

$(x^2 - 4)(x^2 - 9) = 0$ (Factoring)

$x^2 = 4, 9.$ (Equating factors to 0, etc.)

$x = \pm 2, \pm 3.$ (Extracting square roots)

And again the solution may be completed as before in Example 13.

In general, however, *direct substitution from one quadratic equation into another leads to a fourth-degree equation of a more complete type* to be discussed in the sequel to this volume, *Advanced Algebra and Calculus Made Simple.*

Practice Exercise No. 24

(1) $\dfrac{36}{x^2} + \dfrac{1}{y^2} = 18.$ $\dfrac{1}{y^2} - \dfrac{4}{x^2} = 3.$

(2) $2x^2 + y^2 = 6$
　　 $x^2 + 1 = y$

(3) $x^2 = y + 6$
　　 $y^2 = x + 6$

(4) $y^2 = x$
　　 $x^2 = y$

(5) $x^2 + y^2 = 8$
　　 $xy = -4$

Summary

In this chapter we have extended the definitions and principles of Chapters II and III to the new material of Chapter IV with the following major findings . . .

Certain types of **systems involving quadratic equations in two variables** may be *solved by a combination of methods* for solving quadratic equations in one variable (extracting square roots, etc.) and for solving systems of linear equations in two variables (comparison, etc.).

Applied to a defective system consisting of one quadratic equation in two variables, these methods reveal that either variable in such an equation may be a **single-valued function** of the other for all, or for all but one, real value of the latter. However, either variable may also be a **two-valued function** of the other for all real values of the latter. And either may be a **real function** of the other for all, some, one, or no real value(s) of the latter.

Consequently, whereas the relationship between two variables in a linear equation may be depicted graphically by a straight line (Chapter III, page 33), **the relationship between two variables in a quadratic equation** is much more complicated. In different instances it may be represented with respect to a pair of (real) coordinate axes by *a pair of straight lines* (coinciding, in special cases), or by a *parabola*, a *circle*, a *single point*, an *ellipse*, a *hyperbola*, or *no graph at all* (most of these being figures we shall study more precisely, later in analytic geometry, from a geometric point of view).

A determinate system of one quadratic and one linear equation in two variables may therefore yield *two unequal real roots, two equal real roots*, or *two conjugate imaginary roots* — depending upon the various possibilities of intersection, or non-intersection, between a straight-line graph of a linear equation and a conic-section graph of a quadratic equation, illustrated above.

Also, **a determinate system on two quadratic equations in two variables** may yield **two or four roots** in pairs which are either *real and unequal, real and equal*, or *conjugate imaginaries* — depending upon the various possibilities of intersection, or non-intersection of two conic-section graphs, also illustrated above.

However, attempts by the same methods to solve **an indeterminate system involving quadratics** produce the typical result $0 = 0$ when the system is *dependent*, or the typical result $0 = N$ when the system is *inconsistent*. In the former case the graphs coincide. In the latter case the graphs are parallel lines or similarly related figures like concentric circles of different radii, etc.

When methods like those described do not apply, attempts to solve a system involving quadratics lead to higher-degree equations.

Note on Sequence of Study

Readers who have taken the suggested option of beginning Chapters IX and X on analytic geometry before studying the two preceding chapters (page 36) may now return to Chapter XI on Conic Sections in analytic geometry. There they will find typical practical applications of quadratic equations and of systems involving quadratics.

Readers who have not exercised such a choice may also do so now, of course, beginning with Chapter IX and following through to Chapter XIII.

However, if you have not yet departed from the printed order of these chapters, you will also find advantages in continuing to follow the sequence of the text. For, upon completing Chapters VI, VII, and VIII, you should be in a position to study all of Section Three on analytic geometry without referring back for further points of algebraic technique.

CHAPTER VI

LINEAR EQUATIONS IN N VARIABLES

Systems in N Variables

In Chapter II we considered systems of linear equations in only two variables. The methods and findings of that chapter are generalized here for systems of linear equations in n variables, where n is any positive whole-number, 1, 2, 3, 4, etc.

For instance, the elementary methods of comparison and substitution are generalized for solving a determinate system of n linear equations in n variables as follows:

Step One: Compare one equation with all others in the system to obtain n-1 equations in n-1 variables. Or, if it is more convenient, transpose one equation in the system to find the value of one variable in terms of others, and substitute this value in the remaining equations to obtain the same result as above.

Step Two: Repeat this operation to obtain n-2 equations in n-2 variables, n-3 equations in n-3 variables, etc., until you arrive at one equation which gives the solution value of one variable.

Step Three: Substitute solution-values of variables in the above equations in reverse order until you have definite solution-values for all the variables.

Step Four: Check the resulting solution by substitution in the original equations.

EXAMPLE 1: Solve for x, y, z, w, the system,

$$3x + 3y - 2z - w = 40,$$
$$3y - z - w + 2x = 35,$$
$$4z - 2w + 3x + y = 15,$$
$$w - x + y + z = 35.$$

SOLUTION: For convenient comparison, we first rearrange the terms of the equations so that the variables — x, y, z, w, — appear in the same order:

$$3x + 3y - 2z - w = 40, \quad \text{(Equation A)}$$
$$2x + 3y - z - w = 35, \quad \text{(Equation B)}$$

$$3x + y + 4z - 2w = 15, \quad \text{(Equation C)}$$
$$-x + y + z + w = 35. \quad \text{(Equation D)}$$

Step One: Now we note that equation D may conveniently be compared with the others to eliminate the variable, w, as follows:

$$2x + 4y - z = 75 \quad \text{(Adding A and D, obtaining new equation \#1)}$$

$$x + 4y = 70 \quad \text{(Adding B and D, obtaining new equation \#2)}$$

$$x + 3y + 6z = 85 \quad \text{(Adding C and 2-times D, obtaining new equation \#3)}$$

Step Two: Here we have one-fewer equations in one-fewer variables — namely, the three equations, #1, #2, #3, in the three variables, x, y, z. Of these equations, #2 has only the two variables, x and y. Hence this time we may more conveniently use #2 to eliminate another variable by substitution as follows:

$$x = 70 - 4y \quad \text{(Transposing } 4y \text{ in \#2)}$$

$$2(70 - 4y) + 4y - z = 75 \quad \text{(Substitution in \#1)}$$

$$140 - 8y + 4y - z = 75 \quad \text{(Removing parentheses)}$$

$$-4y - z = -65 \quad \text{(Transposing, etc., obtaining new equation, \#4)}$$

$$70 - 4y + 3y + 6z = 85 \quad \text{(Substitution in \#3)}$$

$$-y + 6z = 15 \quad \text{(Transposing, etc., obtaining new equation, \#5)}$$

Again we have one-fewer equations in one-fewer variables — namely, the two equations, #4 and #5, in the two variables, y and z, with the solution:

$$y = 15, \quad z = 5. \quad \text{(As in Chapter II)}$$

Step Three: Now we may find the corresponding solutions of x and w by substitutions in above equations as follows:

$$x = 70 - 4(15) = 70 - 60 = 10$$

$$\text{(Substitution for } y \text{ in the transposed form of \#2 above)}$$

$$-10 + 15 + 5 + w = 35$$

$$\text{(Substitution for } x, y, z, \text{ in equation D)}$$

$$w = 25. \quad \text{(Transposing, etc.)}$$

Answer: $x = 10$, $y = 15$, $z = 5$, $w = 25$; or, as this may be more conveniently written:

$$x, y, z, w = 10, 15, 5, 25.$$

Step Four, Check:

$3(10) + 3(15) - 2(5) - 25 =$
$30 + 45 - 10 - 25 = 40.$ ✓   (Substitution in A)

$2(10) + 3(15) - 5 - 25 =$
$20 + 45 - 5 - 25 = 35.$ ✓   (Substitution in B)

$3(10) + 15 + 4(5) - 2(25) =$
$30 + 15 + 20 - 50 = 15.$ ✓   (Substitution in C)

$-10 + 15 + 5 + 25 = 35.$ ✓   (Substitution in D)

By the method of Chapter II (page 19), it may also be shown that $x, y, z, w = 10, 15, 5, 25$ is the *unique solution* of the above system of equations. Example 1 is therefore an illustration of the general principle that —

A determinate system of n linear equations in n variables has a unique solution consisting of a single set of definite values for its variables.

We can see, moreover, from the steps by which Example 1 is solved that a **defective system** of less than n linear equations in n variables (page 23) is indeterminate in that it has an unlimited (infinite) number of solutions, if any (Example B, Practice Exercise No. 25 below).

From the same steps we can also see that a **redundant system** of more than n linear equations in n variables (page 24) is, in general, indeterminate in that it has no possible solution unless, by coincidence, each sub-set of n equations happens to have the same sub-solution (Problem D, Practice Exercise No. 25 below).

This means that, to be determinate, a verbally stated problem leading to linear equations in n variables must specify n, and in general only n, conditions.

EXAMPLE 1A: When two motors of type x and one motor of type y are working against one motor of type z, the net horsepower generated is 10. When one motor of type x and three motors of type y are working against one motor of type z, the net horsepower generated is 30. But when three motors of type x and three motors of type y are working against

two motors of type z, the net horsepower generated is 25. What is the "horse" of each type motor?

SOLUTION: The three above conditions may be expressed in equations as —

$2x + y - z = 10$		(Equation #1A)
$x + 3y - z = 30$		(Equation #2A)
$3x + 3y - 2z = 25$		(Equation #3A)

Step One:

$4x + 2y - 2z = 20$ (Multiplying #1A by 2, obtaining equation #1A')

$-x + y = 5$ (Subtracting #1A' from #3A, obtaining equation #4A)

$2x + 6y - 2z = 60$ (Multiplying #2A by 2, obtaining equation #2A')

$x - 3y = -35$ (Subtracting #2 from #3, obtaining equation #5A)

Step Two:

$-2y = -30$ (Adding #4A and #5A)

$y = 15$ (Dividing by $-2 = -2$)

Step Three:

$x - 3(15) = -35$ (Substituting in #5A)

$x = -35 + 45 = 10$ (Transposing, etc.)

$2(10) + 15 - z = 10$ (Substituting in #1A)

$-z = -25$ (Transposing, etc.)

$z = 25$ (Multiplying by $-1 = -1$)

Answer: $x, y, z = 10, 15, 25.$

Step Four, Check:

$2(10) + 15 - 25 = 35 - 25 = 10.$ ✓
(Substitution in #1A)

$10 + 3(15) - 25 = 55 - 25 = 30.$ ✓
(Substitution in #2A)

$3(10) + 3(15) - 2(25) = 75 - 50 = 25.$ ✓
(Substitution in #3A)

Thus we see that Example 1A, specifying 3 linear conditions in 3 variables, is determinate in a unique set of definite values for its variables. If this problem had stated fewer than 3 conditions, however, step two of its solution would have arrived at a single linear equation in two or three variables which would have given us an unlimited (infinite) number of solutions, if any. Or, if this problem had stated more than 3 conditions, each sub-set of three equations would, in general, have had a different sub-solution, if any, by the same steps.

Practice Exercise No. 25

A. Show that the solution of Example 1 above is unique. *Hint:* Suppose any other possible solution to be x, y, z, $w = 10 + j$, $15 + k$, $5 + m$, $25 + n$, etc., as on page 19.

B. Excluding equation #4 from the system in Example 1, find x, y, z in terms of w in the remaining three equations, and find the values of x, y, z when $w = 0$, $w = -1$, $w = 5$. *Hint:* In each equation transpose the w-term to the right-hand side of the equality sign and treat it as part of the constant term.

C. Solve the following:

(1) $\quad 2x - 3y - z - w = 1$
$\quad\quad 4x + 3w + 2z = 5$
$\quad -6y + 2x + 4w + 3z = -4$
$\quad\quad 3x + 6y + w + z = 3$

(2) $\quad x - y + z - w = 13$
$\quad\quad 2x + y + 2z + w = 1$
$\quad -x + y + 3z + w = -2$
$\quad\quad 3x + 2y + 5z + 3w = 5$

D. Three motors acting together had a total horsepower of 6. Two of them exactly counterbalance the third. Is there only one answer to the horsepower ratings? What if it is known that each had an integer for its horsepower rating?

Linear Dependence and Independence

If, in a system of n consistent linear equations in n variables, any sub-pair of equations is **dependent** (page 22), steps one and two of the method by which Example 1 is solved will produce a pair of two equivalent linear equations in two variables and the typical algebraic result of an attempt to solve two such equations simultaneously — $0 = 0$ (Chapter II, page 22). In such a case, we can find the values of $n - 1$ of the variables in terms of the remaining variables, but the system is indeterminate in that it has an unlimited (infinite) number of possible solutions.

In this respect, a system of three or more linear equations in the same number of variables is like a system of two linear equations in two variables.

However, a complication arises in the case of systems of linear equations in three or more variables. Even though each sub-pair of equations is separately independent, the system as a whole may be dependent in a different way than that so far defined.

EXAMPLE 2: Solve simultaneously,

$\quad 2x + 4y - z = 75 \quad$ (Equation #1)
$\quad x + 4y \quad = 70 \quad$ (Equation #2)
$\quad 3x + 4y - 2z = 80 \quad$ (Equation #3d)

SOLUTION: Equations #1 and #2 are here the same as in Example 1, and equation #3d is of the same general form as equation #3 in Example 1. It is also clear by inspection that no sub-pair of these equations is equivalent. We proceed, therefore, as in the preceding case to *Step One·*

$x = 70 - 4y$	(Transposing $4y$ in #2)
$2(70 - 4y) + 4y - z = 75$	(Substitution in #1)
$140 - 8y + 4y - z = 75$	(Removing parentheses)
$- 4y - z = -65$	(Transposing, etc., obtaining new equation, #4)
$3(70 - 4y) + 4y - 2z = 80$	(Substitution in #3d)
$210 - 12y + 4y - 2z = 80$	(Removing parentheses)
$- 8y - 2z = -130$	(Transposing, etc., obtaining new equation, #4d)

For *step two*, however, we now find that equations #4 and #4d are dependent (equivalent), since we can obtain #4d from #4 by multiplying by 2 = 2. And, as we already know from Chapter II (page 22), the attempt to solve these by comparison or substitution results only in the truism, $0 = 0$. All we can do, therefore, is to solve the system for any two variables in terms of the third. For instance —

$-4y = z - 65$	(Transposing z in #4)
$y = (65 - z)/4.$	(Dividing by $-4 = -4$)

And,

$x = 70 - 4(65 - z)/4$ (Substitution in #2 transposed as above)

$= 70 - 65 + z = 5 + z$ (Removing parentheses, etc.)

Hence the system has an unlimited (infinite) number of possible solutions corresponding to the unlimited (infinite) number of possible values which we may assign to z in the solution-equations,

$$x = 5 + z \text{ and } y = (65 - z)/4.$$

Why, now, does the system in Example 2 respond to standard solution methods like a dependent system of two linear equations in two variables, even though each sub-pair of its equations is separately independent?

For an answer to this sometimes puzzling question, observe that the equations in Example 2, unlike those in Example 1, are such that we may arrive at any one of them by appropriate operations of comparison upon the other two. For instance —

$4x + 8y - 2z = 150$ (Multiplying #1 by $2 = 2$)

$- x - 4y = -70$ (Multiplying #2 by $-1 = -1$)

_____

$3x + 4y - 2z = 80$ (Equation #3d obtained by adding corresponding sides of the above)

Three or more equations may be related in this way even though each sub-pair is separately independent. And when three or more such equations occur in a system of n linear equations in n variables, steps one and two of the above illustrated method of solution always result eventually in a pair of dependent (equivalent) equations in fewer variables (as in Example 2).

Hence the new *definitions: Three or more equations are said to be* **linearly dependent** *if it is possible to arrive at any one of these equations by comparisons of the remaining equations with each other* (as illustrated above). When they are *not linearly dependent*, a set of equations is said to be **linearly independent**.

Thus the system of equations in Example 2

above is linearly dependent, but the system of equations in Example 1 is linearly independent. And these cases illustrate the general principle that **to be determinate, a system of n linear equations in n variables must be linearly independent.**

From this it follows that **to be determinate, a verbally stated problem leading to a system of n linear equations in n variables must specify n linearly independent conditions.**

For instance, the problem in Example 1A above states 3 linearly independent conditions in 3 variables, and has a unique solution. But consider also —

EXAMPLE 2A: Instead of the third condition in Example 1A, substitute the new condition: When 3 motors of type x and 4 motors of type y are working against 2 motors of type z, the net horsepower generated is 40. Again find the "horse" of each type motor.

SOLUTION: This time the conditions of the problem are expressed by the equations:

$2x + y - z = 10$ (Equation #1A)

$x + 3y - z = 30$ (Equation #2A)

$3x + 4y - 2z = 40$ (Equation #3e)

Step One:

$4x + 2y - 2z = 20$ (Multiplying #1A by $2 = 2$, obtaining equation #1A′)

$x - 2y = -20$ (Subtracting #3e from #1A′, obtaining new equation #4A)

$2x + 6y - 2z = 60$ (Multiplying #2A by $2 = 2$, obtaining equation #2A′)

$- x + 2y = 20$ (Subtracting #3e from #2A′, obtaining equation #4e)

Step Two: Now we have the two equivalent equations, #4A and #4e. By comparing these we can only get the truistic result,

$0 = 0!$ (Adding #4A and #4e)

However, we can solve either equation for one variable in terms of another. For instance,

$x = 2y - 20$ (Transposing $-2y$ in #4A)

Then,

$$2y - 20 + 3y - z = 30 \qquad \text{(Substitution in \#2A)}$$

$$z = 5y - 50 \qquad \text{(Transposing terms, etc.)}$$

Hence the solution of the system for the variables x and z in terms of y is—

$$x, z = 2y - 20, \quad 5y - 50. \qquad \text{(From above.)}$$

This means that the problem has an unlimited (infinite) number of possible solutions corresponding to the unlimited (infinite) number of possible values which we may assign to y.

When each sub-pair of equations in a system is separately independent, the fact that three or more are linearly dependent may be difficult to recognize by inspection. As is illustrated by the above examples, however, the attempt to solve such a system by elementary methods will always result in a pair of equivalent equations and the typical sign of equivalence, $0 = 0$ (Chapter II).

Thus we see that **the concept of linear independence among three or more equations is a generalization of the concept of independence between two equations.** Hence, when we say such a system of equations is **independent** we must mean, not only that each *sub-pair* is *separately independent* as explained in Chapters II and III, but also that the *entire set* is *linearly independent* as explained here.

Practice Exercise No. 26

A. What are the solutions of Example 2 above when $z = 0, -5, 25$? Check each solution.

B. Find the solutions of Example 2A above when $z = 0, 20, 50$. Check each solution.

C. Solve the determinate problems among the following. Find by attempted solution, as in Examples 2 and 2A, which problems are indeterminate because their equations or conditions are linearly dependent.

(1)
$$\begin{aligned}
x + 2y - z + 2w &= -6 \\
2x + y + 3z - 4w &= 15 \\
x + 3y + 2z + 3w &= -10 \\
-3x + 4y + 3z + 2w &= -16
\end{aligned}$$

(2)
$$\begin{aligned}
3x + y - z + w &= 1 \\
4x + y + z - 2w &= 1 \\
2x - y - 3z + w &= 2 \\
x + y + 3z - 2w &= 0
\end{aligned}$$

(3)
$$\begin{aligned}
2x + 3y + z + w &= 5 \\
x + 2y + 2z + 3w &= 7 \\
3x + 4y - z + w &= -4 \\
x + 4y + 8z + 12w &= 26
\end{aligned}$$

(4)
$$\begin{aligned}
x + 2y - 3z + w &= 10 \\
2x + 5y - z - 7w &= -33 \\
x + 4y - 7z + 7w &= 42 \\
x + 3y - 5z + 4w &= 26
\end{aligned}$$

Linear Consistency and Inconsistency

If, in a system of n linearly independent linear equations in n variables, any sub-pair of equations is **inconsistent** (page 21), steps one and two of the method by which Example 1 is solved will produce a pair of inconsistent linear equations in two variables and the typical algebraic result, $N = 0$ (page 21). In such a case the system is indeterminate in that it has no possible solutions (Problem 3 in Practice Exercise No. 27 below).

In this respect, a system of three or more linear equations in the same number of variables is again like a system of two linear equations in two variables.

However, the concept of inconsistency also needs to be generalized for systems of three or more equations.

Definitions: Three or more equations may be said to be **linearly inconsistent (linearly incompatible)** *if any one of these equations is inconsistent (incompatible) with an equation linearly dependent upon the remaining equations.* When they are *not linearly inconsistent*, such equations may be said to be **linearly consistent (linearly compatible).**

When each sub-pair of a system of linear equations is *separately* consistent, the fact that three or more are *linearly* inconsistent may also be difficult to recognize by inspection. However, when you attempt to solve the system by the elementary methods illustrated above, steps one and two will always result in a pair of inconsistent linear equations in two variables and the typical sign of inconsistency, $N = 0!$ (page 21).

EXAMPLE 3: Solve simultaneously,

$$2x + 4y - z = 75 \qquad \text{(Equation \#1)}$$
$$x + 4y \qquad = 70 \qquad \text{(Equation \#2)}$$
$$3x + 4y - 2z = 40 \qquad \text{(Equation \#3i)}$$

ATTEMPTED SOLUTION: *Step One:* As in the solution of Example 2:

$$x = 70 - 4y. \qquad \text{(Transposing } 4y \text{ in \#2)}$$

And by substituting this value of x in #1 we get —

$$-4y - z = -65. \qquad \text{(Steps as in Example 2, obtaining new equation \#4)}$$

However,

$$3(70 - 4y) + 4y - 2z = 40 \qquad \text{(Substitution in \#3i)}$$
$$210 - 12y + 4y - 2z = 40 \qquad \text{(Removing parentheses)}$$
$$-8y - 2z = -170 \qquad \text{(Transposing, etc.)}$$
$$4y + z = 85 \qquad \text{(Dividing by } -2 = -2, \text{ obtaining new equation, \#4i)}$$

For *step two* we now have only the pair of inconsistent equations, #4 and #4i to work with, and the result —

$$0 = -20! \qquad \text{(Adding \#4 and \#4i)}$$

Thus we find that the system of equations in this example is *linearly inconsistent*, and that the system therefore has *no possible simultaneous solution.*

The fact that the system in Example 3 is linearly inconsistent may, indeed, be directly verified by the observation that equation #3i is inconsistent with equation #3d, shown above (page 70) to be linearly dependent upon equations #1 and #2.

Hence, Example 3 illustrates the general principle that, to be determinate, a system of n linear equations in n variables must be both linearly independent and linearly consistent.

This means that to be determinate, a verbally stated problem leading to n linear equations in n variables must specify n conditions which are both linearly independent and linearly consistent.

For instance, the 3 conditions specified in Example 1A above are both linearly independent and linearly consistent. But consider —

EXAMPLE 3A: Instead of the new condition in Example 2A above, substitute the still different condition: When 2 motors of type z are working against 3 motors of type x and 4 motors of type y, the net horsepower generated is 20. Again find the "horse" of each type motor.

ATTEMPTED SOLUTION: Now the conditions of the problem are expressed by the equations:

$$2x + y - z = 10 \qquad \text{(Equation \#1A)}$$
$$x + 3y - z = 30 \qquad \text{(Equation \#2A)}$$
$$-3x - 4y + 2z = 20 \qquad \text{(Equation \#3j)}$$

Step One: As before we get by comparing equations #1A and #2A:

$$x - 2y = -20. \qquad \text{(New equation \#7A)}$$

However,

$$4x + 2y - 2z = 20 \qquad \text{(Multiplying \#2A by } 2 = 2, \text{ obtaining equation \#1A')}$$
$$x - 2y = 40. \qquad \text{(Adding equations 1A' and 3j, obtaining new equation, \#7j)}$$

Step Two: Now we have only the two inconsistent equations, #7A and #7j. By comparing these we can only get the self-contradictory result —

$$0 = 60 \qquad \text{(Adding \#7A and \#7j)}$$

Hence we find that the conditions of the problem are *linearly inconsistent* and that the problem therefore has *no possible simultaneous solution.*

It may be noted, indeed, that equation #3j is inconsistent with equation #3e in Example 2A, shown there to be linearly dependent upon equations #1A and #2A. Hence, equation #3j is seen by direct inspection to be linearly inconsistent with equations #1A and #2A according to the above definition of linear inconsistency.

Hence, when we say that *a system of three or more equations* is consistent we must mean not only that *no sub-pair* is *separately inconsistent* as explained in Chapters II and III, but also that *the entire set* is *not linearly inconsistent* as explained here.

Practice Exercise No. 27

Solve the determinate problems among the following. Find by attempted solution, as in Examples 3 and 3A above, which problems are indeterminate with no possible simultaneous solutions because their equations or conditions are linearly inconsistent.

(1)
$$x + y + w + z = 0$$
$$2x + 3y + 2w + 3z = -3$$
$$-x + 2y - w + 2z = -9$$
$$3x - 2y + 2w - z = 11$$

(2)
$$2x + y - 3w + 3z = -28$$
$$x + 3y - 2w - 4z = 19$$
$$-3x + 4y + w - z = 27$$
$$5x - 2y - 5w - 7z = 11$$

(3)
$$-2x + 5y - 7w + 2z = 1$$
$$x + 4y + 3w - 5z = -3$$
$$3x - 2y + w - z = 7$$
$$x + 3y - 6w + z = 9$$

(4) Three different types of resistances are given, each unknown. The only meter present will read only large numbers, so that a combination of the resistances can be read. The experimenter therefore takes the following readings:

$$x + y + z = 10$$
$$x + 2y + z = 13$$
$$x + y + 2z = 15$$

Find each resistance.

(5) Given any number of groups as in (4), how many measurements have to be made, no matter how delicate the meter?

Geometric Interpretation in Schematic Form

Devices for the geometric interpretation of equations in two variables (Chapter III, pages 30 to 36) can also be extended to cover equations in three variables. For, just as each *pair* of solution values for an equation in *two* variables may be represented geometrically as the *two* coordinates of a point with respect to a system of *two* rectangular coordinate axes (page 33), **each set of *three* solution values for an equation in *three* variables may be represented geometrically as the *three* coordinates of a point with respect to a system of *three* rectangular coordinate axes.**

The latter type of coordinate system is not usually discussed in detail until the student has reached *solid analytic geometry* (Chapters XIV and XV below). It is there shown, however, that just as a (straight) line depicts all the possible solutions of a linear equation in two variables (page 31), **a (flat) plane depicts all the possible solutions of a linear equation in three variables** (page 172). And this graph form is so simple that we may now use it tentatively to illustrate certain general points about linear equations *in schematic form* — without exact numerical values for the coordinates, that is — provided we understand that more exact discussions are to follow in the appropriate place.

Although it is not practicable for graphic solutions of equations in three variables, this device is particularly helpful in further clarifying the sometimes puzzling concepts of linear dependence and linear inconsistency.

As a still further aid to clarification, we shall continue with analogies between the two-variable, or two-dimensional, and the three-variable, or three-dimensional, cases, as follows —

Let e_1, e_1', e_2, etc., represent linear *equations in two variables*, depicted schematically by the lines, l_1, l_1', l_2, etc., in diagrams (*a*) through (*h*) of Figure 21. And let E_1, E_1', E_2, etc., represent linear equations in three variables, depicted schematically by the planes (flat Surfaces), S_1, S_1', S_2, etc., in diagrams (A) through (H) in the same Figure.

For instance, e_1 may be an equation like $x - 2y = -8$ in Example 1 of Chapter II (page 18), and E_1 may be an equation like $2x + 4y - z = 75$ in Example 1 of this chapter (page 67).

From the schematic diagrams in Figure 21, we may now note the following similarities and differences between systems of linear equations in two variables and in three variables:

Just as a pair of equivalent (dependent) linear equations in two variables may be graphed by the same straight line (page 35) as in diagram (*a*) of Figure 21, a pair of equivalent (dependent) linear equations in three varia-

Figure 21. SCHEMATIC GRAPHS OF LINEAR EQUATIONS.

Equations in two variables: e_1, e_1', e_2, e_3, etc., with the respective graph-lines: l_1, l_1', l_2, l_3, etc.

Equations in three variables: E_1, E_1', E_2, E_3, etc., with the respective plane-graphs: S_1, S_1', S_2, S_3, etc.

Case (a) e_1 and e_1' equivalent:

(Lines l_1 and l_1' identical, with all points on l_1 or l_1' representing possible solutions of the defective system.)

Case (A) E_1 and E_1', equivalent:

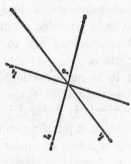

(Planes S_1 and S_1' are identical, with all points on S_1 or S_1' representing possible solutions of the defective system.)

Case (b) e_1 and e_2 inconsistent:

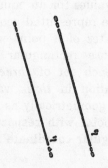

(Lines l_1 and l_4 parallel, with no possible solution for the system.)

Case (B) E_1 and E_2 inconsistent

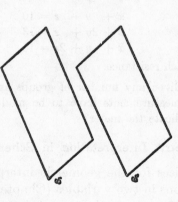

(Planes S_1 and S_2 parallel, with no possible solution for the system.)

Case (f) e_1, e_2, e_3 and e_4 linearly independent and linearly consistent: unique solution:

(The lines intersect by coincidence in the common point, P. The coordinates of P represent the unique solution of the redundant system.)

Case (F) E_1, E_2, E_3 and E_4 linearly independent and linearly consistent:

(The planes intersect in the common point, P. The coordinates of P represent the unique solution of the system.)

Case (G) E_1, E_2, E_3, and E_5 linearly consistent but also linearly dependent:

(The planes intersect in the common line, L. The coordinates of each point on L represent one of the unlimited — infinite — number of possible solutions for the system.)

Case (g) (As for case f)

Case (h)

e_1, e_3, and e_5 such that each sub-pair has a separate solution:

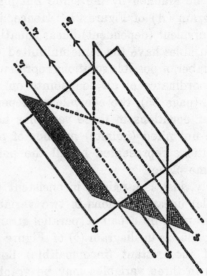

(The lines intersect by pairs in the separate points P_1, P_2, P_3. The redundant system therefore has no simultaneous solution.)

Case (H)

E_1, E_3, and E_5 linearly independent but also linearly inconsistent:

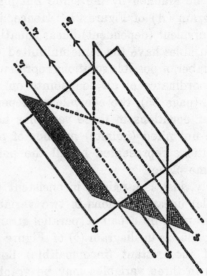

(The planes intersect by pairs in the parallel lines, L_1, L_3, L_5. They therefore have no point in common and the system has no simultaneous solution.)

Case (C) E_1 and E_3 independent and consistent:

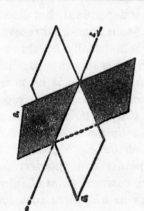

(Planes S_1 and S_3 intersect in the line L. Coordinates of each point on L one of the unlimited — infinite — number of possible solutions for the system.)

Case (c) e_1 and e_3 independent and consistent:

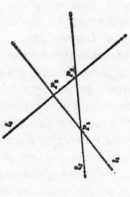

(Lines l_1 and l_2 intersect in the unique point P. Coordinates of P are the unique solution of the system.)

Case (D) E_1 and E_2 inconsistent, but E_1 and E_3 and E_2 and E_3 separately consistent and independent:

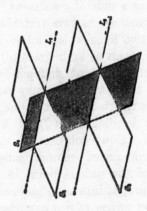

(The planes intersect by pairs in the separate parallel lines, L_1 and L_2. The system has no simultaneous solution.)

Case (d) e_1 and e_2 inconsistent, but e_1 and e_3 and e_2 and e_5 separately consistent and independent:

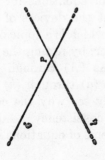

(The lines intersect by pairs in the separate points, P_1 and P_3. The redundant system has no simultaneous solution.)

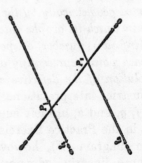

bles may be graphed by the same flat plane as in diagram (*A*) of Figure 21. Hence, just as two equivalent (dependent) linear equations in two variables have the same unlimited (infinite) number of possible solutions represented by the coordinates of all the points on the same line (page 35), two equivalent (dependent) linear equations in three variables have the same unlimited (infinite) number of possible solutions represented by all the points in the same plane.

Similarly, just as a pair of inconsistent (incompatible) linear equations in two variables may be graphed by a pair of parallel straight lines (page 35) as in diagram (*b*) of Figure 21, **a pair of inconsistent (incompatible) linear equations in three variables may be graphed by a pair of parallel planes** as in diagram (*B*) of Figure 21. In both cases, *the fact that a pair of inconsistent linear equations has no common solution corresponds to the geometric fact that their parallel graphs have no point in common.*

Also, just as a pair of consistent, independent linear equations in two variables may be graphed by two lines which intersect in one, and only one, point as in diagram (*c*) of Figure 21, **a pair of consistent, independent linear equations in three variables may be graphed by two planes which intersect in one, and only one, line** as in diagram (*C*) of Figure 21. Whereas the unique solution of the former system is represented geometrically by the coordinates of the unique point in which its straight line graphs intersect, however, *the fact that a defective system of two consistent and independent linear equations in three variables has an unlimited (infinite) number of possible solutions is represented geometrically by the fact that the straight line in which its graphs intersect has an unlimited (infinite) number of points, the coordinates of each point representing a possible simultaneous solution of the defective system.*

Similar geometric interpretations of diagrams *d*, *D*, *e*, *f*, *g*, and *h*, are left as exercises for the student in the Practice Exercise below.

Note now in diagram (*F*), however, that **a system of three linearly independent and** linearly consistent linear equations in three variables is represented geometrically by three planes which intersect in a single point. *The unique solution of such a system is therefore represented geometrically by the coordinates of this unique point of intersection.*

Note also in diagram (*G*), on the other hand, that a system of three linearly consistent, but linearly dependent, linear equations in three variables is represented geometrically by three planes which intersect in a common line. *The unlimited (infinite) number of possible solutions for such a system (page 70) are therefore represented by the coordinates of the unlimited (infinite) number of points which we can select on this line of common intersection. Moreover, the fact that these three planes can intersect in a common line without otherwise coinciding with each other, shows us why three such equations can be linearly dependent while each sub-pair is separately independent!*

Finally, note in diagram (*H*) that **a system of three linearly independent, but linearly inconsistent, linear equations in three variables is represented geometrically by three planes which intersect in three separate parallel lines.** *The fact that such a system has no possible simultaneous solution (page 72), corresponds to the geometric fact that the three separate parallel lines, and consequently the three planes, have no point common to all three. Moreover, the fact that these three planes can intersect in pairs without having any common point of intersection for all three, shows us why three such equations can be linearly inconsistent while each sub-pair is separately consistent!*

Unfortunately, we cannot discuss the geometric interpretation of equations for four or more variables without the devices of higher-dimensional geometry which is a topic of *higher* mathematics. However, by noting the greater complexity of diagrams (*A*) through (*H*) as compared to diagrams (*a*) through (*h*) in Figure 21, you can readily see why the concepts of independence and consistency need to be generalized for all systems of equations in three or more variables.

Practice Exercise No. 28

Analyze practice exercises #27 and #26 in terms of the preceding material.

Summary

The concepts of independence and consistency for systems of two equations may be generalized as the concepts of **linear independence** and **linear consistency** for systems of three or more equations.

To be determinate, a system of linear equations in *n* variables must contain *n*, and in general only *n*, equations which are both linearly independent and linearly consistent. The system then has a **unique solution** which may be found by successive applications of the elementary methods of comparison and substitution.

To be determinate, a physical problem leading to linear equations in *n* variables must specify *n*, and in general only *n*, conditions which are both linearly consistent and linearly independent. The problem then has a **unique solution** which may be found as above.

The graphs of linear equations in three variables are (flat) planes which are identical *for any two dependent equations*, and parallel *for any two inconsistent equations*.

The three graph planes of a system of three linear equations in three variables intersect: (1) **in a single point** *when the equations are linearly independent and linearly consistent;* (2) **in a straight line** *when the equations are linearly consistent but also linearly dependent;* (3) **in two parallel lines** *when one sub-pair of equations is separately inconsistent but the other two sub-pairs are separately consistent and independent;* and (4) **in three parallel lines** *when the equations are linearly independent, but also linearly inconsistent.* These four possibilities correspond respectively to the cases in which the system of equations is (1) *determinate with a unique solution;* (2) *indeterminate with an unlimited (infinite) number of possible solutions;* and (3 or 4) *indeterminate with no possible solution* (just as when each sub-pair of equations is separately inconsistent and their parallel graph-planes do not intersect at all).

CHAPTER VII

DETERMINANTS

A Note on Sequence of Topics

Analytic geometry is often presented without the convenience of determinant methods. This is because some students have not yet studied the topic by the time they are college freshmen.

However, *it is almost universally assumed that the student is familiar with determinants by the time he studies vector analysis. And determinant methods are equally useful in analytic geometry where they have some of their most fascinating applications.*

For these reasons the topic of determinants is introduced here in the present chapter where it can have the greatest practical value in sections immediately following. If the reader has taken suggested options of beginning analytic geometry earlier (pages 36 and 66), he is strongly advised to return to this present chapter as soon as determinants are applied there in the text.

Second Order Determinants

Having derived a general formula for solving quadratic equations in one variable (page 41), we shall now derive general formulas for solving systems of linear equations in two or more variables. As before, the standard procedure is to apply elementary methods of solution to equations written in typical form, as follows:

Any system of two linear equations in two variables can be written in the typical form,

$$a_1x + b_1y = k_1, \quad \text{(Equation \#1)}$$
$$a_2x + b_2y = k_2, \quad \text{(Equation \#2)}$$

where x and y are the two variables and a_1, a_2, b_1, etc., are constants. For instance, in the system of Example 1 in Chapter II (page 18), two particular equations of this type are —

$$x - 2y = -8,$$
$$x + y = 7;$$

and here the values of the constants are —

$$a_1 = 1, \quad b_1 = -2, \quad k_1 = -8,$$
$$a_2 = 1, \quad b_2 = 1, \quad k_2 = 7.$$

When we apply the standard method of comparison to solve the above equations in typical form, we get:

$a_1b_2x + b_1b_2y = k_1b_2$	(Multiplying Equation #1 by b_2, obtaining Equation #1′)
$a_2b_1x + b_1b_2y = k_2b_1$	(Multiplying Equation #2 by b_1, obtaining Equation #2′)
$(a_1b_2 - a_2b_1)x = k_1b_2 - k_2b_1$	(Subtracting Equation #2′ from #1′, obtaining new equation, #3)
$a_1a_2x + a_1b_2y = a_1k_2$	(Multiplying Equation #2 by a_1, obtaining Equation #2″)
$a_1a_2x + a_2b_1y = a_2k_1$	(Multiplying Equation #1 by a_2, obtaining Equation #1″)
$(a_1b_2 - a_2b_1)y = a_1k_2 - a_2k_1$	(Subtracting #1″ from #2″, obtaining new equation, #4)
$x = \dfrac{k_1b_2 - k_2b_1}{a_1b_2 - a_2b_1}$	(Dividing Equation #3 by $a_1b_2 - a_2b_1 = a_1b_2 - a_2b_1 \neq 0$)
$y = \dfrac{a_1k_2 - a_2k_1}{a_1b_2 - a_2b_1}$	(Ditto for Equation #4)

These solutions express the value of the variable x and y in terms of the constant coefficients, a_1, b_1, etc., in the preceding typical equations. Hence they are **solution formulas** for any pair of equations in these typical forms.

If we wish, we can solve any system of two linear equations in two variables simply by substituting appropriate specific values for the constants in these solution formulas. For instance, the system of two equations taken as an example from Chapter II above may be solved by the above illustrated substitutions as follows:

$$x = \frac{(-8)(1) - (-2)(7)}{(1)(1) - (-2)(1)} = \frac{-8 + 14}{1 + 2} = \frac{6}{3} = 2,$$

$$y = \frac{(1)(7) - (-8)(1)}{3} = \frac{7 + 8}{3} = \frac{15}{3} = 5.$$

(Note here that the denominator in the two formulas, being the same, needs to be evaluated only the first time.)

In the above forms, of course, these solution formulas are neither as easy to remember nor as convenient to work with as the previously derived quadratic solution formula (page 41). But suppose we were to rewrite their common denominator in the easier-to-remember form at the left side of the following **formula of definition** —

M1: $\begin{vmatrix} a_1 & b_1 \\ a_2 & b_2 \end{vmatrix} = a_1b_2 - b_1a_2.$

This sort of square array of quantities between two vertical lines is called a **determinant**. The quantities, a_1, b_1, etc., in the determinant are called its **elements**. Since, in this case, there are *two* elements in each row and in each column, the determinant is called a determinant **of the second order**. And since the defined value of the determinant is given by the right hand side of the above formula, the latter expression is called the determinant's **expansion**.

In other words, the *determinant of the second order* on the left side of the above formula is simply a different way of writing the expression involving its elements on the right side of the formula. And the formula itself simply states the following **rule for expanding (evaluating) a determinant of the second order:** *Take the product of the elements on the diagonal from the upper left-hand corner to the lower right-hand corner, and subtract from it the product of the elements on the diagonal from the upper right-hand corner to the lower left-hand corner,* thus —

$$\begin{vmatrix} a_1 & b_1 \\ a_2 & b_2 \end{vmatrix} = a_1b_2 - b_1a_2, \text{ or } a_1b_2 - a_2b_1.$$

For instance, if the elements, a_1, b_1, etc., have the numerical values illustrated further above, the common denominator of the preceding solution formulas may be written and evaluated in determinant form as —

$$\begin{vmatrix} 1 & -2 \\ 1 & 1 \end{vmatrix} = 1(1) - (-2)1 = 1 + 2 = 3.$$
(Substitution)

Moreover, when we define a second order determinant by this expansion rule, we may also write the numerators of the above solution formulas in the same form. Then the formulas become:

$$x = \frac{\begin{vmatrix} k_1 & b_1 \\ k_2 & b_2 \end{vmatrix}}{\begin{vmatrix} a_1 & b_1 \\ a_2 & b_2 \end{vmatrix}}, \quad y = \frac{\begin{vmatrix} a_1 & k_1 \\ a_2 & k_2 \end{vmatrix}}{\begin{vmatrix} a_1 & b_1 \\ a_2 & b_2 \end{vmatrix}}.$$

This is a special case of a more general formula known as **Cramer's Rule** (page 81) after the mathematician, Gabriel Cramer, who first stated it in 1760. The rule's pattern is simple to remember if you note that (1) *the common denominator determinant* — called **the determinant of the system** — *has the same arrangement of a's and b's as the original equations,* and (2) *the numerator determinants have the same arrangement except that k's are substituted for the a-coefficients of x in the x-formula, and for the b-coefficients of y in the y-formula.*

Since pairs of linear equations in two variables are simple to solve in any case, the greatest value of Cramer's rule lies in its extension to systems of linear equations in more variables. Nevertheless, you may be interested to find that, with a little practice, you can apply the determinant method to solve many pairs of linear equations in two variables mentally, or with only a few paper-and-pencil steps.

The "knack" is to multiply along the diagonals mentally, jotting down only the resulting products, or better still, only the difference of these products. The common denominator, of course, needs to be expanded only once.

EXAMPLE 1: Solve by determinants,

$$x - 2y = -8,$$
$$x + y = 7.$$

SOLUTION (with arithmetic steps written out):

$$x = \frac{\begin{vmatrix} -8 & -2 \\ 7 & 1 \end{vmatrix}}{\begin{vmatrix} 1 & -2 \\ 1 & 1 \end{vmatrix}} = \frac{-8 + 14}{1 + 2} = \frac{6}{3} = 2,$$

$$y = \frac{\begin{vmatrix} 1 & -8 \\ 1 & 7 \end{vmatrix}}{3} = \frac{7 + 8}{3} = \frac{15}{3} = 5.$$

Practice Exercise No. 29

Re-work the problems of Practice Exercise No. 2 by determinants. After you have written out a few solutions in full, try doing the remaining ones with as few pencil-on-paper steps as possible.

Determinants of Higher Order

A system of three linear equations in three variables can be written in the typical form,

$$a_1x + b_1y + c_1z = k_1,$$
$$a_2x + b_2y + c_2z = k_2,$$
$$a_3x + b_3y + c_3z = k_3.$$

And when we apply the standard method of comparison for solving these equations (as in Chapter VI, page 67), we get as one of their solution formulas,

$$x = \frac{k_1b_2c_3 + k_2b_3c_1 + k_3b_1c_2 - k_3b_2c_1 - k_2b_1c_3 - k_1b_3c_2}{a_1b_2c_3 + a_2b_3c_1 + a_3b_1c_2 - a_3b_2c_1 - a_2b_1c_3 - a_1b_3c_2},$$

with similar formulas for y and z.

Here again we can solve any system of equations in the above form simply by substituting appropriate specific values for the constants in the solution formulas. As before (page 79), however, the formulas are neither easy to remember nor — what is more important — convenient to apply, when so written. Again, therefore, we re-write their common denom-

inator in the simpler form at the left side of the following definition formula:

M2:
$$\begin{vmatrix} a_1 & b_1 & c_1 \\ a_2 & b_2 & c_2 \\ a_3 & b_3 & c_3 \end{vmatrix} = \begin{aligned} &a_1b_2c_3 + a_2b_3c_1 + a_3b_1c_2 \\ &- a_3b_2c_1 - a_2b_1c_3 - a_1b_3c_2. \end{aligned}$$

Since the left side of this formula is also a square array of elements between two vertical lines, it is by definition a determinant (page 79). But since, in this case, there are *three* elements in each row and column, the determinant is called a determinant of the **third order**.

Obviously, the **expansion rule for a determinant of the third order** is much more complicated than that for a second order determinant (page 79). We may state it most simply if we first supplement the above determinant with its first two columns repeated as below, drawing through it the indicated diagonal lines. With reference to such a dia-

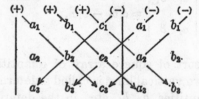

gram, the expansion rule then becomes: *Add the products of the three elements on each of the diagonals sloping down to the right, and subtract from this sum the products of the three elements on each of the diagonals sloping down to the left.*

The preceding solution formula may now be restated in determinant form as —

$$x = \frac{\begin{vmatrix} k_1 & b_1 & c_1 \\ k_2 & b_2 & c_2 \\ k_3 & b_3 & c_3 \end{vmatrix}}{\begin{vmatrix} a_1 & b_1 & c_1 \\ a_2 & b_2 & c_2 \\ a_3 & b_3 & c_3 \end{vmatrix}} = \frac{|k\ b\ c|}{|a\ b\ c|}.$$

The symbols in the numerator and denominator of the second fraction here are simply briefer ways of writing the corresponding typical determinants in the first fraction. This notational device cannot be used, of course, when the elements of a determinant have specific numerical values. But it is very convenient for writing general formulas for determinants expressed in typical form with subscript letters as above.

For instance, using this abbreviated notation we can now write Cramer's rule for the solution of a system of two linear equations in two variables as:

$$x = |\ k\ b\ | \ / \ |\ a\ b\ |,$$
$$y = |\ a\ k\ | \ / \ |\ a\ b\ |.$$

Or, we can write the statement of Cramer's rule for a system of three linear equations in three variables as:

$$x = |\ k\ b\ c\ | \ / \ |\ a\ b\ c\ |,$$
$$y = |\ a\ k\ c\ | \ / \ |\ a\ b\ c\ |,$$
$$z = |\ a\ b\ k\ | \ / \ |\ a\ b\ c\ |.$$

And, since it is shown in *higher* mathematics that this rule is general for a system of n linear equations in n variables, corresponding statements can be made of solution formulas for n = 4, 5, 6, etc. For instance, when n = 4,

$$x = |\ k\ b\ c\ d\ | \ / \ |\ a\ b\ c\ d\ |,$$
$$y = |\ a\ k\ c\ d\ | \ / \ |\ a\ b\ c\ d\ |,$$
$$z = |\ a\ b\ k\ d\ | \ / \ |\ a\ b\ c\ d\ |,$$
$$w = |\ a\ b\ c\ k\ | \ / \ |\ a\ b\ c\ d\ |.$$

The expansion rule for a determinant of the third order is, of course, somewhat complex to be practical to apply in an equation-solving routine. And the expansion rules of determinants of still higher orders are successively more complicated. Consequently, *if* determinants could be evaluated only by expansion rules, they *would* have little practical value.

It happens, however, that there are other ways of evaluating determinants which are both simpler and quicker. We shall therefore discuss these other methods first before applying determinants to the solution of systems of linear equations by Cramer's rule when more than two variables are involved.

Expansion by Minors

The minor of an element in a determinant
is defined as *the determinant of next lower order which is obtained by striking out the row and column in which the given element occurs.* In the case of a third order determinant, for instance, the minor of the element a_1 is

$$\begin{vmatrix} a_1 & b_2 & c_2 \\ a_2 & b_2 & c_2 \\ a_3 & b_3 & c_3 \end{vmatrix} = \begin{vmatrix} b_2 & c_2 \\ b_3 & c_3 \end{vmatrix}.$$

The minor of the element b_2 is

$$\begin{vmatrix} a_1 & b_1 & c_1 \\ a_2 & b_2 & c_2 \\ a_3 & b_3 & c_3 \end{vmatrix} = \begin{vmatrix} a_1 & c_1 \\ a_3 & c_3 \end{vmatrix}.$$

And the minor of the element c_2 is

$$\begin{vmatrix} a_1 & b_1 & c_1 \\ a_2 & b_2 & c_2 \\ a_3 & b_3 & c_3 \end{vmatrix} = \begin{vmatrix} a_1 & b_1 \\ a_3 & b_3 \end{vmatrix}.$$

The reason for so defining the minor of an element is this important general expansion rule: *A determinant is equal to the sum of the products formed by multiplying each element in the N'th row or column by its minor, provided the first such product is given the sign of $(-1)^{N+1}$ and each product thereafter is given an alternate sign.*

Thus, if $N = 1$ or any other odd number, then $(-1)^{N+1} = 1$, and the expansion by minors begins with a plus sign. For instance, if we expand a third order determinant by minors of the *first column*, then $N = 1$ and $(-1)^{N+1} = (-1)^{1+1} = (-1)^2 = 1$. Hence the expansion begins with a plus sign as in the formula,

$$\textbf{M3:}\quad |\ a\ b\ c\ | = \begin{vmatrix} a_1 & b_1 & c_1 \\ a_2 & b_2 & c_2 \\ a_3 & b_3 & c_3 \end{vmatrix}$$

$$= a_1 \begin{vmatrix} b_2 & c_2 \\ b_3 & c_3 \end{vmatrix} - a_2 \begin{vmatrix} b_1 & c_1 \\ b_3 & c_3 \end{vmatrix} + a_3 \begin{vmatrix} b_1 & c_1 \\ b_2 & c_2 \end{vmatrix}.$$

But if $N = 2$ or any other even number, then $(-1)^{N+1} = -1$, and the expansion by minors begins with a minus sign. For instance, if we expand the same third order determinant by minors of the *second row*, then $N = 2$ and $(-1)^{N+1} = (-1)^{2+1} = (-1)^3 = -1$. Hence the expansion begins with a minus sign as in the formula,

M3': $\begin{vmatrix} a & b & c \end{vmatrix} = -a_2 \begin{vmatrix} b_1 & c_1 \\ b_3 & c_3 \end{vmatrix} + b_2 \begin{vmatrix} a_1 & c_1 \\ a_3 & c_3 \end{vmatrix} - c_2 \begin{vmatrix} a_1 & b_1 \\ a_3 & b_3 \end{vmatrix}$.

We must hasten to add at this point that complete general proof for all basic determinant formulas requires complicated reasoning which is ordinarily detailed only in *higher* mathematics beyond the scope or intent of these books. On the *advanced* level the student is ordinarily required only to be able to illustrate the derivation of basic determinant formulas for some representative case — such as that of a typical determinant of the third order.

Nowhere in this chapter, therefore, shall we attempt to give complete general demonstrations. But, assuming that generality can be proven for our theorems by higher methods, we shall simply illustrate the derivation of each principle by typical third order examples.

For instance, we now *illustrate* the derivation of Formula M3 above as follows:

$\begin{vmatrix} a & b & c \end{vmatrix} = a_1b_2c_3 + a_2b_3c_1 + a_3b_1c_2$
$\qquad\qquad - a_3b_2c_1 - a_2b_1c_3 - a_1b_3c_2$ (Formula M2)

$= a_1(b_2c_3 - b_3c_2) - a_2(b_1c_3 - b_3c_1) + a_3(b_1c_2 - b_2c_1)$
                           (Factoring)

$= a_1 \begin{vmatrix} b_2 & c_2 \\ b_3 & c_3 \end{vmatrix} - a_2 \begin{vmatrix} b_1 & c_1 \\ b_3 & c_3 \end{vmatrix} + a_3 \begin{vmatrix} b_1 & c_1 \\ b_2 & c_2 \end{vmatrix}$.

                           (Formula M1)

Although formula M3 permits you to expand a determinant by the minors of any row or column, the question of which row or column to select for expansion depends upon the pattern of the numerical values of the elements. In general, *the choice of a row or column with more zeros or small numbers will shorten the work of calculation.*

EXAMPLE 2: Expand by minors of the first column,
$$D = \begin{vmatrix} 7 & 0 & 9 \\ 32 & 1 & 9 \\ 3 & 0 & 6 \end{vmatrix}.$$

SOLUTION:

$D = 7 \begin{vmatrix} 1 & 9 \\ 0 & 6 \end{vmatrix} - 32 \begin{vmatrix} 0 & 9 \\ 0 & 6 \end{vmatrix} + 3 \begin{vmatrix} 0 & 9 \\ 1 & 9 \end{vmatrix}$ (Formula M3)

$= 7(6-0) - 32(0-0) + 3(0-9)$ (Formula M1)
$= 42 - 0 - 27 = 15$, Answer   (Removing
                                 parentheses)

EXAMPLE 2': Expand D in Example 2 by minors of the second column.

SOLUTION:

$D = -0 + (1) \begin{vmatrix} 7 & 9 \\ 3 & 6 \end{vmatrix} - 0$ (Formula M3)

$= 42 - 27 = 15$, Answer   (Formula M1,
                                    etc.)

The procedure in Example 2' is obviously shorter than that in Example 2 for the same problem. This is because of the more judicious choice of a column by which to expand.

In general, *the higher the order of a determinant, the more important it is to make a judicious choice of the row or column by which to expand.*

EXAMPLE 3: Expand by minors of the first row,
$$D = \begin{vmatrix} 6 & 9 & 3 & 1 \\ 4 & 2 & 7 & 0 \\ 6 & 4 & 2 & 0 \\ 9 & 1 & 9 & 1 \end{vmatrix}.$$

SOLUTION:

$D = 6 \begin{vmatrix} 2 & 7 & 0 \\ 4 & 2 & 0 \\ 1 & 9 & 1 \end{vmatrix} - 9 \begin{vmatrix} 4 & 7 & 0 \\ 6 & 2 & 0 \\ 9 & 9 & 1 \end{vmatrix}$

                                  (Formula M3)

$\quad + 3 \begin{vmatrix} 4 & 2 & 0 \\ 6 & 4 & 0 \\ 9 & 1 & 1 \end{vmatrix} - 1 \begin{vmatrix} 4 & 2 & 7 \\ 6 & 4 & 2 \\ 9 & 1 & 9 \end{vmatrix}$

$= 6 \left(2 \begin{vmatrix} 2 & 0 \\ 9 & 1 \end{vmatrix} - 7 \begin{vmatrix} 4 & 0 \\ 1 & 1 \end{vmatrix} + 0 \right)$

$\quad -9 \left(4 \begin{vmatrix} 2 & 0 \\ 9 & 1 \end{vmatrix} - 7 \begin{vmatrix} 6 & 0 \\ 9 & 1 \end{vmatrix} + 0 \right)$ (Formula M3
                                        again)

$\quad +3 \left(4 \begin{vmatrix} 4 & 0 \\ 1 & 1 \end{vmatrix} - 2 \begin{vmatrix} 6 & 0 \\ 9 & 1 \end{vmatrix} + 0 \right)$

$\quad -1 \left(4 \begin{vmatrix} 4 & 2 \\ 1 & 9 \end{vmatrix} - 2 \begin{vmatrix} 6 & 2 \\ 9 & 9 \end{vmatrix} + 7 \begin{vmatrix} 6 & 4 \\ 9 & 1 \end{vmatrix} \right)$

$= 6[2(2-0) - 7(4-0)]$
$\quad -9[4(2-0) - 7(6-0)]$ (Formula M1)
$\quad +3[4(4-0) - 2(6-0)]$
$\quad -[4(36-2) - 2(54-18) + 7(6-36)]$

$= 6(-24) - 9(-34) + 3(4) - (-146)$

$= 464 - 144 = 320$, Answer.
              (Removing parentheses, etc.)

EXAMPLE 3': Expand D in Example 3 by minors of the fourth column.

SOLUTION:

$$D = -\begin{vmatrix} 4 & 2 & 7 \\ 6 & 4 & 2 \\ 9 & 1 & 9 \end{vmatrix} + 0 - 0 + \begin{vmatrix} 6 & 9 & 3 \\ 4 & 2 & 7 \\ 6 & 4 & 2 \end{vmatrix} \quad \text{(Formula M3)}$$

$$= -\left(4\begin{vmatrix} 4 & 2 \\ 1 & 9 \end{vmatrix} - 2\begin{vmatrix} 6 & 2 \\ 9 & 9 \end{vmatrix} + 7\begin{vmatrix} 6 & 4 \\ 9 & 1 \end{vmatrix}\right) \quad \text{(Formula}$$

$$+ \left(6\begin{vmatrix} 2 & 7 \\ 4 & 2 \end{vmatrix} - 9\begin{vmatrix} 4 & 7 \\ 6 & 2 \end{vmatrix} + 3\begin{vmatrix} 4 & 2 \\ 6 & 4 \end{vmatrix}\right) \quad \text{M3 again)}$$

$$= -[4(36 - 2) - 2(54 - 18) + 7(6 - 36)]$$
$$+ [6(4 - 28) - 9(8 - 42) + 3(16 - 12)]$$

(Formula M1)

$$= -(-146) + 174 = 320, \text{ Answer.}$$

(Removing parentheses, etc.)

Practice Exercise No. 30

Expand the following determinants by minors of a judiciously selected row or column.

(1) $\begin{vmatrix} 1 & 1 & 0 \\ 3 & 3 & 1 \\ -2 & 3 & 1 \end{vmatrix}$

(4) $\begin{vmatrix} a & 0 & 0 \\ 0 & b & 0 \\ 0 & 0 & c \end{vmatrix}$

(2) $\begin{vmatrix} 0 & 1 & -3 & 0 \\ 5 & -2 & 2 & 4 \\ 0 & 1 & 0 & -1 \\ 3 & -1 & 0 & 5 \end{vmatrix}$

(5) $\begin{vmatrix} 0 & 0 & 1 & 1 \\ -1 & 3 & 5 & 0 \\ 2 & 0 & -2 & -1 \\ 1 & -1 & 3 & 0 \end{vmatrix}$

(3) $\begin{vmatrix} 0 & 0 & 3 & -3 \\ 1 & 2 & 0 & 1 \\ 0 & 0 & 1 & 4 \\ 1 & 1 & 5 & 2 \end{vmatrix}$

(6) $\begin{vmatrix} 1 - x & 2 & 0 \\ 3 & 1 & 2 + x \\ 0 & 4 & 0 \end{vmatrix}$

Further Properties of Determinants

The calculation of a determinant's value can usually be greatly shortened if we change its form *before* we expand it.

In other branches of mathematics such as analytic geometry and vector analysis, moreover, the applications of determinants often do not require us to evaluate them fully, but only to show that they can be reduced to a particular form.

For both these reasons it is of great practical importance for us to be able to apply the following general properties of determinants.

Theorem M4: *Changing all columns to rows or all rows to columns, in the same order, leaves the value of a determinant unchanged.* Typical formula:

$$\text{M4:} \quad \begin{vmatrix} a_1 & b_1 & c_1 \\ a_2 & b_2 & c_2 \\ a_3 & b_3 & c_3 \end{vmatrix} = \begin{vmatrix} a_1 & a_2 & a_3 \\ b_1 & b_2 & b_3 \\ c_1 & c_2 & c_3 \end{vmatrix}.$$

Illustration of this theorem follows at once from formula M3 since, according to that formula, it makes no difference whether we expand a determinant by minors of the k'th row or k'th column.

This means that *any theorem concerning the rows of a determinant applies equally to its columns, and vice versa.*

Theorem M5: *Multiplying all the elements of any row (or column) of a determinant by the same quantity multiplies the value of the determinant by that quantity.* Formula:

$$\text{M5:} \quad N\,|\,a\,b\,c\,| = |\,Na\,b\,c\,| = |\,a\,Nb\,c\,| = |\,a\,b\,Nc\,|.$$

Illustration:

$$N\,|\,a\,b\,c\,| = N\left(a_1\begin{vmatrix} b_2 & c_2 \\ b_3 & c_3 \end{vmatrix} - a_2\begin{vmatrix} b_1 & c_1 \\ b_3 & c_3 \end{vmatrix} + a_3\begin{vmatrix} b_1 & c_1 \\ b_2 & c_2 \end{vmatrix}\right)$$

(Formula M3)

$$= Na_1\begin{vmatrix} b_2 & c_2 \\ b_3 & c_3 \end{vmatrix} - Na_2\begin{vmatrix} b_1 & c_1 \\ b_3 & c_3 \end{vmatrix} + Na_3\begin{vmatrix} b_1 & c_1 \\ b_2 & c_2 \end{vmatrix}$$

(Removing parentheses)

$$= |\,Na\,b\,c\,|, \text{ etc.} \quad \text{(Formula M3)}$$

From this theorem it follows that *if we multiply all the elements in any row (or column) of a determinant by the same quantity, and then divide the entire determinant by this quantity, or vice versa, we leave the value of the determinant unchanged.*

$$\text{M5':} \quad |\,a\,b\,c\,| = \frac{1}{N}|\,Na\,b\,c\,| = \frac{1}{N}|\,a\,Nb\,c\,| = \frac{1}{N}|\,a\,b\,Nc\,|$$

$$= N\left|\frac{a}{N}\,b\,c\right| = N\left|a\,\frac{b}{N}\,c\right| = N\left|a\,b\,\frac{c}{N}\right|.$$

These theorems are useful in simplifying the evaluation of determinants, for they can be applied to remove fractions or larger numbers from the rows or columns *before* expansion.

EXAMPLE 4: Evaluate:

$$D = \begin{vmatrix} \frac{1}{4} & 2 & 3 \\ \frac{3}{4} & 0 & 7 \\ \frac{1}{2} & 0 & 5 \end{vmatrix}.$$

SOLUTION: Simplify column 1 by multiplying it by 4, and offset this operation by dividing the entire determinant by 4. Then expand by minors of the second column.

$$D = \frac{1}{4} \begin{vmatrix} (4)\frac{1}{4} & 2 & 3 \\ (4)\frac{3}{4} & 0 & 7 \\ (4)\frac{1}{2} & 0 & 5 \end{vmatrix} = \frac{1}{4} \begin{vmatrix} 1 & 2 & 3 \\ 3 & 0 & 7 \\ 2 & 0 & 5 \end{vmatrix} \quad \text{(Formula M5′)}$$

$$= \frac{1}{4}\left(-2 \begin{vmatrix} 3 & 7 \\ 2 & 5 \end{vmatrix} \right) \quad \text{(Formula M3)}$$

$$= -\tfrac{1}{2}(15 - 14) = -\tfrac{1}{2}. \text{ Ans.} \quad \text{(Formula M1, etc.)}$$

EXAMPLE 5: Evaluate:

$$D = \begin{vmatrix} 7 & 3 & 5 \\ 7 & 0 & 0 \\ 16 & 48 & 32 \end{vmatrix}.$$

SOLUTION: Simplify row 3 by dividing it by 16, and offset this operation by multiplying the entire determinant by 16. Then expand by minors of the second row:

$$D = 16 \begin{vmatrix} 7 & 3 & 5 \\ 7 & 0 & 0 \\ 1 & 3 & 2 \end{vmatrix} \quad \text{(Formula M5′)}$$

$$= 16 \left(-7 \begin{vmatrix} 3 & 5 \\ 3 & 2 \end{vmatrix} \right) \quad \text{(Formula M3)}$$

$$= -112(6 - 15) = 1,008. \text{ Ans.} \quad \text{(Formula M1, etc.)}$$

Practice Exercise No. 31

Simplify and evaluate the following determinants.

(1) $\begin{vmatrix} 3 & 1 & -1 \\ 6 & 2 & 5 \\ 9 & 3 & 7 \end{vmatrix}$ (2) $\begin{vmatrix} 4 & 6 & 1 & 3 \\ 2 & 3 & 0 & -5 \\ 6 & 9 & 1 & 4 \\ 8 & 12 & 2 & 2 \end{vmatrix}$

(3) $\begin{vmatrix} 1 & 7 & 3 & 9 & -2 \\ 0 & 3 & 1 & 6 & 5 \\ 0 & \frac{2}{3} & \frac{1}{2} & 3 & \frac{5}{2} \\ 2 & 0 & 4 & -6 & 8 \\ 1 & 0 & 3 & 5 & -9 \end{vmatrix}$

(4) Show that

$$\begin{vmatrix} 1 & a & -3 \\ -5 & b & 0 \\ 2 & c & -4 \end{vmatrix} + \begin{vmatrix} 1 & d & -3 \\ -5 & e & 0 \\ 2 & f & 4 \end{vmatrix} = \begin{vmatrix} 1 & a+d & -3 \\ -5 & b+e & 0 \\ 2 & c+f & 4 \end{vmatrix}$$

(5) Show that

$$\begin{vmatrix} 1-x & 2 & 0 \\ 3 & 3 & 2+x \\ 0 & 4 & 0 \end{vmatrix} = 0$$

if $x = 1$ or $x = 2$.

(6) Show that if we multiply $\begin{vmatrix} a & b \\ c & d \end{vmatrix}$ by $\begin{vmatrix} e & f \\ g & h \end{vmatrix}$, the correct answer is given by

$$\begin{vmatrix} ae+bg & af+bh \\ ce+dg & cf+dh \end{vmatrix}$$

Theorem M6: *If the elements in any two columns (or rows) of a determinant are proportional, the determinant has the value, 0.* Typical (abbreviated) formula:

M6: $\left| \, a \, b \, Nb \, \right| = 0.$

Illustrated derivation:

$$\left| \, a \, b \, Nb \, \right| = N \left| \, a \, b \, b \, \right| \quad \text{(Formula M5)}$$

$$= N \left(a_1 \begin{vmatrix} b_2 & b_2 \\ b_3 & b_3 \end{vmatrix} - a_2 \begin{vmatrix} b_1 & b_1 \\ b_3 & b_3 \end{vmatrix} + a_3 \begin{vmatrix} b_1 & b_1 \\ b_2 & b_2 \end{vmatrix} \right)$$
$$\text{(Formula M3)}$$

$$= N[a_1(0) - a_2(0) + a_3(0)] = 0. \quad \text{(Formula M1)}$$

From this theorem it follows at once that *the determinant of a system* (defined, page 79) *of n linear equations in n variables must always be 0 if two of its equations are either dependent or inconsistent.* For, the constant coefficients of the variables in any such pair of equations are proportional (Chapters II and VI above).

From the theorem it also follows that, if *you were to attempt to apply Cramer's rule* (page 81) *to solve a system of n linear equations in n variables with two of the equations dependent,* then *the numerator determinants of Cramer's rule would also be 0.* For, all the constants in any such pair of equations are proportional (Chapters II and VI).

The significance of these facts is discussed further below under the heading of *Exceptions to Cramer's Rule* (page 88). But first we shall apply theorem M6 to certain further important properties of determinants in general.

Practice Exercise No. 32

Verify by formula M6 that the determinants of all the following problems or systems of equations are 0:

> Exercise No. 2, problems 3, 5.
> Exercise No. 20, problems 1, 4.
> Exercise No. 21, problem 1.

Theorem M7: *If each of the elements in a column (or row) of a determinant can be expressed as the sum of two quantities, the determinant itself can be expressed as the sum of two determinants according to the typical (abbreviated) formula:*

M7: $\quad |\, m{+}n \ b \ c\,| = |\, m \ b \ c\,| + |\, n \ b \ c\,|.$

Illustration follows directly from expansion of the above determinants by minors of the first column of each (Practice Exercise No. 33, Problem A below).

This theorem, together with M6, enables us to establish the more important one —

Theorem M8: *If all the elements in any column (or row) of a determinant are multiplied by the same constant and the resulting products are added to the corresponding elements in another column (or row), the value of the determinant is unchanged.* Typical (abbreviated) formula:

M8: $\quad |\, a \ b \ c{+}Na\,| = |\, a \ b \ c\,|.$

Illustrated derivation:

$$|\, a \ b \ c{+}Na\,| = |\, a \ b \ c\,| + |\, a \ b \ Na\,| \quad \text{(Formula M7)}$$
$$= |\, a \ b \ c\,| + 0 = |\, a \ b \ c\,|.$$
$$\text{(Formula M6)}$$

The great practical importance of this theorem is that *it may be applied repeatedly to reduce all, or all but one, of the elements in a selected row or column to 0.* Thus it may greatly expedite the labor of computing the determinant's value.

EXAMPLE 6: Apply theorem M8 to evaluate the determinant in Example 3 (page 82):

$$D = \begin{vmatrix} 6 & 9 & 3 & 1 \\ 4 & 2 & 7 & 0 \\ 6 & 4 & 2 & 0 \\ 9 & 1 & 9 & 1 \end{vmatrix}.$$

SOLUTION:

$$D = \begin{vmatrix} 6 & 9 & 3 & 1 \\ 4 & 2 & 7 & 0 \\ 6 & 4 & 2 & 0 \\ 3 & -8 & 6 & 0 \end{vmatrix} \qquad \begin{array}{l}\text{(Subtracting}\\ \text{row 1 from}\\ \text{row 4)}\end{array}$$

$$= - \begin{vmatrix} 4 & 2 & 7 \\ 6 & 4 & 2 \\ 3 & -8 & 6 \end{vmatrix} + 0 - 0 + 0 \qquad \begin{array}{l}\text{(Applying M3}\\ \text{to minors of}\\ \text{column 4)}\end{array}$$

$$= - \begin{vmatrix} 4 & 2 & 7 \\ -2 & 0 & -12 \\ 15 & 0 & 10 \end{vmatrix} \qquad \begin{array}{l}\text{(Adding } -2\\ \text{times row 1}\\ \text{to row 2, and}\\ \text{2 times row 2}\\ \text{to row 3)}\end{array}$$

$$= - (-2) \begin{vmatrix} -2 & -12 \\ 15 & 10 \end{vmatrix} \qquad \begin{array}{l}\text{(Applying M3}\\ \text{to minors of}\\ \text{column 2)}\end{array}$$

$$= 2(-20 + 180) = 320. \qquad \text{(Formula M1)}$$

EXAMPLE 7: Show without expanding that

$$D = \begin{vmatrix} b & c & 1 \\ \dfrac{a}{2} & 0 & 1 \\ \dfrac{a+b}{3} & \dfrac{c}{3} & 1 \end{vmatrix} = 0.$$

SOLUTION:

$$D = \begin{vmatrix} b & c & 1 \\ \dfrac{a}{2} & 0 & 1 \\ 0 & 0 & 0 \end{vmatrix} \qquad \begin{array}{l}\text{(Adding } -\tfrac{1}{3}\text{ row}\\ 1 \text{ and } -\tfrac{2}{3}\text{ row 2}\\ \text{to row 3 by}\\ \text{formula M8)}\end{array}$$

$$= 0 - 0 + 0 = 0. \qquad \begin{array}{l}\text{(Formula M3 ap-}\\ \text{plied to minors}\\ \text{of row 3)}\end{array}$$

Practice Exercise No. 33

A. Illustrate the derivation of formula M7 by expanding its determinants.

B. Illustrate the derivation of formula M8 in the variant form,

$$|\, a{+}Nc \ b \ c\,| = |\, a \ b \ c\,|.$$

C. Evaluate the following determinants, expediting each step by means of formula M8:

(1) $\begin{vmatrix} 1 & 3 & 2 \\ 5 & 7 & 10 \\ 7 & 9 & 14 \end{vmatrix}$ (2) $\begin{vmatrix} 1 & 6 & 1 & 2 \\ -3 & -5 & 2 & -2 \\ 5 & 13 & -3 & 6 \\ -2 & -6 & 7 & -1 \end{vmatrix}$

D. Show, without expanding, that each of the following determinants = 0.

(1) $\begin{vmatrix} 3 & -3 & 2 & 4 \\ 6 & -6 & 4 & 8 \\ -2 & 1 & 3 & 6 \\ 6 & -5 & 1 & 2 \end{vmatrix}$ (2) $\begin{vmatrix} 1 & 5 & 3 & 4 & 1 \\ 3 & 12 & 6 & 9 & -2 \\ 5 & 8 & -2 & 3 & 1 \\ -2 & 1 & 3 & -2 & 7 \\ 2 & -1 & -5 & 3 & 14 \end{vmatrix}$

(3) $\begin{vmatrix} 1 & a & b+c \\ 1 & b & a+c \\ 1 & c & a+b \end{vmatrix}$ (4) $\begin{vmatrix} 1 & 5 & 3 & 2 \\ 3 & -2 & -8 & 7 \\ -2 & 6 & 10 & 1 \\ 2 & 1 & -3 & 6 \end{vmatrix}$

Applications of Cramer's Rule

Having introduced Cramer's rule earlier to show how and why determinants are defined (pages 79 and 81), and having already applied the rule to the solution of determinate systems of linear equations in two variables (page 80), we may now use the preceding theorems to *extend the application of Cramer's rule to the solution of determinate systems of linear equations in three or more variables.*

One of the many conveniences of solving systems of linear equations by determinants is that the method can be applied, when appropriate, to find the value of only one variable without the need of solving the entire system for all variables as when more elementary methods are used, (for instance, in Chapter VI).

EXAMPLE 8: Solve for z only, the system of equations in Example 1 of Chapter VI (page 67):

$$3x + 3y - 2z - w = 40$$
$$2x + 3y - z - w = 35$$
$$3x + y + 4z - 2w = 15$$
$$-x + y + z + w = 35$$

SOLUTION:

$|a\,b\,c\,d| = \begin{vmatrix} 3 & 3 & -2 & -1 \\ 2 & 3 & -1 & -1 \\ 3 & 1 & 4 & -2 \\ -1 & 1 & 1 & 1 \end{vmatrix}$ (Substitution)

$= \begin{vmatrix} 3 & 6 & 1 & 2 \\ 2 & 5 & 1 & 1 \\ 3 & 4 & 7 & 1 \\ -1 & 0 & 0 & 0 \end{vmatrix}$ (Adding column 1 to columns 2, 3, and 4 by formula M8)

$= -(-1)\begin{vmatrix} 6 & 1 & 2 \\ 5 & 1 & 1 \\ 4 & 7 & 1 \end{vmatrix}$ (Formula M3 applied to minors of row 4)

$= \begin{vmatrix} -2 & -13 & 0 \\ 1 & -6 & 0 \\ 4 & 7 & 1 \end{vmatrix}$ (Adding -1 times row 3 to row 2 and -2 times row 3 to row 1 by formula M8)

$= \begin{vmatrix} -2 & -13 \\ 1 & -6 \end{vmatrix} = 12 - (-13) = 25.$ (Formulas M3 and M1)

$|a\,b\,k\,d| = 5\begin{vmatrix} 3 & 3 & 8 & -1 \\ 2 & 3 & 7 & -1 \\ 3 & 1 & 3 & -2 \\ -1 & 1 & 7 & 1 \end{vmatrix}$ (Substitution. Then divide column 3 by 5 and multiply the determinant by 5 according to formula M5′)

$= 5\begin{vmatrix} 2 & 4 & 15 & 0 \\ 1 & 4 & 14 & 0 \\ 1 & 3 & 17 & 0 \\ -1 & 1 & 7 & 1 \end{vmatrix}$ (Adding row 4 to rows 1 and 2, and twice row 4 to row 3 by formula M8)

$= 5\begin{vmatrix} 2 & 4 & 15 \\ 1 & 4 & 14 \\ 1 & 3 & 17 \end{vmatrix}$ (Formula M3 applied to minors of column 4)

$= 5\begin{vmatrix} 0 & -2 & -19 \\ 0 & 1 & -3 \\ 1 & 3 & 17 \end{vmatrix}$ (Adding -1 times row 3 to row 2, and -2 times row 3 to row 1 by formula M8)

$= 5\begin{vmatrix} -2 & -19 \\ 1 & -3 \end{vmatrix}$ (Formula M3 applied to minors of column 1)

$= 5(6 + 19) = 125.$ (Formula M1)

$z = |a\,b\,k\,d| / |a\,b\,c\,d|$ (Cramer's rule)

$= 125/25 = 5.$ Answer.   (Substitution)

Practice Exercise No. 34

(1) An airplane flies with a tailwind 750 miles in two hours. It returns against a wind twice as great and flies 900 miles in three hours. How fast does the plane fly?

(2) In a typical printer's metal there is 80% lead, 10% zinc, 5% tin, 5% antimony, If we start with 1000 lbs. of such a mixture how much of each should be added to produce a mixture which is slightly softer 80% lead, 10% zinc, 4.6% tin, and 5.2% antimony?

(3) Solve the following system of equations:

$$\begin{aligned} x + y + z + w &= -2 \\ 2x + y - z - w &= 1 \\ 3x - 2y + z + 2w &= 2 \\ x + 2y - z - 3w &= 6 \end{aligned}$$

(4) If you were told that from a system of equations, the solution for x was exactly the following:

$$x = \frac{\begin{vmatrix} 1 & 2 & -1 & 1 \\ 5 & 3 & 2 & 5 \\ 7 & 4 & 3 & -6 \\ 2 & 6 & 2 & 7 \end{vmatrix}}{\begin{vmatrix} 1 & 2 & -1 & 1 \\ 3 & 3 & 2 & 5 \\ 5 & 4 & 3 & -6 \\ -7 & 6 & -2 & 7 \end{vmatrix}},$$

write the solutions for the other unknowns, and write down what the original equations must have been.

Properties of Determinants, Continued

From previously derived determinant formulas we may derive many more which express still other interesting properties of determinants. For instance,

Theorem M10: *Moving a column (or row) of a determinant over n other columns (or rows) leaves the value of the determinant unchanged if n is an even number, and changes only the sign of the determinant if n is an odd number.* Typical (abbreviated) **formula:**

M10: $| a\ b\ c | = - | b\ a\ c | = + | b\ c\ a |$, etc.

Illustrated derivation follows from direct expansion of the above determinants by formula (Practice Exercise No. 35, Problem A). Although this particular theorem is not often applied in the solution of systems of linear equations or in analytic geometry, it is extremely useful in vector analysis.

The final determinant formula which we shall consider here requires a preliminary definition.

From the previously stated definition of linearly dependent equations (page 70), it is clear that any three or more linear equations in any number of variables,

$$\begin{aligned} a_1 x + b_1 y + c_1 z + \cdots &= k_1 \\ a_2 x + b_2 y + c_2 z + \cdots &= k_2 \\ &\vdots \\ a_i x + b_i y + c_i z + \cdots &= k_i \end{aligned}$$

are **linearly dependent** if

$$\begin{aligned} a_i &= M a_1 + N a_2 + \cdots + Q a_{i-1}; \\ b_i &= M b_1 + N b_2 + \cdots\ Q b_{i-1}; \\ c_i &= M c_1 + N c_2 + \cdots + Q c_{i-1};\ \text{etc.} \end{aligned}$$

By analogy, therefore, we may define any three of more sets of quantities — $a_1, b_1, \ldots k_1$; $a_2, b_2, \ldots k_2$; $\ldots$; $a_i, b_i, \ldots k_i$ — as **linearly dependent** if they are related by a set of quantities, M, N, $\ldots$ Q, etc., as above. Thus in the same way that the concept of linear dependence among three or more equations is a generalization of the concept of dependence between a pair of equations (page 71) *the concept of linear dependence among three or more sets of quantities is a generalization of the concept of proportionality between two sets of quantities.* And by means of this generalization we may now state a more generalized form of theorem M6 above (page 84); namely:

Theorem M9: *If the elements of any three or more rows (or columns) of a determinant are linearly dependent, the determinant has the value, 0.* Typical (abbreviated) **formula:**

M9: $| a\ b\ Ma{+}Nb | = 0.$

Although we shall apply this theorem mainly to the rows of a determinant, we illustrate its formula and derivation in terms of columns in order to take advantage of the convenience of abbreviated notation. This makes no difference, since we know from theorem M4 that any theorem applying to the columns of a determinant applies equally to its rows:

$$| a\,b\;Ma+Nb\,| = |\,a\;b\;Ma\,| + |\,a\;b\;Nb\,|$$
(Theorem M7)
$$= M\,|\,a\;b\;a\,| + N\,|\,a\;b\;b\,|$$
(Theorem M5)
$$= M(0) + N(0) = 0$$
(Theorem M6)

From this theorem it follows at once that *the determinant of a system* (defined, page 79) *of n linear equations in n variables must always be 0 if three or more of its equations are linearly dependent or linearly inconsistent.* For, the constant coefficients of the variables in any such sets of equations are linearly dependent (Chapter VI).

From the theorem it also follows that, *if you were to attempt to apply Cramer's rule* (page 81) *to solve a system of n linear equations in n variables with three or more of the equations linearly dependent,* then *the numerator determinants of Cramer's rule* would *also be 0.* For, the sets of constants in any such system of equations are linearly dependent by definition.

A discussion of the significance of these facts follows next after —

Practice Exercise No. 35

A. Illustrate the derivation of formula M10 as suggested in the text above.

B. Illustrate the derivation of formula M9 in the variant form:

$$|\;Mb+Nc\;\;b\;\;c\;| = 0.$$

C. Verify by formula M9 that the determinant of the following system of equations is 0: Exercise No. 27, problem (1).

$$2x - 3y + z = 6$$
$$x + 5y + 2z = 7$$
$$x - 21y - 4z = 15$$

D. Apply formula M9 to Example 7 in the text above (page 86). What are the values of M and N by which we may show row 3 to be linearly dependent upon rows 1 and 2 in this example?

E. Show that the following determinant has the value, 0.

$$\begin{vmatrix} 1 & 2 & 0 & 3 & 1 \\ 5 & 3 & -7 & -2 & -2 \\ -3 & -2 & 8 & 6 & 3 \\ 6 & -5 & -17 & -1 & 1 \\ 7 & 4 & -10 & 7 & 0 \end{vmatrix}$$

F. Show that the following is a second degree equation in x, and use the determinant properties to show that the roots are ± 4.

$$\begin{vmatrix} x & 1 & x \\ 11 & x & 12 \\ 1 & x & 4 \end{vmatrix} = 0$$

Exceptions to Cramer's Rule

To the statements of Cramer's rule above (page 81), it is usually added that the denominator determinant, called **the determinant of the system, must not be zero.** In algebraic form:

$$|\,a\,b\,| \neq 0, \quad |\,a\,b\,c\,| \neq 0, \quad |\,a\,b\,c\,d\,| \neq 0, \quad \text{etc.}$$

The reason for this important restriction is that *we can assign no definite numerical value to division by 0, or to the supposed fraction, K/0 by applying our usual rules for division by non-zero quantities.*

One may say, for instance, that "6/2 = 3" because "2 goes into 6 three times, and 6 divided by 2 is therefore 3." But zero will "go into" any non-0 quantity "forever and forever without end." In commonsense terms: no matter how many times you eat *none* of your cake, you still have all of it left. In mathematical terms: no actual number, however large, is big enough to be the quotient of a presumed division of any definite number, K, by 0.

Consequently, we say that $K/0$ *is undefined;* or, in another (equivalent) mode of speech, we say that $K/0$ *does not exist,* since a combination of symbols must first be defined in order *to be* the subject matter of mathematical treatment.

The fact that $K/0$ *is an impossible demand for an indefinitely large number* is sometimes indicated by the **equation-like expression,**

$$K/0 = \infty.$$

This may be read: "*K*-over-0 is indefinitely large;" or: "*K*-over-0 is larger than any pre-assignable quantity;" or even: "*K*-over-0 is infinite," meaning by "infinite" only "indefinitely large" or "larger than any pre-assignable quantity." But the expression should never be read: "*K/0* *equals* infinity." This is because "equals" has a very different meaning in connection with "infinity"—a meaning to be discussed again later in calculus (*Advanced Algebra and Calculus Made Simple*) where the problem of dealing with such expressions is more often encountered.

At this stage of your mathematical studies, therefore, *you must always be careful not to divide by 0* even when the cipher appears in some disguised form like $a_1b_2 - b_1a_2$ with a possible value of 0 as in Cramer's rule. At the same time, however, *it is important to be able to interpret the mathematical situations in which you may inadvertently chance upon this undefined case of division.*

As has already been seen in connection with the discussion of theorems M6 and M9 above (page 84 and page 87), **Cramer's rule produces a result in the form, $K/0$,** if we attempt to apply it to an inconsistent system of equations. For instance—

EXAMPLE 9: Apply Cramer's rule (legitimately?) to the inconsistent system of equations in Example 2 of Chapter II (page 20):

$$x - 2y = -8,$$
$$-3x + 6y = -6.$$

ATTEMPTED SOLUTION:

$$x = \frac{\begin{vmatrix} -8 & -2 \\ -6 & 6 \end{vmatrix}}{\begin{vmatrix} 1 & -2 \\ -3 & 6 \end{vmatrix}} = \frac{-48 - 12}{6 - 6} = \frac{-60}{0} = \infty\,!$$

(Cramer's rule except that $|a\ b| = 0$)

$$y = \frac{\begin{vmatrix} 1 & -8 \\ -3 & -6 \end{vmatrix}}{0} = \frac{-6 - 24}{0} = \frac{-30}{0} = \infty\,!$$

This example illustrates the general fact that **a result in the excluded form, $K/0$, from Cramer's rule is a sign that the system of equations is inconsistent and therefore has no possible simultaneous solution.** The example has, moreover, the interesting geometric interpretation that the parallel lines which graph the equations of such a system (page 34) can intersect only in a point with "infinite coordinates"—that is to say, in no actual point at all.

But the fact that the determinant of a system of linear equations is 0 does not necessarily mean that the system has no simultaneous solution. We have also seen in connection with the discussion of theorems M6 and M9 above, that Cramer's rule produces a result in the form, 0/0 if we attempt to apply it to a dependent system of equations. For instance—

EXAMPLE 10: Apply Cramer's rule (legitimately?) to find x in the linearly dependent system of equations of Example 2, Chapter VI (page 69):

$$2x + 4y - z = 75,$$
$$x + 4y = 70,$$
$$3x + 4y - 2z = 80.$$

ATTEMPTED SOLUTION:

$$x = \frac{\begin{vmatrix} 75 & 4 & -1 \\ 70 & 4 & 0 \\ 80 & 4 & -2 \end{vmatrix}}{\begin{vmatrix} 2 & 4 & -1 \\ 1 & 4 & 0 \\ 3 & 4 & -2 \end{vmatrix}} = \frac{0}{0}$$

(By theorem M9, since the elements in the three rows of each determinant are linearly dependent)

It may seem to you at first glance that 0/0 is a possible exception to the rule against division by 0. You may perhaps think that, "since

$$1/1 = 2/2 = \tfrac{1}{2}/\tfrac{1}{2} = n/n, \text{ when } n \neq 0,$$

it may also be reasonable to say that $0/0 = 1$."

But let us put this thought to the test of consistency with other mathematical definitions and rules. Suppose we do let

$$0/0 = 1, \qquad \text{just as} \quad 2/2 = 1.$$

Then, according to Axiom 3 (page 16),

$$0 = 0(1), \qquad \text{just as} \quad 2 = 2(1).$$

So far this may appear all right.

Notice, however, that *only* the number 1 will do as the coefficient of 2 in the righthand equation above. In other words, 1 is the unique value of the fraction 2/2. But *any number whatsoever* may be substituted for 1 in the lefthand equation. It is equally true, for instance, that

$$0 = 0 \ (999{,}999{,}999).$$

This means that the **fraction-like expression, 0/0**, *may be assigned any value at all.* Thus it illustrates the inappropriateness of division by 0 even more strikingly than the fraction-like expression, $K/0$. For, it is not merely indeterminately large, but *absolutely indeterminate.*

Consequently, the main lesson of Example 10 is its illustration of the general fact that a result in the excluded form, 0/0, from Cramer's rule is a sign that the system of equations is dependent and therefore has infinitely many solutions.

Practice Exercise No. 36

Verify by actual expansion that the determinants in the attempted solution of Example 10, page 89 are 0.

Summary

When the determinant of a system of n linear equations in n variables is not equal to zero, the system is determinate and its unique solution is given by **Cramer's rule.**

The typical formula of Cramer's rule for the case where **n = 3** is —

$x = |\ k\ b\ c\ |/|\ a\ b\ c\ |,$

$y = |\ a\ k\ c\ |/|\ a\ b\ c\ |,$ provided $|\ a\ b\ c\ | \neq 0.$

$z = |\ a\ b\ k\ |/|\ a\ b\ c\ |.$

Exclusion of the case where $|\ a\ b\ c\ | = 0$ in Cramer's rule is an instance of the more general fact that *division by zero is an undefined mathematical operation.* This is because, to be consistent with other mathematical definitions, we must regard the fraction-like expression, $K/0$, as "infinite," which means "larger than

any pre-assignable quantity"; and the fraction-like expression, 0/0, could be assigned any value whatsoever if it were to be admitted.

However, when an excluded application of Cramer's rule would lead to *a result in the form, $K/0$,* this is a sign that the system of equations is **inconsistent.** And when an excluded application of Cramer's rule would lead to *a result in the form, 0/0,* this is a sign that the system of equations is **dependent.**

Determinants are defined and evaluated by formulas like those in the following —

Table of Determinant Formulas

M1: $|\ a\ b\ | = a_1 b_2 - a_2 b_1.$

Direct expansion of a second-order determinant.

M2: $|\ a\ b\ c\ | = a_1 b_2 c_3 + a_2 b_3 c_1 + a_3 b_1 c_2 - a_3 b_2 c_1 - a_2 b_1 c_3 - a_1 b_3 c_2.$

Direct expansion of a third-order determinant.

(Third-order examples hereafter)

M3: $|\ a\ b\ c\ | = a_1 \begin{vmatrix} b_2 & c_2 \\ b_3 & c_3 \end{vmatrix} - a_2 \begin{vmatrix} b_1 & c_1 \\ b_3 & c_3 \end{vmatrix} + a_3 \begin{vmatrix} b_1 & c_1 \\ b_2 & c_2 \end{vmatrix}$

$= -b_1 \begin{vmatrix} a_2 & c_2 \\ a_3 & c_3 \end{vmatrix} + b_2 \begin{vmatrix} a_1 & c_1 \\ a_3 & c_3 \end{vmatrix} - b_3 \begin{vmatrix} a_1 & c_1 \\ a_2 & c_2 \end{vmatrix}$, etc.

Expansion by minors.

M4: $\begin{vmatrix} a_1 & b_1 & c_1 \\ a_2 & b_2 & c_2 \\ a_3 & b_3 & c_3 \end{vmatrix} = \begin{vmatrix} a_1 & a_2 & a_3 \\ b_1 & b_2 & b_3 \\ c_1 & c_2 & c_3 \end{vmatrix}.$

Equivalence of columns and rows.

M5: $N|\ a\ b\ c\ | = |\ Na\ b\ c\ | = |\ a\ Nb\ c\ | = |\ a\ b\ Nc\ |.$

M5′: $|\ a\ b\ c\ | = N \left|\ \dfrac{a}{N}\ b\ c\ \right| = \dfrac{1}{N} |\ Na\ b\ c\ |$, etc.

Multiplication or division of a determinant by a constant, N.

M6: $|\ a\ b\ Nb\ | = 0.$

A condition that a determinant = 0.

M7: $|\ m+n\ b\ c\ | = |\ m\ b\ c\ | + |\ n\ b\ c\ |.$

The sum of two determinants.

M8: $|\ a\ b\ c\ | = |\ a+Nb\ b\ c\ | = |\ a+Nc\ b\ c\ |$, etc.

Equivalence of two determinants.

M9: $| a \ b \ Ma+Nb | = 0.$

    Generalization of Formula M6.

M10: $| a \ b \ c | = - | b \ a \ c | = + | b \ c \ a |$
$$= - | c \ b \ a | \text{, etc.}$$

    Changes of sign by interchange of columns or rows.

The practical routine for evaluating determinants of higher orders is to simplify their form by formulas M5 and M8 so that they can be reduced to lower orders by formula M3 until finally expanded by formula M1.

CHAPTER VIII

TRIGONOMETRIC FUNCTIONS AND EQUATIONS

From elementary mathematics the reader should already be familiar with the basic trigonometric functions — $\sin x$, $\cos x$, $\tan x$, etc. — defined for angles from 0° to 90°. But for advanced mathematics, including certain applications in analytic geometry and vector analysis, a further knowledge of this topic is required.

Trigonometric Equations

Since trigonometric functions are useful in stating many basic scientific principles, they are often encountered in equations.

When only *constant values of trigonometric functions* appear in an equation, these constants may be treated like any others and no new mathematical problem arises from their occurrence.

For instance, $\sin 30°$ is a constant $= \frac{1}{2}$ (*MMS*, Chapter XV). Hence an equation like

$$y = x \sin 30°$$

is in every way equivalent to the already familiar equation,

$$y = \tfrac{1}{2}x, \qquad \text{(substituting } \sin 30° = \tfrac{1}{2}),$$

and so may be treated as in Chapter II above.

When *variable trigonometric functions* appear in an equation, however, the equality is of a basically different type. Then it is called a **trigonometric equation** because it cannot be fully solved by algebraic methods alone.

For instance, $\sin x$ is a variable with a definite value corresponding to each possible value of x (Chapter III, page 27). Hence an equality like

$$y = \tfrac{1}{2} \sin x$$

is by definition a *trigonometric equation*. And it will be seen throughout this chapter that such equations require, for their *complete* solution, a kind of mathematical treatment not yet discussed here.

Preliminary Algebraic Treatment

From a purely *algebraic* point of view, trigonometric equations may be treated as *linear equations* or as *quadratic equations*, etc., not in variables like x and y, etc., but rather in the given trigonometric functions of these variables. From this point of view, for instance, the equation,

$$2 \tan x - 1 = 0,$$

is a *linear equation*, not in x, but *in* $\tan x$. And from the same point of view the equation,

$$2 \cos^2 x + \cos x - 1 = 0,$$

is a *quadratic equation*, not in x, but *in* $\cos x$.

Consequently, we can solve such equations by the algebraic methods already discussed, not for values of a variable like x, but only for values of trigonometric functions of these variables, like $\tan y$, $\cos x$, etc.

The simplest way to do this is first to substitute other arbitrary letters for the trigonometric functions — $u = \tan x$, etc. But the device may be dropped as soon as you get used to thinking of trigonometric functions as the variables for which trigonometric equations may first have to be solved algebraically.

EXAMPLE 1a: Solve for $\tan x$:
$$2 \tan x - 1 = 0.$$

SOLUTION (algebraic): Substituting $u = \tan x$, we obtain the equivalent linear equation in u,
$$2u - 1 = 0.$$
Then, as in Chapter II,

$$2u = 1, \qquad \text{(Transposing } -1)$$
$$u = \tfrac{1}{2}. \qquad \text{(Dividing by } 2 = 2)$$

Hence our *answer* is:
$$\tan x = \tfrac{1}{2}. \qquad \text{(Resubstituting } u = \tan x)$$

Check:
$$2(\tfrac{1}{2}) - 1 = 1 - 1 = 0. \ \checkmark \ \text{(Substitution)}$$

EXAMPLE 2a: Solve for $\cos x$:
$$2 \cos^2 x + \cos x - 1 = 0. \qquad \text{(See Note)}$$

(NOTE: Recall here that $\cos^2 x$ means $(\cos x)^2$, the exponent applying to the cosine of x, and not to x itself. The expression is therefore to be read: "cosine x ... squared," or better: "cosine-squared x," meaning: "the square of the cosine of x." But it is **never** to be read as "cosine ... x-squared," for this would be the different quantity, $\cos x^2$. See *MMS*, Chapter XV.)

SOLUTION (algebraic): As in the preceding example, we may substitute $u = \cos x$ to obtain the equivalent quadratic equation, $2u^2 + u - 1 = 0$, etc. Recognizing the quadratic form of the equation in $\cos x$, however, we may also write directly:

$$\cos x = \frac{-1 \pm \sqrt{1^2 - 4(2)(-1)}}{2(2)}$$ (The quadratic formula with $a = 2, b = 1, c = -1$)

$$= \frac{-1 \pm \sqrt{9}}{4}$$ (Removing parentheses, etc.)

$$= \frac{-1 \pm 3}{4} = \tfrac{1}{2}, -1.$$ (Extracting roots, etc.)

Check:
$2(\tfrac{1}{2})^2 + \tfrac{1}{2} - 1 = \tfrac{1}{2} - \tfrac{1}{2} = 0.$ ✓ (Substituting $\cos x = \tfrac{1}{2}$)

$2(-1)^2 + (-1) - 1 = 2 - 2 = 0.$ ✓ (Substituting $\cos x = -1$)

When a system of trigonometric equations is **defective** — has fewer equations than trigonometric variables (page 23) — it may often be made non-defective if the number of variables is first reduced by means of the basic trigonometric formulas — those numbered 49 to 53 in the Review Table (Chapter I, page 16). Since this step usually raises the degree of the equations, you must be particularly careful to check the solution to make sure that extraneous roots have not been introduced (as when solving radical equations equivalent to quadratics, Chapter IV, page 44).

EXAMPLE 3a: Solve for $\sin x$ and $\csc x$:
$$2 \sin x + 2 \csc x - 5 = 0.$$

SOLUTION (algebraic): Here we have two trigonometric variables and only one equation. Consequently, if we set $u = \sin x$, $v = \csc x$, we shall have the defective system of one linear equation in two variables, $2u + 2v - 5 = 0$, which cannot be solved for a unique solution (Chapter II). We may recall, however, that

$$\csc x = 1/\sin x.$$ (Formula R49)

Hence we may rewrite our equation as

$$2 \sin x + 2/\sin x - 5 = 0.$$ (Substitution)

And then,

$$2 \sin^2 x - 5 \sin x + 2 = 0,$$ (Multiplying by $\sin x = \sin x$)

$$\sin x = \frac{-(-5) \pm \sqrt{(-5)^2 - 4(2)2}}{2(2)}$$ (The quadratic formula)

$$= \frac{5 \pm 3}{4} = 2, \tfrac{1}{2}.$$ (Removing parentheses, etc.)

$$\csc x = 1/2, 1/\tfrac{1}{2} = \tfrac{1}{2}, 2.$$ (Substitution in Formula R49)

Answer (algebraic):
$\sin x, \csc x = 2, \tfrac{1}{2}; \tfrac{1}{2}, 2.$ (But see discussion below)

Check (algebraic only):
$2(2) + 2(\tfrac{1}{2}) - 5 = 5 - 5 = 0.$ ✓ (Substituting $2, \tfrac{1}{2}$)

$2(\tfrac{1}{2}) + 2(2) - 5 = 5 - 5 = 0.$ ✓ (Substituting $\tfrac{1}{2}, 2$)

Equations involving trigonometric functions may, of course, be indeterminate for the same general *algebraic* reasons as equations involving other kinds of variables (Chapter II, page 26; Chapter V, page 66). Such equations may also be indeterminate, however, because of the special *trigonometric* relationships between their variables.

For instance, the root $\sin x, \csc x = 2, \tfrac{1}{2}$ checks just as well *algebraically* in the equation of Example 3a above as the root $\sin x, \csc x = \tfrac{1}{2}, 2$. But we shall see presently that, although there are values of x which satisfy the latter *trigonometrically*, there are no possible values of x which satisfy the former *trigonometrically*. Consider, moreover —

EXAMPLE 4a: Solve for $\sin x$ and $\cos x$:
$$\sin^2 x + \cos^2 x - 1 = 0.$$

ATTEMPTED SOLUTION: Although this equation has two trigonometric variables, it may be reduced as in the solution of Example 3a to an equation in one such variable. For,

$$\cos^2 = 1 - \sin^2 x. \qquad \text{(Formula R53)}$$

Then, however,

$$\sin^2 x + (1 - \sin^2 x) - 1 = 0. \qquad \text{(Substitution)}$$

And we get the typical result of mathematical indeterminacy (see references above), —

$$0 = 0! \qquad \text{(Removing parentheses, etc.)}$$

Explanation: The given trigonometric equation is simply a different way of writing Review Formula 53 which is an **identity** — an equation which holds good for *all* values of its variables, as in the non-trigonometric cases of $x = x$ or $x + y = y + x$. Hence standard methods for solving equations, when applied to such an equality, will produce only the above mathematical truism (page 22). As in non-trigonometric cases, however, we can always verify such an equality for particular values of its variables. For instance, when $x = 30°$, $\sin x = \frac{1}{2}$, $\cos x = \sqrt{3}/2$, (*MMS*, Chapter XV), and

$$(\tfrac{1}{2})^2 + (\sqrt{\tfrac{3}{2}})^2 - 1 = \qquad \text{(Substitution)}$$
$$\tfrac{1}{4} + \tfrac{3}{4} - 1 = 0. \checkmark \qquad \text{(Removing parentheses, etc.)}$$

Practice Exercise No. 37

Solve the following equations algebraically for values of their trigonometric functions as above. Check your results algebraically. But do not assume that all results which do so check will necessarily hold good under further analysis made later in this chapter.

(1) $\sin x + \cos x = 1$
(2) $\sin x - \cos x = 1$
(3) $\cos^2 x - \sin^2 x = 2$
(4) $\tan x - \cot x = 2$
(5) $\tan x = \cot x$
(6) $\csc x + 2 \sin x = 1$
(7) $\sin^2 x = 4$
(8) $\tan x + \cot x = 1$

Inverse Trigonometric Functions

By the *algebraic* methods illustrated in the above examples, we have thus far solved trigonometric equations only for values of their trigonometric variables which satisfy their *algebraic* conditions. Hence our answers have taken the form,

$$\tan x = \tfrac{1}{2}. \qquad \text{(Example 1a)}$$
$$\cos x = \tfrac{1}{2}, -1. \qquad \text{(Example 2a)}$$
$$\sin x, \csc x = 2, \tfrac{1}{2}; \ \tfrac{1}{2}, 2. \qquad \text{(Example 3a)}$$

Although simpler than the original equations, these algebraic solutions are still trigonometric equations. For most practical problems, they still need to be solved *trigonometrically* for those values of the variable which satisfy their *trigonometric* conditions. This brings up the topic of *inverse trigonometric functions*.

The inverse of any function $y = f(x)$ has already been defined as the equivalent function, $x = g(y)$ (Chapter III, page 29). Hence for a trigonometric function like

$$y = \sin x,$$

the **inverse trigonometric function**, stated in words, is

x equal to the angle whose sine function is y.

The customary way of writing this more briefly, however, is:

$$x = sin^{-1} y;$$

or sometimes:

$$x = \arcsin y.$$

Either of these notations may be read as the verbal statement preceding, or as: "$x = arcsine\ y$." It is important, however, never to confuse $\sin^{-1} y$ with the minus-one-power of a trigonometric function. For this reason, the minus-one-power of a function like $\sin x$ is **never** properly written with the exponent "-1" like the exponent "2" in $\sin^2 x$. Rather, it is always written with a parenthesis as in $(\sin x)^{-1}$; or else it is expressed by some equivalent function such as $1/\sin x$ or $\sec x$.

In terms of inverse trigonometric functions, we may now rewrite the algebraic solution of Example 1a above as

$$x = \tan^{-1} \tfrac{1}{2}.$$

This means: "x is the angle whose tangent is $\frac{1}{2}$."

Or, we can rewrite the algebraic solution of Example 2a above as

$$x = \cos^{-1} \tfrac{1}{2}, -1.$$

This means: "x is the angle whose cosine is $\frac{1}{2}$, and x is also the angle whose cosine is -1."

Finally, we can rewrite the algebraic solution of Example 3a above as

$$x = \sin^{-1} 2, \csc^{-1} \tfrac{1}{2}; \ \sin^{-1} \tfrac{1}{2}, \csc^{-1} 2.$$

This means: "x is the angle whose sine is 2, and whose cosecant is $\frac{1}{2}$; and x is also the angle whose sine is $\frac{1}{2}$, and whose cosecant is 2."

In order to complete the solution of our original trigonometric equations in these examples, therefore, we must next find and check solutions for the resulting **inverse trigonometric equations** — wherever that is possible.

The student who has his elementary trigonometry fresh in mind can, of course, find from a table of natural trigonometric functions, by interpolation if necessary, that $\frac{1}{2} = \tan 26°34'$ to the nearest minute (*MMS*, Chapter XV). Hence he may correctly conclude that, to the nearest minute, one possible *trigonometric solution* of the equation in Example 1a is

$$x = \tan^{-1} \tfrac{1}{2} = 26°34'.$$

Recalling also that $\frac{1}{2} = \cos 60° = \sin 30°$, and that $2 = \csc 30°$, you may similarly conclude that one possible *trigonometric solution* of the equation in Example 2a is

$$x = \cos^{-1} \tfrac{1}{2} = 60°;$$

and one possible *trigonometric solution* of the equation in Example 3a is

$$x = \sin^{-1} \tfrac{1}{2}, \csc^{-1} 2 = 30°.$$

Check in the last case, for instance, is:

$2 \sin 30° + 2 \csc 30° - 5 = \qquad$ (Substituting
$2(\tfrac{1}{2}) + 2(2) - 5 = 5 - 5 = 0.$ ✓ $\quad x = 30°$)

In no standard table of trigonometric functions (*MMS*, Chapter XV), however, can you find values of x for which $x = \cos^{-1} -1$, for which $x = \sin^{-1} 2$, or for which $x = \csc^{-1} \tfrac{1}{2}$, as required by the remaining algebraic solutions of Examples 2a and 3a. The questions therefore arise:

Are there valid trigonometric solutions for our equations corresponding to these algebraic solutions? And also, *are there perhaps other trigonometric solutions than those already found corresponding to the previously considered algebraic solutions?*

Scanning a brief table of trigonometric functions (*MMS*, Chapter XV), you can readily satisfy yourself that there are no other possible values of x and y *from 0° to 90°* for which $y = \tan^{-1} \tfrac{1}{2}$, etc. But we shall presently see that the above questions require a more complete answer than one restricted to the trigonometric functions of angles within this range.

Practice Exercise No. 38

Restate as inverse trigonometric equations your algebraic solutions of the problems in Practice Exercise No. 37. Find the values of x and/or y from 0° to 90° which satisfy these equations trigonometrically. Use a table of natural trigonometric functions where necessary (*MMS*, Chapter XV).

Angles of Any Magnitude

Only positive angles are usually considered in elementary mathematics, and these only from 0° to 360° (*MMS*, Chapter XIV). But in advanced mathematics, this limited concept of the amount of opening between two lines may be generalized to include angles of any real magnitude, positive or negative.

For this purpose, it is customary to take the positive x-axis of an x,y rectangular coordinate system as the standard reference line.

Imagine a wheel in the plane of such a system of coordinates, with its hub at the origin O, and with its outer rim intersecting the positive x-axis at the fixed point Q.

If this wheel is set so that a given spoke OP lies along the x-axis in its positive direction, the point P at the outer end of the spoke will coincide with the fixed point Q on the x-axis.

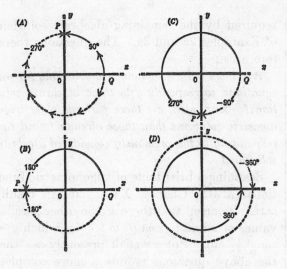

Fig. 22

Then, since there is no opening between the two lines OQ and OP, we may say that the angle $QOP = 0°$.

If, now, our wheel is rotated about the origin so that the spoke OP moves in a *counter-clockwise direction* — like the hands of a clock moving backwards — the generated angle QOP is regarded as **positive**. But if the wheel is rotated in the opposite, *clockwise direction* — like the hands of a clock moving normally — the generated angle QOP is regarded as **negative**.

Thus the angle between the positive x-axis and the positive y-axis is $(+)90°$ measured counter-clockwise, but is $-270°$ measured clockwise, as in diagram A of Figure 22, etc. And the angle between the positive x-axis and itself, which we previously noted to be $0°$, is also $360°$ measured through one complete revolution about the origin in a counter-clockwise direction, but is $-360°$ measured through one complete revolution about the origin in a clockwise direction, as in diagram D of Figure 22.

As our imaginary wheel makes more than one complete rotation about its hub at the origin, counter-clockwise or clockwise, the spoke OP will of course duplicate its positions the first time around. Suppose, then, that the wheel come to rest on its last partial rotation so that the positive x-axis and the spoke OP

form the positive angle θ, or the negative angle $-\phi = \theta - 360°$. (Note: θ and ϕ are the Greek letters, *theta* and *phi*, pronounced *thay-ta* with a "th" as in "*thin*," and *fee* or *figh*, respectively.)

If, now, our mathematical purposes require us to keep track of the fact that this is a second, third, or n'th revolution of OP about O, we may do so by defining the angle QOP as $= \theta + 360°$ for a second counter-clockwise revolution, or as $= \theta + (n - 1)360°$ for an n'th counter-clockwise revolution, and as $= -\phi - 360°$ for a second clockwise revolution, or as $= -\phi - (n - 1)360°$ for an n'th clockwise revolution.

In Figure 23, for instance, OP is shown coinciding with the positive y-axis on its second revolution so that $\theta = 90°$, and $\phi = 90° - 360° = -270°$. Hence the angle $QOP = 90° + 360° = 450°$ measured counter-clockwise, or $QOP = -270° - 360° = -630°$ measured clockwise. The next time around the corresponding measures of the angle QOP would be $90° + 2(360°) = 810°$ for a third counter-clockwise revolution, or $-270° - 2(360°) = -990°$ for a third clockwise revolution, etc.

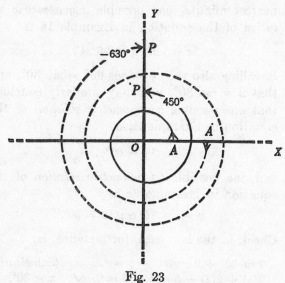

Fig. 23

Practice Exercise No. 39

(1) What are all the possible values from $-1,000°$ to $+1,000°$ of the angle between the

positive and negative directions of the x-axis?

(2) Between the positive x-axis and the negative y-axis?

(3) Between the positive x-axis and itself?

(4) Between the positive x-axis and the graph of $y = x$ in the first quadrant?

(5) The same in the third quadrant?

SUGGESTION:

Sketch diagrams similar to those of Figures 22 and 23 to illustrate your answers.

Radian Measure of Angles

While discussing *the magnitude of angles* it is convenient to introduce *a different unit of measure* — that of **radians** rather than of degrees, This unit is essential in the calculus and is therefore often used in other branches of mathematics as well.

Suppose the imaginary wheel described above to have a radius $= OP = r$, and suppose the wheel to be turned so that the arc $PQ = r$, as in Figure 24. **An angle of one radian** is then defined as an angle equal to POQ. In more formal language meaning the same thing: *an angle of one radian is an angle subtended by an arc of a circle equal in length to the radius of the circle.*

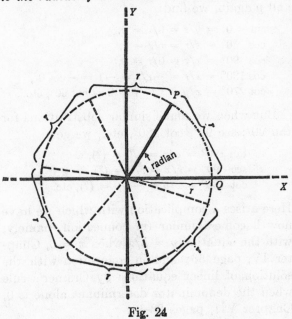

Fig. 24

Since the circumference of a circle is $2\pi r = 2(3.14\ldots)r = (6.28\ldots)r$ (*MMS*, Chapter XIV), we can measure off $6.28\ldots$ arcs of length r on the circumference of a circle of radius r. As shown in Figure 24, therefore,

2π radians = 6.28 ⋯ radians = **360°.**

Dividing by 2π,

1 radian = $360°/2\pi$ = **57.3°**, approximately.

Or dividing by 360,

1° = $2\pi/360$ = **.017453 radians**, approximately.

Normally, however, angles measured in radians are expressed in terms of π, thus:

$\pm \pi$ radians $= \pm 180°$,   (Div. by 2 = 2)
$\pm \pi/2$ " $= \pm\ 90°$,   (Div. by 4 = 4)
$\pm \pi/4$ " $= \pm\ 45°$,   (Div. by 8 = 8, etc.)
$\pm 4\pi$ " $= \pm 720°$, etc. (Mult. by 2 = 2)

Consequently, when the measure of an angle is given as "$n\pi$" this is understood to mean $n\pi$ *radians*

Practice Exercise No. 40

A. How many radians are there in the following angles?

(1) 30°   (2) −60°   (3) 235°   (4) −225°
(5) 450°   (6) −720°   (7) 10°   (8) −2°

B. How many degrees are there in the following angles?

(9) $\pi/4$ (10) $-\pi/3$ (11) $\pi/6$ (12) -3π
(13) 4π (14) -9π (15) .1 radians
(16) $-.2$ radians

Trigonometric Functions of Any Angle

Although trigonometric functions are usually defined in elementary mathematics only for positive angles from 0° to 90°, in advanced mathematics these definitions also are extended. When generalized, they apply to angles of any magnitude or either sign.

Returning to the imaginary wheel described above, let its radius again be r so that Q is the fixed point $(r, 0)$ on the positive x-axis (page 95). Let the point P at the outer end of the spoke OP be designated by coordinates as $P(x,y)$ so that $r = \sqrt{x^2 + y^2}$ (formula R40, page 16). And let the angle QOP again be designated θ. Then, regardless of the magnitude or sign of θ, the generalized definitions for the six basic trigonometric functions of θ are:

$$\begin{array}{ll} \sin\theta = y/r, & \csc\theta = r/y, \\ \cos\theta = x/r, & \sec\theta = r/x, \\ \tan\theta = y/x, & \cot\theta = x/y. \end{array}$$

Suppose $P(x,y)$ to assume any first quadrant position $P_1(a,b)$ so that θ is the acute angle θ_1 shown in the first quadrant of Figure 25. The above definitions are then equivalent to the familiar elementary ones for functions of an acute angle in a right triangle. For here, r is the hypotenuse; x, equal to a, is the side adjacent to the acute angle θ_1; and y, equal to b, is the side opposite θ_1 (*MMS*, Chapter XV). Hence, substituting $x = a$, $y = b$, we get:

$$\sin\theta_1 = y/r = b/r,$$
$$\cos\theta_2 = x/r = a/r, \text{ etc.}$$

And the other trigonometric functions of θ_1 have their usual values, listed in the Quadrant I column of the Table below Figure 25.

But suppose $P(x,y)$ to assume the position $P_2(-a,b)$ so that θ is the **corresponding second-quadrant angle**,

$$\theta_2 = 180° - \theta_1 = \pi - \theta_1;$$

or the position $P_3(-a, -b)$ so that θ is the **corresponding third-quadrant angle**,

$$\theta_3 = 180° + \theta_1 = \pi + \theta_1;$$

or the position $P_4(a, -b)$ so that θ is the **corresponding fourth-quadrant angle**,

$$\theta_4 = 360° - \theta_1 = 2\pi - \theta_1.$$

Now we no longer have a right triangle of which θ_2, θ_3, or θ_4, is an acute angle. Nevertheless, the generalized definitions above still give us *the same absolute values* — numerical values without regard to sign — for the trigonometric functions of these angles. But *in each quadrant the signs of certain functions*

change as the signs of x and y change while r, being the length of a line, always remains positive.

In the *second quadrant*, for instance, by substituting $y = b$, $x = -a$, we find:

$$\sin\theta_2 = y/r = b/r = \sin\theta_1,$$
$$\cos\theta_2 = x/r = -a/r = -\cos\theta_1,$$
$$\tan\theta_2 = y/x = b/-a = -b/a = -\tan\theta_1, \text{ etc.}$$

Or, in the *third quadrant*, by substituting $x = -a$, $y = -b$, we find:

$$\sin\theta_3 = y/r = -b/r = -\sin\theta_1,$$
$$\cos\theta_3 = x/r = -a/r = -\cos\theta_1,$$
$$\tan\theta_3 = y/x = -b/-a = b/a = \tan\theta_1, \text{ etc.}$$

All other values and changes of sign for the trigonometric functions of angles in different quadrants are shown in the four Quadrant columns of the Table below Figure 25, and may be verified as above.

Suppose $P(x,y)$ next to assume any of the positions on the coordinate axes: $Q(r,0)$ so that $\theta = 0°$, or $P_1'(0,r)$ so that $\theta = 90° = \pi/2$, or $P_2'(-r,0)$ so that $\theta = 180° = \pi$, etc., as also shown in Figure 25. All entries for the sin and cos functions in the appropriate (alternate) columns of the accompanying Table, and most entries for the other functions in these columns may still be verified in the same way. For instance, by appropriate substitutions for x and y again, we find:

$$\begin{array}{l} \sin \ \ 0° = y/r = 0/r = 0, \\ \cos \ \ 0° = x/r = r/r = 1, \\ \cos \ 90° = x/r = 0/r = 0, \\ \cos 180° = x/r = -r/r = -1 = -\cos 0°, \\ \cot 270° = x/y = 0/-r = 0 = \cot 90°, \text{ etc.} \end{array}$$

But when we make similar substitutions for tan 90°, sec 90°, cot 180°, etc., we get:

$$\begin{array}{l} \tan \ \ 90° = y/x = r/0 = (?), \\ \sec \ \ 90° = r/x = r/0 = (?), \\ \cot 180° = x/y = -r/0 = (?), \text{ etc.} \end{array}$$

Here arises a complication with which we have now become familiar (in connection, namely, with the equation $y = k/x$ when $x = 0$, Chapter IV, page 56; and in connection with the solution of linear equations by Cramer's rule when the denominator determinant alone is 0, Chapter VII, pages 88–9).

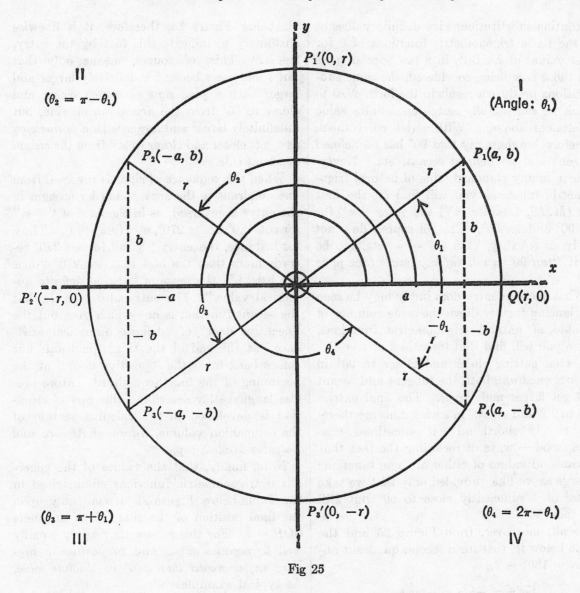

Fig 25

TABLE: Typical Values of Trigonometric Functions

Quad't		I		II		III		IV	
θ	0	θ_1	$\pi/2$	θ_2	π	θ_3	$3\pi/2$	θ_4	2π
P (x,y)	Q $(r,0)$	P_1 (a,b)	P'_1 $(0,r)$	P_2 $(-a,b)$	P'_2 $(-r,0)$	P_3 $(-a,-b)$	P'_3 $(0,-r)$	P_4 $(a,-b)$	Q $(r,0)$
$Sin\ \theta$	0	b/r	1	b/r	0	$-(b/r)$	-1	$-(b/r)$	0
$Cos\ \theta$	1	a/r	0	$-(a/r)$	-1	$-(a/r)$	0	a/r	1
$Tan\ \theta$	0	b/a	$\pm\infty$	$-(b/a)$	0	b/a	$\pm\infty$	$-(b/a)$	0
$Cot\ \theta$	$\mp\infty$	a/b	0	$-(a/b)$	$\mp\infty$	a/b	0	$-(a/b)$	$\mp\infty$
$Sec\ \theta$	1	r/a	$\mp\infty$	$-(r/a)$	-1	$-(r/a)$	$\mp\infty$	r/a	1
$Csc\ \theta$	$\mp\infty$	r/b	1	r/b	$\pm\infty$	$-(r/b)$	-1	$-(r/b)$	$\mp\infty$

Routine substitutions give definite values of all the basic trigonometric functions of θ for most values of θ. But, in a few special cases like those now being considered, the same substitutions produce a result in the form $N/0$ to which we can usefully assign no definite value (references above). With strict correctness, therefore, we must say that 90° has no defined tangent and no defined secant, etc. Nevertheless, in any standard table of natural trigonometric functions you will find on the first line (*MMS*, Chapter XV) the entry "∞" for tan 90° and sec 90°. This, of course, does not imply an equality, "tan 90° = ∞, etc." to be read: "tan 90° *equals infinity*, etc." (See page 89).

What such an entry does imply may be seen by glancing further down the same columns of a table of natural trigonometric functions. There you will find that for values of $\theta = 87°$, 88°, etc., getting closer and closer to 90° in the *first quadrant*, both the tangent and secant of θ get larger and larger. The final entries for tan 90° and sec 90° in such a table are therefore simply shorthand — if sometimes misunderstood — ways of recording the fact that we can find values of either of these functions as large as we like, provided only that we take values of θ sufficiently close to 90° but still less than 90°.

Recall, moreover, from Figure 25 and the Table below it, that for a second quadrant angle $\theta_2 = 180° - \theta_1$,

$$\tan \theta_2 = -b/a = -\tan \theta_1,$$

and

$$\sec \theta_2 = -b/r = -\sec \theta_1.$$

This means that for values of $\theta = 93°$, 92°, etc., getting closer and closer to 90° from the *second quadrant* side, the tan and sec of θ have similar sets of larger and larger values, but prefixed by a *minus* sign. In other words, we can find values of either function less than any pre-assigned quantity, provided only that we take values of θ sufficiently close to 90° but still more than 90°. When tabulating typical values of trigonometric functions as in the Ta-

ble below Figure 25, therefore, it is likewise customary to indicate this fact by the entry, "$\pm \infty$." This, of course, means only that tan θ and sec θ become indefinitely larger and larger with a *plus sign* as θ gets closer and closer to 90° from the *first quadrant* side, but indefinitely larger and larger with a *minus sign* as θ gets closer and closer to 90° from the *second quadrant* side.

When this sequence of signs is reversed from one quadrant to the next, the order of signs in the entry is reversed, as in the case of "$\mp \infty$" for cot 180° or sec 270°, etc. (page 99). Thus, for instance, the entry "$\mp \infty$" for sec 270° records more than the fact that sec 270° "does not exist" in the sense of "has no definite numerical value." The entry also tells us that the secant function is one which goes out the "negative door" of "definite numerical existence" at the end of the third quadrant, but comes back in at the "positive door" at the beginning of the fourth quadrant. More precise language for describing this sort of situation is developed in the calculus section of the companion volume, *Advanced Algebra and Calculus Made Simple*.

Note, finally, that the values of the generalized trigonometric functions summarized in the Table below Figure 25 depend only upon the final position of the side OP of an angle $QOP = \theta$. For this reason they apply equally well to *negative angles* and to *positive or negative angles greater than 360° in absolute value*. As typical examples:

$$\sin -\theta_1 = y/r = -b/r = -\sin \theta_1, \text{ etc.}$$

And

$$\sin (\theta_1 \pm 2\pi n) = y/r = b/r = \sin \theta_1, \text{ etc.}$$

In view of these generalized definitions, the fact that tables of trigonometric functions range only from 0° to 90° (*MMS*, Chapter XV) may at first seem puzzling. The explanation is, however, that *the absolute values of these functions are repeated in each quadrant* somewhat, although not exactly, as the values of common logarithmic mantissas are repeated between each pair of consecutive integral powers of 10 (*MMS*, Chapter XII). Hence a table of natural trigonometric functions contains only first quadrant entries for much the same reason that a table of common logarithms contains mantissas only for numbers from 10 to 100 or from 100 to 1,000. *Anyone who uses a trigonometric table is expected to be able to apply it to angles in other quadrants in the same way that anyone who uses a logarithmic table is expected to be able to apply it to numbers with other decimal places.*

The steps for finding a trigonometric function of any angle θ are as follows:

1. Find the corresponding first quadrant angle θ_1.

2. Find the required function of θ_1.

3. Determine whether the required function of θ has the same sign or a minus sign.

In *step one*, if θ is a negative angle or an angle greater than 360°, we can always find the equivalent first, second, third, or fourth quadrant angle — θ_1, θ_2, θ_3, or θ_4, respectively — as previously explained (page 96). If θ or its equivalent angle is then a first quadrant angle, this is θ_1 and there is no further problem of step one. But if θ or its equivalent angle is in the second, third, or fourth quadrant, we can always find θ_1 by substitution in the above formulas for corresponding angles (page 98), transposed and combined as follows:

$$\theta_1 = 180° - \theta_2 = \theta_3 - 180° = 360° - \theta_4.$$

Step 2 next follows as in elementary mathematics (*MMS*, Chapter XV).

And *step* 3 may be concluded simply by inspection of Figure 25, or of its sketched or mentally pictured equivalent. Although standard trigonometry texts give many rules and formulas for this purpose, all are easy to confuse and are derived in the first place from diagrams like that in Figure 25 anyway. Rather than try to learn these rules by rote and run the risks of misrecollection, therefore, you will do better to sketch out such a diagram whenever necessary until its plan becomes so firmly fixed in your mind's eye that you can mentally formulate the textbook rules for yourself.

EXAMPLE 5: Find the sin, cos, and tan of $\theta = -150°$

SOLUTION: *Step One:* The positive angle equivalent to θ is:

$$360° + \theta = 360° - 150° = 210° = \theta_3. \quad \text{(Page 100)}$$

Hence the corresponding first quadrant angle is:

$$\theta_1 = \theta_3 - 180° = 210° - 180° = 30°. \quad \text{(Formula above)}$$

Step Two: From a table of natural logarithms,

$$\sin 30° = .5000$$
$$\cos 30° = .8660 \qquad (MMS,$$
$$\tan 30° = .5774 \qquad \text{Chapter XV)}$$

Step Three: But from Figure 25 and its Table:

$$\sin \theta_3 = -b/r = -\sin \theta_1 \quad \text{(Page 99)}$$
$$\cos \theta_3 = -a/r = -\cos \theta_1,$$
$$\tan \theta_3 = -b/-a = b/a = \tan \theta_1.$$

Hence our answer is:

$$\sin -150° = -.5000,$$
$$\cos -150° = -.8660, \qquad \text{(Substitu-}$$
$$\tan -150° = .5774. \qquad \text{tion)}$$

EXAMPLE 6: Find the same functions of $\theta = 510°$.

SOLUTION: *Step One:* The angle less than 360° equivalent to θ is

$$\theta - 360° = 510° - 360° = 150° = \theta_2. \quad \text{(Page 100)}$$

Hence the corresponding first quadrant angle is

$$\theta_1 = 180° - \theta_2 = 180° - 150° = 30°. \quad \text{(Formula above)}$$

Step Two is now the same as in Example 5.

Step Three: In this second-quadrant case, however, we find from Figure 25 and its Table that

$$\sin \theta_2 = b/r = \sin \theta_1,$$
$$\cos \theta_2 = -a/r = -\cos \theta_1,$$
$$\tan \theta_2 = b/-a = -b/a = -\tan \theta_1.$$

Hence this time our answer is:

$$\sin 510° = .5000,$$
$$\cos 510° = -.8660, \qquad \text{(Substitution)}$$
$$\tan 510° = -.5774.$$

When θ is an exact multiple of 90°, of course, Step Two is not necessary and the required function values can be read directly from the diagram as in the Table below Figure 25.

EXAMPLE 7: Find the six basic trigonometric functions of $\theta = 990°$.

SOLUTION: *Step One:* The angle less than 360° equivalent to θ is

$$\theta - 2(360°) = 990° - 720° = 270°.$$

Steps Two and Three: Hence the trigonometric functions of θ are those listed in the $3\pi/2$ column of the Table under Figure 25.

$$\sin 990° = -1, \text{ etc.}$$

Practice Exercise No. 41

A. Verify all the entries in the Table below Figure 25, page 99, by substituting appropriate values for x and y in the generalized definitions of the trigonometric functions.

B. Referring to that Table, complete the following table of signs for the trigonometric functions of angles in the several quadrants:

Quadrant	I	II	III	IV
Sin	+	+	−	
Cos	+	−		
Tan	+			
Cot				
Sec				
Csc				

C. For which pairs of functions in the above table are the sign-entries the same? Why?

D. Referring only to Figure 25, write formulas like the one below for $\tan \theta_1$.

$$sin\ \theta_1 = sin\ (\pi - \theta_1) = -sin\ (\pi + \theta_1) = -sin\ (2\pi - \theta_1).$$

E. Find the values of the following trigonometric functions:

(1) $\sin 120°$ (5) $\sin -240°$
(2) $\tan 135°$ (6) $\sin 210°$
(3) $\cot 765°$ (7) $\csc 330°$
(4) $\csc -330°$ (8) $\sec 1140°$

Applications of
Generalized Trigonometric Functions

To understand *why* the definitions of the trigonometric functions have been generalized as above, it is helpful to consider a typical use to which they are put in this form.

As we shall see in more detail in vector analysis and in calculus, many basic physical quantities — like forces, displacements, velocities, accelerations, etc, — may be represented with respect to a set of rectangular coordinate axes by a line which changes in length and direction as these quantities change. For the mathematical treatment of these quantities, therefore, it is important to have precise measures of:

(1) How far such a line extends in the direction of the positive x-axis. This is called the line's *x-component*.

(2) How far the line extends in the direction of the positive y-axis. This is called its *y-component*.

(3) The "lie" or "set" of the line with respect to the coordinate system as measured by the ratio of its y-component over its x-component. This is called the line's *slope.*

For instance, suppose such a line to be OP_1, of known length r, forming the known first quadrant angle θ_1 with the positive x-axis, as in Figure 26. If we designate the (at first) unknown coordinates of P_1 as (x_1, y_1), the x-component of OP_1 is, by definition, x_1; the y-component of OP_1 is, by definition, y_1; and the slope of OP_1, designated S_1, is by definition the ratio y_1/x_1. But also by definition,

$\sin \theta_1 = y_1/r,$	(Substituting $y = y_1$)
$\cos \theta_1 = x_1/r,$	(" $x = x_1$)
$\tan \theta_1 = y_1/x_1.$	(" the same)

Hence in terms of the known quantity r and trigonometric functions of the known angle θ_1,

$y_1 = r \sin \theta_1$	(Multiplying by $r = r$)
$x_1 = r \cos \theta_1$	(Ditto)
$S_1 = \tan \theta_1.$	(Substitution)

Suppose a similar line, however, to be OP_2, also of known length r, but forming the known second quadrant angle $\theta_2 = 180° - \theta_1$ with the positive x-axis, as in Figure 26. It is clear from the diagram that $y_2 = y_1$, and that OP_2 therefore has the same y-component as OP_1. But it is also clear from the diagram that, although x_2 has the same length as x_1, it extends in the opposite (negative) direction along the x-axis. In order mathematically not to confuse the differently directed x-components of OP_1 and OP_2, therefore, we should assign the

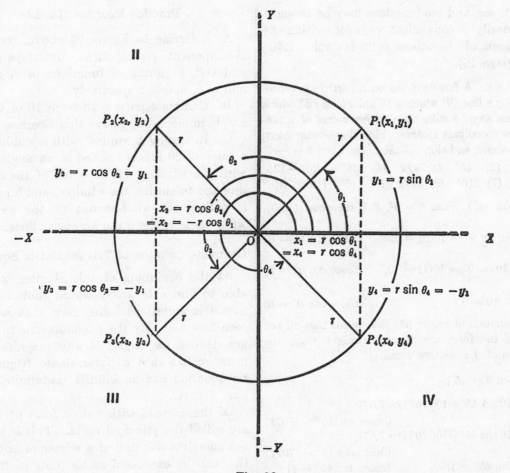

Fig. 26

latter a value equal to the *negative* of the former. Likewise, not to mathematically confuse the obviously different although related slopes of OP_1 and OP_2, we should assign S_2 a value equal to the *negative* of S_1.

But this is precisely what the generalized definitions of the trigonometric functions do automatically if we regard the above equalities, not as special first quadrant cases, but as general formulas. For instance, if we substitute the subscript 2 for the subscript 1 in these equalities, we find —

$$y_2 = r \sin \theta_2 = r \sin \theta_1 = y_1,$$
$$\text{(Since } \sin \theta_2 = \sin \theta_1\text{)}$$
$$x_2 = r \cos \theta_2 = r(-\cos \theta_1) = -x_1,$$
$$\text{(Since } \cos \theta_2 = -\cos \theta_1\text{)}$$
$$S_2 = \tan \theta_2 = -\tan \theta_1 = -S_1,$$
$$\text{(Since } \tan \theta_2 = -\tan \theta_1\text{)}$$

all as required by the above discussion. And similar results may be verified for lines OP forming angles in the third or fourth quadrants.

This means that if a line OP, of length r and forming an angle θ of any magnitude with the positive x-axis, represents a physical quantity:

1. The sine of θ is the function by which we must multiply r to find the physical quantity's y-component, $y = r \sin \theta$.

2. The cos of θ is the function by which we must multiply r to find the physical quantity's x-component, $x = r \cos \theta$.

3. The tan of θ is the function which directly gives the slope of the line along which this physical quantity acts, $S = \tan \theta$.

When such physical interpretations are made,

the cot, sec, and csc functions may be thought of primarily as convenient ways for writing the reciprocals of the others as in formulas R49–R51 (page 16).

EXAMPLE 8: A force exerted on a particle is represented by a line OP which is 10 units long and which forms an angle θ with the positive x-axis of a rectangular coordinate system. Find the y-component, x-component, and slope of this force when $\theta = -$

(1) $0°$ (2) $45°$ (3) $\pi/2$ (4) $135°$ (5) $-135°$
(6) π (7) $210°$ (8) $3\pi/2$ (9) $315°$ (10) $510°$

SOLUTION: (1) When $\theta = 0°$, P is the point $(10,0)$, and

$$y = 10 \sin 0° = 10(0) = 0, \qquad \text{(Since } \sin 0° = 0\text{)}$$

$$x = 10 \cos 0° = 10(1) = 10, \qquad \text{(Since } \cos 0° = 1\text{)}$$

$$S = 0/10 = 0. \qquad \text{(Since } \tan 0° = 0\text{)}$$

A combination of values like this means that all ten units of the force are directed straight along the direction of the positive x-axis.

(2) When $\theta = 45°$,

$$y = 10 \sin 45° = 10(.7071) = 7.071,$$
$$\text{(Since } \sin 45° = .7071\text{)}$$
$$x = 10 \cos 45° = 10(.7071) = 7.071,$$
$$\text{(Since } \cos 45° = .7071\text{)}$$
$$S = \tan 45° = 1. \qquad \text{(Since } \tan 45° = 1\text{)}$$

(3) When $\theta = \pi/2$, P is the point $(0,10)$ and

$$y = 10 \sin \pi/2 = 10(1) = 10,$$
$$\text{(Since } \sin \pi/2 = 1\text{)}$$
$$x = 10 \cos \pi/2 = 10(0) = 0, \quad \text{(Since } \cos \pi/2 = 0\text{)}$$
$$S = \tan \pi/2 = \infty. \qquad \text{(See page 100)}$$

A combination of values like this means that all ten units of the force are directed straight upward along the direction of the positive y-axis.

(4) When $\theta = 135°$,

$$y = 10 \sin 135° = 10 \sin 45°$$
$$\text{(Since } \sin 135° = \sin 45°\text{)}$$
$$= 10(.7071) = 7.071, \qquad \text{(As above)}$$
$$x = 10 \cos 135° = 10 (-\cos 45°)$$
$$\text{(Since } \cos 135° = -\cos 45°\text{)}$$
$$= 10 (-.7071) = -7.071, \qquad \text{(As above)}$$
$$S = \tan 135° = -\tan 45° = -1.$$
$$\text{(Since } \tan 135° = -\tan 45°\text{)}$$

Etc.

Practice Exercise No. 42

A. Referring to Figure 26 above, write the x-component, y-component, and slope of OP_3 and OP_4 in terms of functions of θ_3 and θ_1, and of θ_4 and θ_1 respectively.

B. Complete parts 5 through 10 of the Example immediately above this Exercise.

C. In closing a window with a window pole a force of 20 lbs. is exerted at an angle of 30° with the window. How much of the force is directed to closing the window, and how much is pulling the window out of the wall? At what angle is there the greatest efficiency?

Complete Solution of Trigonometric Equations

At the beginning of this chapter we were able to find *only first-quadrant* roots of trigonometric equations. But now that we have seen how and why the trigonometric functions are defined for angles of any magnitude, we must realize that a determinate trigonometric equation has an infinite (indefinite) number of roots.

Of these, roots with values from 0° to 360° are called the principal roots. This is because the complete solution of a trigonometric equation can be expressed as its roots in the four basic quadrants, plus exact multiples of 360° or 2π, namely: $n(360°)$ or $2\pi n$, where $n = 0,1,2,3, \ldots$ etc.

EXAMPLE 1: Solve completely the equation in Example 1a above:
$$2 \tan x - 1 = 0. \qquad \text{(Page 92)}$$

SOLUTION: We have already found the algebraic solution of this equation to be
$$\tan x = \tfrac{1}{2}. \qquad \text{(Page 92)}$$
And we have already found its first-quadrant trigonometric solution to be
$$x = \tan^{-1} \tfrac{1}{2} = 26°34'. \qquad \text{(Page 95)}$$
But since
$$\tan x = \tan (180° + x), \qquad \text{(Page 99)}$$
it follows that another (third-quadrant) root is
$$x = \tan^{-1} \tfrac{1}{2} = 180° + 26°34' = 206°34'.$$
Hence the *principal roots* of the equation are
$$x = 26°34', 206°34'.$$
And since
$$\tan x = \tan [x \pm n(360°)] \qquad \text{(Page 100)}$$

the *complete solution* of the equation is

$$x = 26°34' \pm n(360°), \; 206°34' \pm n(360°),$$

for $n = 0,1,2,3, \cdots$ etc., hereafter understood.

Check:

$$2 \tan 26°34' \; - 1 = 2(\tfrac{1}{2}) - 1 = 0. \; \checkmark \quad \text{(Substitu-}$$
$$2 \tan 206°34' - 1 = 2(\tfrac{1}{2}) - 1 = 0. \; \checkmark \quad \text{tion)}$$

For the same reason that it is possible to express the complete solution of a trigonometric equation in terms of its principal roots, it is of course necessary to check only these principal roots as above.

EXAMPLE 2: Solve completely the equation in Example 2a above:

$$2 \cos^2 x + \cos x - 1 = 0. \qquad \text{(Pages 92–3)}$$

SOLUTION: We have already found the algebraic solution of this equation to be the pair of simplified equations,

$$\cos x = \tfrac{1}{2}, -1. \qquad \text{(Same reference)}$$

And we have already found the first-quadrant trigonometric solution for the first of these simplified equations to be

$$x = \cos^{-1} \tfrac{1}{2} = 60°. \qquad \text{(Page 95)}$$

But since

$$\cos x = \cos (360° - x), \qquad \text{(Page 99)}$$

it follows that another (fourth quadrant) root is

$$x = \cos^{-1} \tfrac{1}{2} = 360° - 60° = 300°.$$

Moreover, since

$$\cos 180° = -1, \qquad \text{(Table, page 99)}$$

still another root is

$$x = \cos^{-1} -1 = 180°.$$

Hence the *three principal roots* of the equation are

$$x = 60°, \; 180°, \; 300°,$$

and the *complete solution* is

$$x = 60° \pm n(360°),$$
$$180° \pm n(360°),$$
$$300° \pm n(360°).$$

Check:

$$2 \cos^2 60° + \cos 60° - 1 =$$
$$2(\tfrac{1}{2})^2 + \tfrac{1}{2} - 1 = 1 - 1 = 0. \; \checkmark$$
$$\text{(Substitution)}$$
$$2 \cos^2 180° + \cos 180° - 1 =$$
$$2(-1)^2 - 1 - 1 = 2 - 2 = 0. \; \checkmark \; \text{Etc.}$$

In the above examples, all results which previously checked algebraically have been found to lead to roots which also check trigonometrically. In other instances, however, *results which check at the preliminary algebraic stage of solution may be found to lead to inverse trigonometric equations for which there are no possible trigonometric solutions.*

EXAMPLE 3: Solve completely the equation in Example 3a above:

$$2 \sin x + 2 \csc x - 5 = 0. \qquad \text{(Page 93)}$$

SOLUTION: We have already found, and checked, the algebraic solution of this equation as the two pairs of simplified trigonometric equations,

$$\sin x, \csc x = 2, \tfrac{1}{2}; \; \tfrac{1}{2}, 2. \qquad \text{(Same reference)}$$

And we have already found the first-quadrant trigonometric solution of the *second* of these pairs of equations to be

$$x = \sin^{-1} \tfrac{1}{2}, \csc^{-1} 2 = 30°. \qquad \text{(Page 95)}$$

But since

$$\sin x, \csc x = \sin (180° - x), \csc (180° - x),$$

it follows that another (second quadrant) root is

$$x = \sin^{-1} \tfrac{1}{2}, \csc^{-1} 2 = 180° - 30° = 150°.$$

However, there is **no angle, x, in any quadrant** for which

$$x = \sin^{-1} 2, \csc^{-1} \tfrac{1}{2} \; (?). \qquad \text{(Figure 25 and Table,}$$
$$\text{page 99)}$$

Hence the *principal solution* of the equation consists solely of the *two principal roots*,

$$x = 30°, \; 150°,$$

and the *complete solution* is

$$x = 30° \pm n(360°), \; 150° \pm n(360°).$$

In a case like the above, an algebraic result which leads to no trigonometric roots may be said to be **trigonometrically invalid** or **extraneous** even though the result checks algebraically in the preliminary algebraic phase of solution (page 94).

When a trigonometric equation has no roots at all for this reason, it may be said to be **trigonometrically indeterminate** (page 19). Such is the case, for instance, in Example A1, 2, 3, 6, 7, 8, of the Exercise below.

Practice Exercise No. 43

A. Complete the solutions of the equations in Practice Exercise No. 37.

GRAPHS OF TRIGONOMETRIC FUNCTIONS

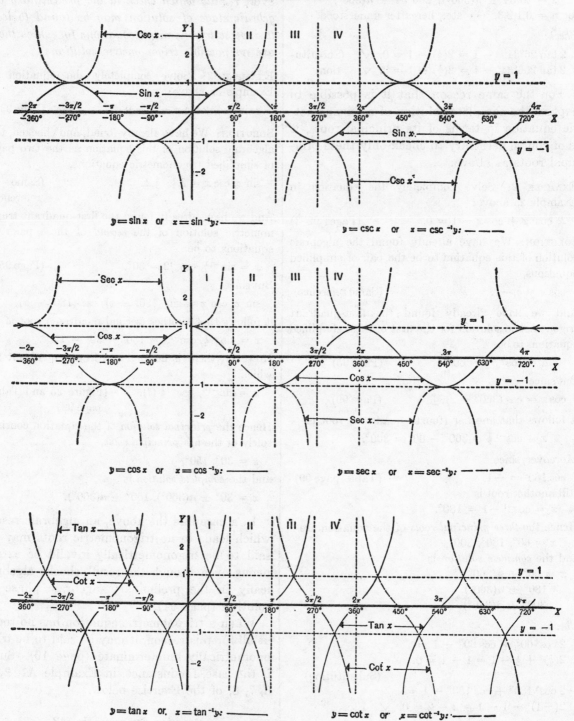

Fig. 27

B. Solve:

$$2 \sin^2 x + 3 \sin x - 2 = 0.$$

Graphs of Trigonometric Functions

Using typical values like those computed in preceding examples and exercises, it's a simple matter to plot the trigonometric functions of a variable angle x with respect to an x,y set of rectangular coordinate axes.

The graph of $y = \sin x$ is the regular wave-like curve of indefinite length drawn with an unbroken contour in the upper diagram of Figure 27. Note how it fluctuates back and forth between the horizontal lines $y = \pm 1$, lying above the x-axis in quadrants I and II, lying below the x-axis in quadrants III and IV, and alternating in sign every 180° along the x-axis on either side of these four quadrants.

Distinguished from the sine curve in the same diagram by broken-line drawing, the graph of $y = \csc x$ is a series of open-ended U-shaped curves, every other one inverted, but always on the same side of the x-axis as the sine curve, and "touching" it every 180° at the points where $\csc x = \sin x = \pm 1$. Note how it changes in value from an indicated $+ \infty$, down to $+1$, back up to an indicated $+ \infty$, in the first two quadrants; next "jumps" to an indicated $- \infty$, increases to -1, and returns to an indicated $- \infty$, in the next two quadrants. This is, of course, the pattern we should expect from the arithmetic of the formula, $\csc x = 1/\sin x$ (R49, page 16).

The Cos and Sec Columns in a table of natural trigonometric functions, you should recall, are merely the Sin and Csc Columns read in reverse order. This typographical saving of space is possible for the mathematical reason that $\cos x = \sin (90° - x)$ and $\sec x = \csc (90° - x)$ (*MMS*, Chapter XV; and Figure 25, page 99 above). Hence the graph of $y = \cos x$ and the graph of $y = \sec x$, as shown in the middle diagram of Figure 27, are identical with the sin and csc graphs in the upper diagram except that they are shifted one quadrant, or $90° = \pi/2$, to the left along the x-axis.

The graph of $y = \tan x$, finally, is the series of upward sweeping curves shown intersecting the x-axis every 180° in the lower diagram of Figure 27. And the graph of $y = \cot x$ is the series of downward sweeping curves in the same diagram. Note how the tan curves pass in value every 180° from an indicated $- \infty$, to 0, on up to an indicated $+ \infty$, typically in the second and third quadrants. And note how the cot curves pass in value every 180° from an indicated $+ \infty$, to 0, on down to an indicated $- \infty$, typically in the first and second quadrants.

Check the patterns of values depicted by these graphs with the diagram and table of Figure 25 (page 99). Try reconstructing them for yourself until you have their contours firmly pictured in your mind's eye. For then you may be able to use them, perhaps even more conveniently than diagrams like that of Figure 25, as an aid in improvising formulas (page 101), in finding the functions of angles outside the first quadrant (page 98), or in finding the complete solutions of trigonometric equations (page 104).

Suppose, for instance, that you need to find trigonometric functions of $x = -150°$, as in Example 5 above (page 101). Glance down the vertical line, $x = -150°$, on the diagrams of Figure 27. If necessary, use the right edge of a vertically placed ruler — preferably a transparent plastic ruler — to guide your eye. From the intersection of this line with the several graphs, and from the pattern of these graphs in each 180° interval of the x-axis, you should be able to see at once that

$$\sin -150° = -\sin 30°, \; \tan -150° = \tan 30°, \text{ etc.}$$

Or suppose that you need to find all the angles, x, such that $x = \sin^{-1} \frac{1}{2}, 2$, as in completing the solution of the trigonometric equation in Example 4 above (page 93). Glance across the horizontal lines, $y = \frac{1}{2}$ and $y = 2$, in the upper diagram of Figure 27. From the intersections of the line $y = \frac{1}{2}$ with the sine curve, you should be able to see at once that

$$x = \sin^{-1}\tfrac{1}{2} = 30° \pm n(360°), \; 150° \pm n(360°).$$

And from the failure of the line $y = 2$ to intersect the sine curve at any point, you can also see at once that there is no angle $x = \sin^{-1}2$.

Practice Exercise No. 44

A. Find graphically where $\tan x$ and $\cos x$ are equal.

B. Find graphically where $\cos x$ and x are equal, it being understood that x is measured in radians.

Variations of Trigonometric Functions

Each of the graphs in Figure 27 repeats a pattern of values, called its **cycle**, at a regular interval, called its **period**. For this reason the trigonometric functions are classed as **periodic functions**.

Half the difference between the largest and smallest values which such a function attains in each cycle, is called its **amplitude**. The horizontal distance between corresponding points in the cycles of any two periodic functions is called their **phase difference** or **phase displacement**. When two such functions have a phase difference of 0, they are said to be **in phase** with each other; otherwise they are said to be **out of phase**.

Thus, for instance, the sin and cos functions have basically similar *cycles* with a *period* of 360° or 2π and an *amplitude* of 1. But there is a *phase difference* of 90° or $\pi/2$ between them, since that is the distance which it would be necessary to move (displace) either along the x-axis in order to bring it *into phase* with the other.

Some idea of the physical meaning of these terms is suggested by the fact that the movement of molecules of matter in sound waves can be depicted by modified sine or cosine curves, and have lower-pitched sounds when the *periods* of these curves are longer, and have louder sounds when the *amplitudes* of these curves are greater.

Here, however, we are primarily concerned with interpreting the mathematical devices by which these variations in periodic behavior are expressed in periodic functions.

As in the case of other functions, *the graph of any trigonometric function is simply raised or lowered c units by the addition of a constant $\pm c$ term.* By tabulating and plotting values of $y = \sin x + c$, for instance, we obtain a graph identical to that of the sine curve in Figure 27, except that it is raised three units higher with respect to the x-axis.

The addition of a constant term, $\pm c\pi$, to the

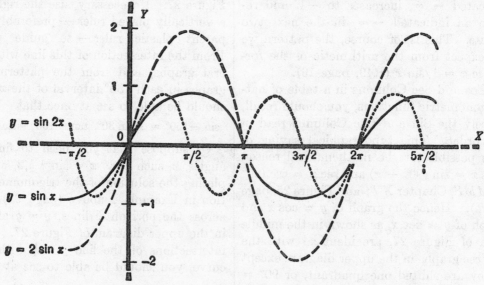

Fig. 28

independent variable angle of a trigonometric function, however, displaces its phase that number of units to the left or right with respect to the vertical y-axis. By tabulating and plotting values of $y = \sin (x + \pi/2)$, for instance, we obtain the graph of the cos x curve in Figure 27 which has already been observed to be identical to that of the sin x curve except that it is displaced $\pi/2$ units to the left (page 107).

Moreover, *when the trigonometric functions are multiplied by a constant c, their amplitude is multiplied by the same constant. But when their independent variable angle is multiplied by c, their period is divided by this constant.* For instance, the graph of $y = 2 \sin x$, shown in Figure 28, has twice the amplitude, but the same period, as that of $y = \sin x$. On the other hand, the graph of $y = \sin 2x$ has the same amplitude, but half the period, of that of $y = \sin x$.

When a trigonometric function is multiplied by a variable cofactor, however, the resulting product is no longer strictly a periodic function. But its pattern of variation may nevertheless reflect the periodicity of its trigonometric factor. For instance, the graph of $y = \frac{1}{2}x \sin x$, as shown in Figure 29, intersects the x-axis every 180° like the graph of $y = \sin x$. But its amplitude within each interval of 180° changes with the magnitude of its cofactor, $\frac{1}{2}x$.

Practice Exercise No. 45

Sketch at least one cycle of the graphs of the following functions. What is their phase and amplitude?

1. $y = -\sin x$, and $y = -\sin 2x$.
2. $y = \sin x$, and $y = -\sin (x - \pi/2)$.
3. $y = \frac{1}{2} \cos x$, and $y = \cos \frac{1}{2}x$.
4. $y = \cos 3x$, and $y = 3 \cos x$.
5. $y = \sin x + \cos x$, and $y = \sin x - \cos x$.
6. $y = \sin^2 x$, and $y = \sin^2 x + \cos^2 x$.
7. $y = \frac{1}{2} \tan \frac{1}{2}x$.
8. $y = \frac{1}{2} \sec x - 1$.

Regarding the functions in Examples 1 to 6 above as pairs of simultaneous trigonometric equations, what are their graphic solutions?

Summary

A **trigonometric equation** may be *solved algebraically*, by methods previously discussed, only for those values of its trigonometric functions which satisfy its *algebraic conditions*. To complete the solution of such an equation, we must treat its algebraic roots as **inverse trigonometric equations** and solve the latter for those values of the variable angles which satisfy the *trigonometric conditions* of the original equation.

When the **definitions of the trigonometric functions are generalized** to apply to angles of

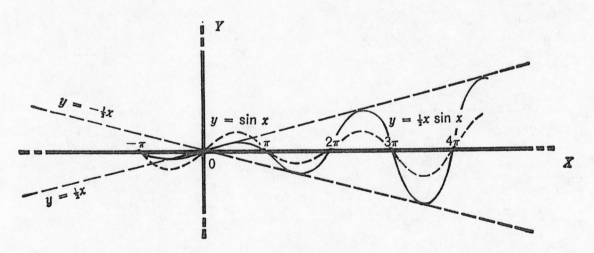

Fig. 29

any magnitude (as illustrated in Figure 25 and its Table, page 99), these functions become **periodic** (as depicted in the graphs of Figure 27, page 106).

The **complete solution** of a determinate trigonometric equation therefore consists of *an unlimited (infinite) number of roots*. But those roots — x_1, x_2, etc. — which occur in the four basic quadrants are distinguished as the **principal roots** because the complete solution of such an equation can always be expressed in terms of them as $x_1 \pm n(360°)$, $x_2 \pm n(360°)$, etc., where $n = 0,1,2,3, \ldots$ etc.

In this chapter we have also illustrated graphically certain variations in the periodicity of trigonometric functions. This was in anticipation of applications to be made later on in calculus.

CHAPTER IX

POINTS, DISTANCES, AND SLOPES

What is Analytic Geometry?

The kind of geometry usually taught in high school is called **synthetic,** which literally means: "put-together."

This name is appropriate since **the method of synthetic geometry** is to put geometric facts together — one upon the other like building stones of a temple on the ancient Acropolis. Its primary *definitions, axioms,* and *postulates* (*MMS,* Chapter XIV) are foundations. Its long sequences of *theorems, constructions,* and *corollaries* are superstructures. And to reach any one of its higher *propositions,* you are required to follow a step-by-step path of reasoning all the way up from the base.

The original invention of such geometry by the ancient Greeks was, of course, a brilliant cultural achievement of broader importance than its mathematical usefulness. Although limited to the simple subject matter of space relationships, it was one of the first great historical efforts of man to think clearly — with logical precision, rather than wishfully, with misplaced poetic imagination.

Its definitions are attempts to clarify ideas which might carelessly be supposed to be "obvious." Its axioms and postulates are attempts to identify the assumptions which underlie these ideas. Its sequences of propositions are attempts to organize the same systematically in terms of their logical dependence upon each other. And its entire method expresses a civilized concern to solve problems rationally.

As it was developed by authors like Euclid, Apollonius, and Archimedes, synthetic geometry also has a kind of imperishable classic beauty, far outclassing that of Hellenic architecture. And if we have a taste for intellectual gymnastics, we can experience in it the exhilaration of a fascinating intellectual game.

For more advanced mathematical applications, however, *the synthetic method in geometry* has certain *practical disadvantages*. One is that it requires you to keep constantly in mind a very large number of previously demonstrated propositions. Another is that it often requires elaborate constructions and indirect methods of deduction through many intermediate steps. Consequently, it often involves the time-consuming, trial-and-error sort of reasoning one would normally apply to a puzzle of the crossword or picture-cut-out type.

The kind of geometry usually taught in first college courses, on the other hand, is called **analytic,** which literally means: "loosening-up" in the sense of "disentangling."

This name also is appropriate since **the method of analytic geometry** is to separate out the essential elements in each new problem by stating them in the form of equations, and then to resolve the geometric question by solving these equations algebraically.

An immediate advantage of such a procedure is that, to solve most practical problems, you need to keep in mind only a few basic formulas. The greatest **advantages of the analytic method,** however, are that it is *more direct, quicker,* and *much more powerful*. Relatively little progress was made in synthetic geometry for two thousand years after its techniques were brought to their classic perfection by the ancient Greeks. But the invention of the analytic method by mathematicians like René Descartes in the seventeenth century soon led to the discovery of calculus and all the related tools of modern mathematical science.

If you have carefully studied the preceding algebra section of this book, the easiest way for you to approach this new method now is to think of it as *graphic-interpretation-in-reverse*.

Hitherto we have studied various types of mathematical functions by plotting the relationships between variables in equations graphically. Thus we were able to find *geometric interpretations for algebraic processes* and, in certain instances, even *geometric solutions for algebraic problems*.

But now we shall proceed the other way around. For **the geometric conditions in a problem** we shall first seek **analytic-equivalents** — corresponding relationships between variables in equations. Then, by treating these equations algebraically, we shall obtain *algebraic solutions for geometric problems*.

To apply the method systematically, however, we must first derive certain **basic formulas** which will then be our *work-kit of analytic tools* for solving other geometric problems. To distinguish these from previously stated review formulas (pages 14–6) and algebra formulas (Chapters II through VIII), we number them, G1, G2, G3, etc., in the text which follows.

(*Suggestion:* If you do not have the introductory ideas of Chapter III fresh in mind, it would be well to review pages 27 to 36 before proceeding with what now follows.)

Positions of Points

We have previously seen that any pair of real values for a pair of variables — $x = x_1$, $y = y_1$ — can be depicted geometrically with respect to a system of rectangular coordinate axes by the point $P_1(x_1, y_1)$. As in Figure 30 further below, P_1 is then the point whose position is defined by the intersection of its abscissa, $x = x_1$, and its ordinate, $y = y_1$ (Chapter III, page 33).

This method of graphic representation, however, can be a two-way process. Beginning at the other end, let us *start* with the *geometric position* of the point P_1, defined as before by its distances, x_1 and y_1, from the y-axis and x-axis in the appropriate directions.

Corresponding to the geometric position of P_1, we now have the *unique pair of x,y values*, $x = x_1$ and $y = y_1$. For, by the definitions of the system, this pair of values can correspond to no other point. And any other point — say P_2 in Figure 30 — must have the different

corresponding pair of x,y values, $x = x_2$ and $y = y_2$.

Or, in terms defined a bit further above (this page), **the geometric position of the point P_1 is expressed analytically by the pair of simultaneous linear equations,**

G1: $x = x_1,$ $y = y_1.$

Although very elementary, this first simple step in analytic geometry illustrates half the method of the entire subject. For *it translates a geometric concept* — that of the position of a point — *into the corresponding algebraic terms of an analytic equivalent* — here, a pair of simultaneous linear equations.

To illustrate the other half of the same method, let us now see how Formula G1 can be applied to the algebraic solution of a simple geometric problem.

Suppose we need to find the point P which divides a given straight-line segment P_1P_2 in a given ratio m/n — that is to say, the point P on P_1P_2 such that

$$\frac{P_1P}{PP_2} = \frac{m}{n}.$$

To solve this problem in *synthetic* geometry we need, in each instance, to construct a divided-off auxiliary line and a set of parallel transversals between this auxiliary line and the given line (*MMS*, Chapter XIV).

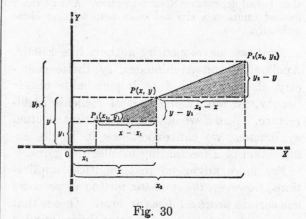

Fig. 30

Proceeding *analytically*, however, we may first apply Formula G1 to designate the known positions of P_1 and P_2, with reference to a set of rectangular coordinate axes, by the known pairs of coordinates (x_1, y_1) and (x_2, y_2) respectively, as in Figure 30. And we may represent the as-yet unknown point P with reference to the same set of coordinate axes, by the as-yet unknown pair of coordinates (x, y).

From the fact that the two shaded triangles in the diagram are similar, we now see that

$$\frac{P_1P}{PP_2} = \frac{x - x_1}{x_2 - x}. \qquad \text{(MMS, Chapter XIV)}$$

Hence the x-coordinates of P, P_1, and P_2, are related algebraically by the equation,

$$\frac{x - x_1}{x_2 - x} = \frac{m}{n}. \qquad \text{(Substitution from above)}$$

This can be solved for x as follows:

$$nx - nx_1 = mx_2 - mx, \qquad \begin{array}{l}\text{(Multiplying by}\\ n(x_2 - x) = \text{etc.)}\end{array}$$

$$(m + n)x = nx_1 + mx_2, \qquad \begin{array}{l}\text{(Transposing}\\ \text{and factoring)}\end{array}$$

$$x = \frac{nx_1 + mx_2}{m + n}. \qquad \begin{array}{l}\text{(Dividing by}\\ m + n = m + n)\end{array}$$

And since it is also seen that

$$\frac{P_1P}{PP_2} = \frac{y - y_1}{y_2 - y}, \qquad \text{(Figure 30)}$$

it likewise follows that

$$y = \frac{ny_1 + my_2}{m + n}. \qquad \text{(Steps as above)}$$

Consequently, we have for **the point P which divides a given line P_1P_2 in a given ratio m/n**, the formulas,

$$\textbf{G2:} \quad x = \frac{nx_1 + mx_2}{m + n}, \quad y = \frac{ny_1 + my_2}{m + n}.$$

An important special case is that in which $m = n = 1$. For then $P_1P = PP_2$, and **the midpoint M of P_1P_2 is given by the formulas,**

$$\textbf{G2':} \quad x = \frac{x_1 + x_2}{2}, \quad y = \frac{y_1 + y_2}{2}. \qquad \begin{array}{l}\text{(Substituting}\\ m = n = 1)\end{array}$$

Whenever we are given the coordinates of the end-points of a line segment, we can always find the coordinates of the point which divides it in any given ratio, by direct substitution into Formulas G2 or G2'.

Example 1: A truss for a construction job is to be built by connecting girders with steel pins through holes near their ends. Suppose a plan for the truss to be laid out with respect to a pair of x,y coordinate axes as in Figure 31, with the points P_1, P_2, etc.,

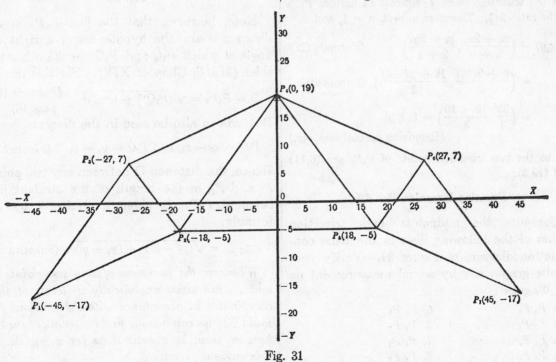

Fig. 31

representing the centers of the connecting pins. Find the midpoint M of the line P_1P_2.

SOLUTION:

$$M(x,y) = \left(\frac{x_1 + x_2}{2}, \frac{y_1 + y_2}{2}\right) \quad \text{(Formula G2')}$$

$$= \left(\frac{-45 + (-27)}{2}, \frac{-17 + 7}{2}\right) \text{(Substitution)}$$

$$= \left(\frac{-72}{2}, \frac{-10}{2}\right) = (-36, -5).$$
(Removing parentheses, etc.)

Hence, M is the point whose x-coordinate is -36, and whose y-coordinate is -5.

EXAMPLE 2: Find the trisection points of the line P_4P_5 in the same diagram.

SOLUTION: The trisection point T_1 nearest P_4, divides P_4P_5 in the ratio 1:2 (*MMS*, Chapter VI). Therefore $m = 1$, $n = 2$, and

$$T_1(x,y) = \left(\frac{2x_4 + x_5}{1 + 2}, \frac{2y_4 + y_5}{1 + 2}\right) \quad \text{(Formula G2)}$$

$$= \left(\frac{2(0) + 18}{3}, \frac{2(19) - 5}{3}\right) \quad \text{(Substitution)}$$

$$= \left(\frac{18}{3}, \frac{38 - 5}{3}\right) = (6, 11).$$
(Removing parentheses, etc.)

But the trisection point T_2 nearest P_5 divides P_4P_5 in the ratio 2:1. Therefore $m = 2$, $n = 1$, and

$$T_2(x,y) = \left(\frac{x_4 + 2x_5}{2 + 1}, \frac{y_4 + 2y_5}{2 + 1}\right) \quad \text{(Formula G2)}$$

$$= \left(\frac{0 + 2(18)}{3}, \frac{19 + 2(-5)}{3}\right) \quad \text{(Substitution)}$$

$$= \left(\frac{36}{3}, \frac{19 - 10}{3}\right) = (12, 3)$$
(Removing parentheses, etc.)

Hence the two trisection points of P_4P_5 are (6,11) and (12,3).

Practice Exercise No. 46

Compute the midpoints and trisection points of the following lines in the truss construction diagram of Figure 31. Verify your results graphically by actual measurement on the diagram:

1. $P_1 P_2$ 5. $P_4 P_6$
2. $P_2 P_3$ 6. $P_5 P_6$
3. $P_3 P_4$ 7. $P_5 P_7$
4. $P_3 P_4$ 8. $P_6 P_7$

9. Given the points $P_1(3,2)$ and $P_2(5,7)$, find the point Q such that $P_1P_2/P_2Q = 2/3$.

Distances or Lengths

The distance between any two points is, by definition, the length of the straight line drawn between them. Thus, for instance, the distance D between the points P_1 and P_2 is the length of the straight line P_1P_2 drawn between them in Figure 32.

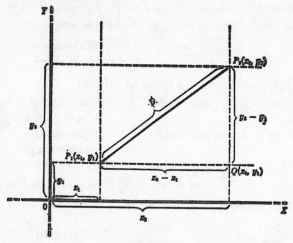

Fig. 32

Note, however, that the line P_1P_2 in the diagram is also the hypotenuse of a right triangle of which P_1Q and P_2Q are the other two sides (*MMS*, Chapter XIV). For this reason,

$$D = P_1P_2 = \sqrt{(P_1Q)^2 + (P_2Q)^2} \quad \text{(Formula R40, page 16)}$$

But, as can also be seen in the diagram,

$$P_1Q = x_2 - x_1, \text{ and } P_2Q = y_2 - y_1. \quad \text{(Figure 32)}$$

Hence, the distance D between any two points P_1 and P_2, or the length of the straight line segment P_1P_2, is given analytically by the formula,

$$\text{G3: } D = \sqrt{(x_2 - x_1)^2 + (y_2 - y_1)^2}. \text{ (Substitution)}$$

Whenever the positions of any two points P_1 and P_2 are given analytically in terms of their coordinates in accordance with the previous formula G1, we can always find the length of the line between them by substitutions for x_1, y_1, etc., in this present formula.

EXAMPLE 3: Find D in Figure 32, assuming $(x_1,y_1) = (4,4)$ and $(x_2,y_2) = (8,7)$.

SOLUTION: Making the given substitutions in Formula G3, we get

$$D = \sqrt{(8-4)^2 + (7-4)^2} = \sqrt{4^2 + 3^2}$$
(Removing parentheses)

$$= \sqrt{16 + 9} = \sqrt{25} = 5, \text{ Ans.} \quad \text{(Squaring, etc.)}$$

Since the quantities $x_2 - x_1$ and $y_2 - y_1$ are squared in Formula G3, it makes no difference in which order we consider any two points to find the distance between them by means of the formula. For instance, if we take points P_1 and P_2 of Example 3 in the opposite order, we still find the distance between them to be

$$D = \sqrt{(4-8)^2 + (4-7)^2} = \sqrt{(-4)^2 + (-3)^2}$$

$$= \sqrt{16 + 9} = \sqrt{25} = 5. \quad \text{(As above)}$$

For similar reasons, Formula G3 always gives the same, correct result regardless of which quadrants P_1 and P_2 are in, and regardless of the fact that both may sometimes lie on a line parallel to one of the axes.

EXAMPLE 4: Find the distance between P_1 and P_2 in Figure 33 below:

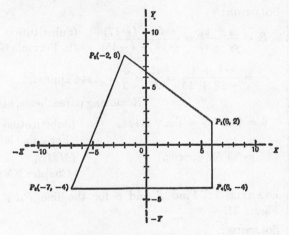

Fig. 33

SOLUTION: By substitution in Formula G3,

$$D = \sqrt{(-2-6)^2 + (8-2)^2} = \sqrt{(-8)^2 + 6^2}$$

$$= \sqrt{64 + 36} = \sqrt{100} = 10, \text{ Ans.}$$

This may be verified by direct measurement on the diagram against the scale marked on the axes.

EXAMPLE 5: Find the distance between P_3 and P_4 in Figure 33.

SOLUTION: Substituting the subscript 3 for the subscript 1, and the subscript 4 for the subscript 2, in Formula G3, we get

$$D = \sqrt{(x_4 - x_3)^2 + (y_4 - y_3)^2} \quad \text{(As explained)}$$

$$= \sqrt{[6 - (-7)]^2 + [-4 - (-4)]^2}$$
(Substitution)

$$= \sqrt{13^2 + 0^2} = 13, \text{ Ans.}$$
(Removing parentheses, etc.)

Practice Exercise No. 47

A. Find the lengths of the following lines in Figure 33:

(1) P_2P_3 (2) P_1P_4 (3) P_1P_3

B. Find the lengths of the following lines in the truss diagram of Figure 31.

(4) P_1P_2 (5) P_1P_3 (6) P_2P_4 (7) P_2P_3
(8) P_6P_6 (9) P_4P_6 (10) P_6P_7

Slopes

Many of the most fundamental axioms and propositions of elementary geometry concern parallel lines and relationships between angles at which non-parallel lines meet. In analytic geometry the useful content of these is translated into equivalent algebraic terms by means of the linking concept of *a line's slope*.

In common sense terms, the *slope* of anything is its "pitch" or "grade" — how much it goes up or down as it goes out along a given horizontal line of reference. Thus we speak of a roof having a "four-in-five pitch," meaning that it goes up four feet every five feet measured along the attic floor below. Or we speak of a road having a "twelve percent downgrade," meaning that it goes down twelve feet every 100 feet measured along a line of level altitude in the direction of its descent.

Similarly, when a straight line is referred to a system of x,y rectangular coordinates, its slope is *defined* as the ratio of (1) the amount its y-coordinate increases or decreases between any two points, over (2) the amount its x-coordinate increases or decreases between these same two points.

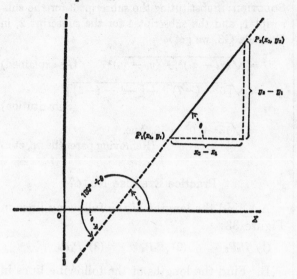

Fig. 34

In Figure 34, for instance, the y-coordinate of a point on the straight-line segment P_1P_2 increases from y_1 to y_2 between the points P_1 and P_2, while its x-coordinate increases from x_1 to x_2 between the same two points. Hence the slope S of P_1P_2 is by definition:

$$S = \frac{y_2 - y_1}{x_2 - x_1}.$$

Observe in Figure 34 that the line P_1P_2, extended, forms the angle θ with the x-axis in the latter's positive direction. This is called the line's **angle of inclination** with the x-axis. And it is equal to the corresponding angle (*MMS*, Chapter XIV) which P_1P_2 makes with the horizontal line drawn through the point P_1 parallel to the x-axis. From the diagram, moreover,

$$\tan\theta = \frac{y_2 - y_1}{x_2 - x_1}. \qquad \text{(By definition)}$$

Hence the slope S of the line segment P_1P_2 is given by the formula,

G4: $\quad S = \dfrac{y_2 - y_1}{x_2 - x_1} = \tan\theta.$

And **the angle of inclination** θ of P_1P_2 is given by the inverse of this same formula,

G4′: $\quad \theta = tan^{-1}S = tan^{-1}\dfrac{y_2 - y_1}{x_2 - x_1}.$

(*Note*: Readers who have begun analytic geometry without having studied Chapter VIII or its equivalent, should refer briefly to page 94 above for definitions of symbols and terms in this last statement.)

In order better to understand a wide variety of applications to be made shortly, you should first realize how this formula works out numerically in different geometric situations.

If the line segment P_1P_2 lies so that its y-coordinate increases with an increase in its x-coordinate, as in Figure 34, then its *slope* is *positive* and is *greatest when the angle of incidence θ is greatest*. The student who has studied Chapter VIII or its equivalent before beginning analytic geometry will recognize this as the case in which θ is a first quadrant angle, and in which $\tan\theta$ is therefore positive and increasing as θ increases towards $90°$ (page 107).

EXAMPLE 6: Find the slope S and the angle of inclination θ of the line P_1P_2 in the truss diagram of Figure 31 above.

SOLUTION:

$$S = \frac{y_2 - y_1}{x_2 - x_1} = \frac{-5 - (-17)}{-18 - (-45)} \quad \text{(Substitution in Formula G4)}$$

$$= \frac{-5 + 17}{-18 + 45} = \frac{12}{27} = \frac{4}{9} = .4444 \text{ approx.}$$

(Removing parentheses, etc.)

$$\theta = \tan^{-1}S = \tan^{-1}.4444 \quad \text{(Substitution in Formula G4′)}$$

$$= 23°58' \text{ approx.} \quad \text{(\textit{MMS}, Chapter XV)}$$

EXAMPLE 7: Find S and θ for the line P_1P_3 in Figure 31.

SOLUTION:

$$S = \frac{y_2 - y_1}{x_3 - x_1} = \frac{7 - (-17)}{-27 - (-45)} \quad \text{(Substitution in G4)}$$

$$= \frac{7 + 17}{-27 + 45} = \frac{24}{18} = \frac{4}{3} = 1.3333 \text{ approx.}$$

$$\theta = \tan^{-1}1.3333 \quad \text{(Substitution in G4′)}$$

$= 53°08'$ approx. *(MMS, Chapter XV)*

Note in Examples 6 and 7 that the slope of P_1P_3 is substantially greater than that of P_1P_2. This is because P_1P_3 is "pitched up more steeply" to the right in the diagram.

Formula G4 gives the same numerical value for the slope of a line segment P_1P_2 regardless of the end from which we consider it to extend. This is because, algebraically,

$$\frac{y_2 - y_1}{x_2 - x_1} = \frac{-(y_1 - y_2)}{-(x_1 - x_2)} = \frac{y_1 - y_2}{x_1 - x_2}.$$

(Formulas R9 and R13, page 14)

The student who has completed Chapter VIII should recognize from Figure 34, moreover, that *the angle of inclination of P_2P_1 is $180° + \theta$* which has the same tangent function as θ.

EXAMPLE 8: Find S and θ for the line P_2P_1 in Figure 31.

SOLUTION: As for the line P_1P_2 in Example 6, we find

$$S = \frac{y_1 - y_2}{x_1 - x_2} = \frac{-17 - (-5)}{-45 - (-18)}$$ (Substitution in G4)

$$= \frac{-12}{-27} = \frac{12}{27} = .4444 \text{ approx.}$$

(Formula R13, page 14)

But since the angle of inclination is now in the third quadrant, we also find:

$$\theta = \tan^{-1} S = \tan^{-1} .4444$$ (Substitution in G4')

$$= 180° + 23°58' = 203°58' \text{ approx.}$$ (Page 101)

Comparing the results of Example 8 and Example 6, we see that the lines P_1P_2 and P_2P_1 have the same slope, but angles of inclination which differ by 180°.

However, *if a line lies so that its y coordinate decreases as its x coordinate increases, then its slope is negative and greatest in negative value when its angle of incidence is closest to 90° or 270°.* The student who has studied Chapter VIII will recognize this as the case in which θ is a second or fourth quadrant angle, and in which $\tan \theta$ is therefore negative (page 107).

EXAMPLE 9: Find S and θ for the line P_5P_7 in Figure 31.

SOLUTION: As before,

$$S = \frac{y_7 - y_5}{x_7 - x_5} = \frac{-17 - (-5)}{45 - 18}$$ (Substitution in G4)

$$= \frac{-17 + 5}{27} = -\frac{12}{27} = -.4444 \text{ approx.}$$

But since θ is now a fourth quadrant angle,

$$\theta = \tan^{-1} -.4444$$ (Substitution in G4')

$$= 360° - \tan^{-1} .4444$$ (Page 101)

$$= 360° - 23°58' = 336°02' \text{ approx.}$$

(As above)

In the special case when a line P_1P_2 is parallel to the x-axis, then $y_2 = y_1$, and $\theta = 0°$. Hence $y_2 - y_1 = 0$, $\tan 0° = 0$, and the slope $S = 0$.

In the other special case when a line P_1P_2 is parallel to the y-axis, then $x_2 = x_1$, and $\theta = 90°$. Hence the formula gives $S = (y_2 - y_1)/0 = \tan 90°$ for which we can usefully define no definite value (page 100). We sometimes hear that "the slope of a line parallel to the y-axis is infinite, or $\pm \infty$." But this means merely what has already been said, plus the fact that the slope of a line becomes indefinitely larger as the line becomes more nearly parallel to the y-axis, with a plus sign when the line's angle of incidence is in the first or third quadrants, and with a minus sign when the line's angle of incidence is in the second and fourth quadrants, (reference above).

Practice Exercise No. 48

Find the slopes of the lines between the points:

(1) $(2,1)$, $(-3,5)$

(2) $(-3,2)$, $(4,-7)$

(3) $(3,1)$, $(3,4)$

(4) $(3,1)$, $(2,1)$

(5) The line which makes an angle of 30° with the y-axis.

(6) The line which has twice the angle of inclination of the line in (1).

(7) The line which makes a 60° angle with the line in (2).

Angles Between Lines

Let l_1 and l_2 be any two non-parallel lines with slopes S_1 and S_2, and with angles of incidence θ_1 and θ_2 respectively. From the diagram of Figure 35 — in which an auxiliary line has been drawn through the intersection of l_1 and l_2, parallel to the x-axis — it is clear that the angle θ, measured counter-clockwise (page 96) from l_1 to l_2, is given by the formula,

G5: $\theta = \theta_2 - \theta_1 = tan^{-1}S_2 - tan^{-1}S_1.$

(See G4′, page 106)

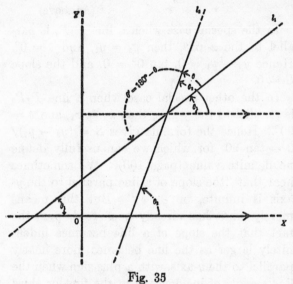

Fig. 35

Consequently, *we can always find the counter-clockwise angle θ between any two lines of known slopes by* (1) *finding their angles of incidence, and* (2) *substituting these values in Formula G5.*

EXAMPLE 10: Find the angle $P_2P_1P_3$ in the truss diagram of Figure 31, page 103.

SOLUTION: We have already found the angle of incidence for P_1P_2 to be

$\theta_1 = 23°58'.$ (Example 6)

And we have already found the angle of incidence for P_1P_3 to be

$\theta_2 = 53°08'.$ (Example 7)

Hence the counter-clockwise angle, $P_2P_1P_3$, is

$\theta = \theta_2 - \theta_1$ (Formula G5)

$= 53°08' - 23°58' = 29°10'$ (Substitution)

There are cases, however, in which we may

not wish to bother finding the angles of incidence, θ_1 and θ_2. To handle these we may derive a related formula as follows:

$\tan \theta = \tan (\theta_2 - \theta_1).$ (From formula G5)

But from trigonometry,

$$\tan (\theta_2 - \theta_1) = \frac{\tan \theta_2 - \tan \theta_1}{1 + \tan \theta_1 \tan \theta_2},$$

$$= \frac{S_2 - S_1}{1 + S_1 S_2}.$$ (Formula G4)

Hence, for the counter-clockwise angle between any two lines with slopes S_1 and S_2 respectively, we have the alternative formula,

G5′: $\theta = tan^{-1} \dfrac{S_2 - S_1}{1 + S_1 S_2}.$ (Substitution, G4′)

EXAMPLE 11: Find the angle in Example 10 directly.

SOLUTION: We have already found the slopes of P_1P_2 and P_1P_3 to be, respectively,

$S_1 = 4/9,$ and $S_2 = 4/3.$ (Examples 6 and 7)

Hence

$\theta = \tan^{-1} \dfrac{4/3 - 4/9}{1 + (4/3)(4/9)}$ (Substitution in G5′)

$= \tan^{-1} \dfrac{8/9}{43/27}$ (Adding fractions)

$= \tan^{-1} 24/43$ (Simplifying the fraction)

$= \tan^{-1} .55814 = 29°10',$ Ans. (*MMS*, Chapter XV)

As expected, this answer agrees with that of Example 10 above.

When Formula G5 produces a *negative result,* this is because the *angle θ is between 90° and 180°,* and tan θ is therefore negative (page 99).

For instance, interchanging the values of S_1 and S_2 in Example 11 above, we get

$\theta' = \tan^{-1} \dfrac{4/9 - 4/3}{1 + (4/9)(4/3)} = \tan^{-1} - .55814$

$= 180° - \tan^{-1} .55814$ (Page 101)

$= 180° - 29°10' = 150°50'.$

This result is also precisely what we should expect from the fact that the angle θ', measured counter-clockwise from P_1P_3 to P_1P_2 (extended back through the point P_1), is

$\theta' = 180° - \theta$ (Figure 35)

The relationship is also shown schematically in the diagram of Figure 35 above where θ' is identified as the supplement of θ.

In applying Formula G5, therefore, you must always be careful to keep track of the relative counter-clockwise positions of your lines. Then you will not mistakenly accept a result which gives the supplement of the angle for which you are actually looking.

Practice Exercise No. 49

A. Find the counter-clockwise measured angles between the following pairs of lines:

(1) L_1 from $(3,5)$ to $(-2,1)$, and L_2 from $(4,3)$ to $(1,-5)$.

(2) L_1 from $(2,-3)$ to $(4,-2)$, and L_2 from $(3,1)$ to $(4,-1)$.

B. Find the following angles in the truss diagram of Figure 31:

(3) $P_1 P_2 P_3$

(4) $P_2 P_1 P_3$

(5) $P_3 P_4 P_2$

Problem-Solving Technique

With as few tools in our analytic work-kit as formulas G1 through G5, we may now begin the analytic treatment of certain types of geometric problems covered by those formulas.

In general **the analytic method of solving geometric problems** may be divided into the following steps:

Step One: *Set up an analytic diagram by relating the geometric figure to a set of coordinate axes.* For the present, this means applying formulas G1 and, possibly, G2 or G2'.

When all the conditions in your problem are reasonably simple, as in the case of Examples 12, 13, and 14 below, it is usually best to designate all fixed points in the generalized form, $P_1(x_1,y_1)$, $P_2(x_2,y_2)$, etc. This tends to keep both the equations and their solutions symmetric, thus simplifying your algebraic operations and helping you more readily to identify or interpret your results.

But if many fixed points are specified the resulting large number of x_n's and y_n's may be awkward to manipulate algebraically. Also, it is sometimes convenient to make use of the origin or the perpendicular x- and y-axes as part of the figure. In problems such as those in Examples 15, 16, and 17 below, it is therefore best to forego the advantages of symmetry in your equations and locate your diagram with respect to the axes in such a way that as many as possible of the fixed-point coordinates become 0, or have the same absolute value other than 0.

Caution: It is absolutely essential, however, that you always set up the diagram in such a way as not to specialize the figure. Thus, if "a triangle" is specified you may locate its vertices, P_1, P_2, and P_3, in any of the four ways shown in the diagrams of Figure 36:

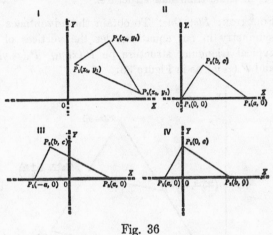

Fig. 36

But only the special case of a specified "*isosceles* triangle" may be set up as in the diagrams of Figure 37:

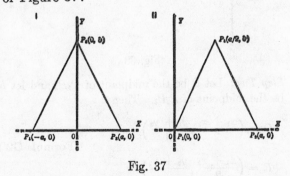

Fig. 37

Step Two: *State the other conditions of the problem analytically by expressing them in* *equation form.*   For the present this means applying formulas G2 through G5.

Step Three: *Anticipate the required analytic form of the geometric solution in terms of the required outcome of the above formulated algebraic equations.*

Step Four: *Develop the equations formulated in Step Two by whatever algebraic steps which may be necessary in order to verify whether the result anticipated in step three is obtainable.*

EXAMPLE 12: A number of triangular-shaped structures are to be strengthened by adding braces to the midpoints of pairs of adjacent sides.   Determine what the length and slope of each brace will be as compared to the length and slope of the third side of the same triangular structure.

SOLUTION: *Step One:* To obtain the advantages of symmetry in our equations, let the vertices of a typical triangular structure be $P_1(x_1,y_1)$, $P_2(x_2,y_2)$, and $P_3(x_3,y_3)$ as in Figure 38:

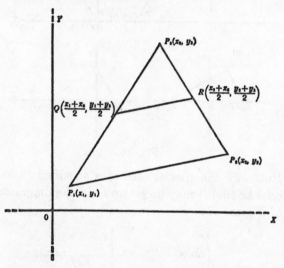

Fig. 38

Step Two: Let Q be the midpoint of P_1P_3 and let R be the midpoint of P_2P_3.   Then

$$Q = \left(\frac{x_1 + x_3}{2}, \frac{y_1 + y_3}{2}\right),$$
$$R = \left(\frac{x_2 + x_3}{2}, \frac{y_2 + y_3}{2}\right).$$
(Formula G2′)

Moreover, letting S and D be the slope and length of P_1P_2,

$$S = \frac{y_2 - y_1}{x_2 - x_1}$$
(Formula G4)

$$D = \sqrt{(x_2 - x_1)^2 + (y_2 - y_1)^2}.$$
(Formula G3)

Step Three: We are now required to find S' and D', the slope and length of the brace QR, in terms of S and D respectively.

Step Four:

$$S' = \frac{\dfrac{y_2 + y_3}{2} - \dfrac{y_1 + y_2}{2}}{\dfrac{x_2 + x_3}{2} - \dfrac{x_1 + x_2}{2}}$$
(Substitution, Formula G4)

$$= \frac{\dfrac{y_2 + y_3 - y_1 - y_3}{2}}{\dfrac{x_2 + x_3 - x_1 - x_3}{2}}$$
(Adding fractions)

$$= \frac{y_2 - y_1}{x_2 - x_1} = S.$$
(Simplifying and substitution)

$$D' = \sqrt{\left(\frac{x_2 + x_3}{2} - \frac{x_1 + x_3}{2}\right)^2 + \left(\frac{y_2 + y_3}{2} - \frac{y_1 + y_3}{2}\right)^2}$$
(Formula G3)

$$= \sqrt{\left(\frac{x_2 + x_3 - x_1 - x_3}{2}\right)^2 + \left(\frac{y_2 + y_3 - y_1 - y_3}{2}\right)^2}$$
(Adding fractions)

$$= \tfrac{1}{2}\sqrt{(x_2 - x_1)^2 + (y_2 - y_1)^2} = \tfrac{1}{2}D.$$
(Simplifying and substitution)

Answer: Each brace has the same slope as the base and $\frac{1}{2}$ the length of the base.

EXAMPLE 13: A number of quadrilateral-shaped structures are to be strengthened by pairs of braces joining the midpoints of opposite sides.   Verify that each pair of braces should be bored at their midpoints so that they may be riveted at their intersection.

SOLUTION: *Step One:* To obtain the advantages of symmetry again, let the vertices of the quadrilateral be P_1, P_2, P_3, and P_4, as in Figure 39:

Step Two: Let M_1 be the midpoint of P_1P_2, etc.   Then

$$M_1 = \left(\frac{x_1 + x_2}{2}, \frac{y_1 + y_2}{2}\right), \quad M_2 = \left(\frac{x_2 + x_3}{2}, \frac{y_2 + y_3}{2}\right),$$
$$M_3 = \left(\frac{x_3 + x_4}{2}, \frac{y_3 + y_4}{2}\right), \quad M_4 = \left(\frac{x_4 + x_1}{2}, \frac{y_4 + y_1}{2}\right).$$
(Formula G2′)

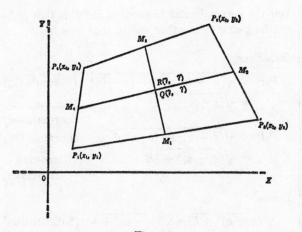

Fig. 39

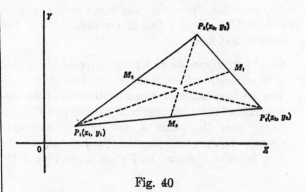

Fig. 40

Step Three: We are now required to verify that Q and R, the midpoints of M_1M_3 and M_2M_4 respectively, coincide.

Step Four: First,

$$Q = \left(\frac{\frac{x_1 + x_2}{2} + \frac{x_3 + x_4}{2}}{2}, \frac{\frac{y_1 + y_2}{2} + \frac{y_3 + y_4}{2}}{2} \right)$$
(Formula G2′)

$$= \left(\frac{x_1 + x_2 + x_3 + x_4}{4}, \frac{y_1 + y_2 + y_3 + y_4}{4} \right)$$
(Simplifying)

But also,

$$R = \left(\frac{\frac{x_2 + x_3}{2} + \frac{x_4 + x_1}{2}}{2}, \frac{\frac{y_2 + y_3}{2} + \frac{y_4 + y_1}{2}}{2} \right)$$
(Formula G2′)

$$= \left(\frac{x_1 + x_2 + x_3 + x_4}{4}, \frac{y_1 + y_2 + y_3 + y_4}{4} \right) = Q.$$
(Simplifying and Substitution)

Note: A line which joins the mid-points of opposite sides of a quadrilateral is called a *median* of the quadrilateral. The above steps therefore prove the proposition that the medians of a quadrilateral bisect each other.

EXAMPLE 14: Prove analytically that the medians of a triangle intersect in a point which divides each median in the ratio 2:1.

SOLUTION: *Step One:* Let the vertices of the triangle be $P_2(x_1, y_1)$, etc., and the other extremities of the corresponding medians be M_1, M_2, and M_3, as in Figure 40:

Step Two:

$$M_1 = \left(\frac{x_2 + x_3}{2}, \frac{y_2 + y_3}{2} \right),$$

$$M_2 = \left(\frac{x_3 + x_1}{2}, \frac{y_3 + y_1}{2} \right), \quad \text{(Formula G2′)}$$

$$M_3 = \left(\frac{x_1 + x_2}{2}, \frac{y_1 + y_2}{2} \right).$$

Step Three: We are now required to verify that Q_1, Q_2, and Q_3, the points which divide the medians P_1M_1, etc., in the ratio 2:1, coincide.

Step Four:

$$Q_1 = \left(\frac{x_1 + 2\frac{x_2 + x_3}{2}}{2 + 1}, \frac{y_1 + 2\frac{y_2 + y_3}{2}}{2 + 1} \right) \text{(Formula G2)}$$

$$= \left(\frac{x_1 + x_2 + x_3}{3}, \frac{y_1 + y_2 + y_3}{3} \right). \quad \text{(Simplifying)}$$

$$Q_2 = \left(\frac{x_2 + 2\frac{x_3 + x_1}{2}}{2 + 1}, \frac{y_2 + 2\frac{y_3 + y_1}{2}}{2 + 1} \right) \text{(Formula G2)}$$

$$= \left(\frac{x_1 + x_2 + x_3}{3}, \frac{y_1 + y_2 + y_3}{3} \right). \quad \text{(Simplifying)}$$

$$Q_3 = \left(\frac{x_3 + 2\frac{x_1 + x_2}{2}}{2 + 1}, \frac{y_3 + 2\frac{y_1 + y_2}{2}}{2 + 1} \right) \text{(Formula G2)}$$

$$= \left(\frac{x_1 + x_2 + x_3}{3}, \frac{y_1 + y_2 + y_3}{3} \right). \quad \text{(Simplifying)}$$

Hence,

$$Q_1 = Q_2 = Q_3. \quad \text{(Substitution)}$$

Note: You will find it interesting to compare the directness and brevity of the above demonstration with the many steps of indirect reasoning usually required in *synthetic* proofs of the same proposition.

EXAMPLE 15: Prove that any point on the perpendicular bisector of a line is equidistant from the ends of that line.

SOLUTION: *Steps One and Two:* In problems of this type we may make use of the fact that the coordinate axes are perpendicular to each other. Let the extremities of the given line be $P_1(-a,0)$ and $P_2(a,0)$. Then, since the y-axis is now the perpendicular bisector of P_1P_2, we may let "any point on the perpendicular bisector" be $P_3(0,b)$, as in Figure 41:

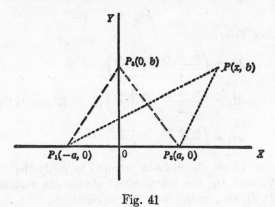

Fig. 41

Step Three: Now all that needs to be proven is that $P_1P_3 = P_2P_3$.

Step Four:

$$P_1P_3 = \sqrt{[0 - (-a)]^2 + (b - 0)^2} \qquad \text{(Formula G3)}$$
$$= \sqrt{a^2 + b^2}. \qquad \text{(Removing parentheses)}$$

$$P_2P_3 = \sqrt{(a - 0)^2 + (0 - b)^2} \qquad \text{(Formula G3)}$$

$$= \sqrt{a^2 + b^2}. \qquad \text{(Removing parentheses)}$$

Hence,

$P_1P_3 = P_2P_3$, as required. (Substitution)

The **converse of a proposition** is *a second proposition which has the given condition of the first as its conclusion, and the conclusion of the first as its given condition.*

EXAMPLE 16: Prove the converse of the proposition in Example 15.

SOLUTION: *Steps One and Two:* Using the same diagram (Figure 41), let the point $P(x,b)$ be any point such that

$$P_1P = P_2P. \qquad \text{(The given condition)}$$

Step Three: The analytic condition that $P(x,b)$ be "on the perpendicular bisector of P_1P_2" is that it be on the y-axis — in other words, that $x = 0$.

Step Four:

$$P_1P = \sqrt{[x - (-a)]^2 + (b - 0)^2} \qquad \text{(Formula G3)}$$
$$= \sqrt{(x + a)^2 + b^2}. \qquad \text{(Removing parentheses)}$$

$$P_2P = \sqrt{(x - a)^2 + (b - 0)^2} \qquad \text{(Formula G3)}$$
$$= \sqrt{(x - a)^2 + b^2}. \qquad \text{(Removing parentheses)}$$

Hence

$$\sqrt{(x + a)^2 + b^2} = \sqrt{(x - a)^2 + b^2} \quad \text{(Substitution)}$$
$$x^2 + 2ax + a^2 + b^2 = x^2 - 2ax + a^2 + b^2 \quad \text{(Squaring both sides)}$$
$$4ax = 0 \qquad \text{(Transposing)}$$
$$x = 0/4a = 0. \qquad \text{(Dividing by } 4a \neq 0\text{)}$$

And so

$$P(x,b) = P(0,b), \qquad \text{(Substitution)}$$

which means that P must be on the Y-axis, the perpendicular bisector of P_1P_2.

EXAMPLE 17: Find the angle at which the diagonals of a square intersect.

SOLUTION: *Step One:* Again we may make use of the fact that our coordinate axes meet at right angles — this time by letting the vertices of our square be $A(0,0)$, $B(b,0)$, $C(b,b)$, $D(0,b)$, as in Figure 42:

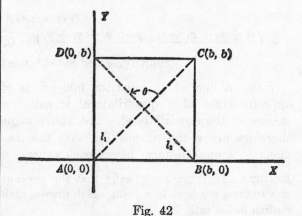

Fig. 42

Steps Two and Three: Let the diagonal AC be the line l_1, let the diagonal BD be the line l_2, and let θ be the angle at which l_1 and l_2 intersect. Then

$$S_1 = \frac{b - 0}{b - 0} = 1,$$

(Formula G4)

$$S_2 = \frac{b - 0}{0 - b} = -1,$$

and

$$\theta = \tan^{-1} \frac{S_2 - S_1}{1 + S_1 S_2}$$

(Formula G5')

Step Four: Hence,

$$\theta = \tan^{-1} \frac{-1 - 1}{1 + 1(-1)}$$

(Substitution)

$$= \tan^{-1} - 2/0$$

(Removing parentheses)

$$= \tan^{-1} \infty = 90°.$$

(Page 100)

Note that the above steps, in effect, prove the general proposition that the diagonals of a square intersect each other at right angles.

Practice Exercise No. 50

1. Rework Example 12 above, placing the diagram as in Case II of Figure 36. Do the same placing the diagram as in Cases III and IV of Figure 36. Note in all three instances that your algebraic work involves fewer constants, but that your equations and steps are less symmetrical in form.

2. Rework Example 13 above, placing the vertices of the quadrilateral at points $P_1(0,0)$, $P_2(b,0)$, $P_3(c,d)$, and $P_4(e,f)$. Does the reduction in the number of constants compensate for the loss of symmetry in the equations?

3. Rework Example 14 as in Problem 1, above.

4. P is the midpoint of the side BC of triangle ABC. Show that

$$AB^2 + AC^2 = 2AP^2 + BP^2 + PC^2$$

5. Prove that the line joining the midpoints of the non-parallel sides of a trapezoid has the same slope as the bases and length = half the sum of the lengths of the bases. *Suggestion:* Let the vertices be $(0,0)$, $(a,0)$, (b,c) and (d,c).

6. Prove that if a triangle is isosceles the medians to its equal sides are also equal. *Suggestion:* Set up the diagram as in diagram II, Figure 36 above.

7. Prove the converse of the preceding theorem. *Suggestion:* Apply the reasoning of Example 16 above.

8. Prove that if the diagonals of a parallelogram are equal, the figure is a rectangle. *Suggestion:* Let the vertices be $(0,0)$, $(a,0)$, (x,b), and $(x + a,b)$. Why?

Summary

Analytic geometry is *the technique of treating geometric problems by algebraic means.*

The *basic analytic tools* are formulas for translating geometric concepts and conditions into equivalent algebraic expressions and equations. Thus far we have derived the following —

G1: $x = x_1,\ y = y_1$.
Position of the point $P_1(x_1,y_1)$.

G2: $P(x,y) = \left(\dfrac{nx_1 + mx_2}{m + n}, \dfrac{ny_1 + my_2}{m + n} \right)$.
Position of the point P which divides the line P_1P_2 in the ratio $m:n$.

G2': $M(x,y) = \left(\dfrac{x_1 + x_2}{2}, \dfrac{y_1 + y_2}{2} \right)$.
Position of the mid-point M of the line P_1P_2.

G3: $D = \sqrt{(x_2 - x_1)^2 + (y_2 - y_1)^2}$.
The distance D between the points P_1 and P_2, or the length P_1P_2.

G4: $S_1 = \dfrac{y_2 - y_1}{x_2 - x_1} = \tan \theta_1$.
The slope S_1 of the line P_1P_2 with angle of incidence θ_1.

G4': $\theta_1 = \tan^{-1} S_1$.
The angle of incidence of P_1P_2 with slope S_1.

G5: $\theta = \theta_2 - \theta_1 = \tan^{-1} S_2 - \tan^{-1} S_1$.

G5': $\theta = \tan^{-1} \dfrac{S_2 - S_1}{1 + S_1 S_2}$.
The counter-clockwise angle between l_1 and l_2 with slopes S_1 and S_2 respectively.

In general, the analytic method is to:

1. *Set up a diagram which relates the given geometric figure to a set of coordinate axes.*

2. Express the other geometric conditions of the problem in equations.

3. Anticipate the corresponding analytic form which the solution must take.

4. Treat the equations of step 2 algebraically to determine whether the solution anticipated in step 3 is obtainable.

When the *fixed points of a diagram* have the generalized form, $P_1(x_1,y_1)$, $P_2(x_2,y_2)$, etc., our equations are more symmetrical, and our algebraic steps are easier to perform and to interpret.

When the *fixed points of a diagram* are strategically located with respect to the origin and axes, our equations are less symmetrical and less easy to interpret. But then we have fewer constants to manipulate, and we may take advantage of the geometric properties of the coordinate axes themselves.

In all cases, however, *we must be careful never to set up our diagram so as to make the figure less general than it should be under the conditions of the problem.*

CHAPTER X

STRAIGHT LINES

Problems solved in the preceding chapter have involved only **segments of straight lines** — for instance, P_1P_2 which begins at the point P_1 and ends at the point P_2. But other types of geometric problems involve **lines of indefinite length** — extending as far as required in either direction.

To develop analytic techniques for dealing with the latter is the purpose of this present chapter.

Point-Slope Line Formulas

An axiom of elementary geometry states that: *Through a given point only one straight line can be drawn parallel to a given line* (*MMS*, Chapter XIV). For instance, through the point P_1 in Figure 43 only the line L can be

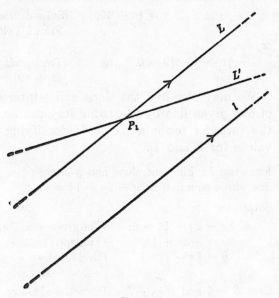

Fig. 43

drawn parallel to the given line l; and any other line L' through P must eventually intersect l if extended far enough (same reference).

How can we now express analytically — in terms of equations, that is — the fact that a

point P is on the line L parallel to the given line l, and not on some different line L'?

For a solution we first refer the diagram, Figure 43 above, to a set of rectangular coordinate axes as in Figure 44 below. Since l is a *given* line, we may here designate its angle of incidence with the x-axis as the *known* quantity θ (page 116). Since L is specified to be parallel to l, we may designate its corresponding angle of incidence by the same quantity θ (*MMS*, Chapter XIV). Since the point P_1 is *given*, we may designate its coordinates by the *known* quantities (x_1, y_1). But since the point P is specified only as "anywhere on the line L," we must designate its coordinates by the *variable* quantities (x, y).

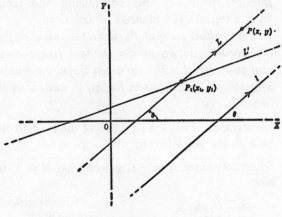

Fig. 44

Observe that the *slope* of each of the lines, l and L, is also a known quantity; namely —

$$S = \tan \theta. \qquad \text{(Formula G3)}$$

This is because the analytic equivalent of the above postulate from elementary geometry is: *Through a given point there can be only one line with the same slope.* Or, to word the same principle a little differently: *A straight line is determined by a point and a slope.*

However, since the point $P(x, y)$ is specified

125

to be on the line L, the latter's slope can also be written, by Formula G3, as $(y-y_1)/(x-x_1)$. Hence we have the new formula,

G6: $\dfrac{y-y_1}{x-x_1} = S.$ (Substitution)

This is called **the point-slope formula of a straight line.** For it is clear from the above that, if L is the line through $P_1(x_1,y_1)$ with slope S, then the coordinates of any point $P(x,y)$ on L must satisfy this equation; and conversely, if the coordinates of any point $P(x, y)$ satisfy this equation, then P must be on the line L.

The same formula may also be written in **determinant form** as:

G6m: $\begin{vmatrix} x & y & 1 \\ x_1 & y_1 & 1 \\ 1 & S & 0 \end{vmatrix} = 0.$

Readers who have already studied Chapter VII may verify this by expanding the determinant (page 80) and comparing the result with Formula G6 cleared of fractions.

Using either form of Formula G6, we may find the linear equation of the straight line through any given point with any given slope, by making appropriate substitutions for x_1, y_1 and S in the formula.

EXAMPLE 1: Write the equation of the straight line through the point $(-1,3)$ with slope $=2$.

SOLUTION: Since $x_1 = -1$, $y_1 = 3$, and $S = 2$, we have

$\dfrac{y-3}{x-(-1)} = 2,$ (Substitution in G6)

$y - 3 = 2(x+1),$ (Multiplying by $x+1 = x+1$)

$y - 3 = 2x + 2,$ (Removing parentheses)

$-2x + y - 5 = 0.$ Answer.   (Transposing $2x + 2$)

Or, using the determinant form:

$\begin{vmatrix} x & y & 1 \\ -1 & 3 & 1 \\ 1 & 2 & 0 \end{vmatrix} = 0,$ (Substitution in G6m)

$-2x + y - 5 = 0.$ Answer.   (Expanding by minors of row 1)

An important special case of these formulas is that in which the fixed point P_1 is $(0,B)$ on the y-axis. The ordinate B of P_1 is then called the **y-intercept** of the line L, and Formula G6 becomes,

$\dfrac{y-B}{x-0} = S,$ (Substituting $x_1 = 0, y_1 = B$)

$y - B = Sx,$

or:

G6': $y = Sx + B.$ (Transposing $-B$)

Called **the y-intercept-slope formula of a straight line,** this has the **determinant form,**

G6'm: $\begin{vmatrix} x & y & 1 \\ 0 & B & 1 \\ 1 & S & 0 \end{vmatrix} = 0.$ (The same substitutions)

Using either form of Formula G6', we may find the linear equation of the straight line with any given slope and y-intercept.

EXAMPLE 2: Write the equation of the line with slope $= \frac{1}{2}$ and y-intercept $= 19$.

SOLUTION: Since here $S = \frac{1}{2}$, $B = 19$, we get

$y = \tfrac{1}{2}x + 19,$ (Substitution in Formula G6)

or,

$-\tfrac{1}{2}x + y - 19 = 0.$ Ans.   (Transposing $\tfrac{1}{2}x + 19$)

We may also find the slope and y-intercept of any given line by re-writing its equation in the form of formula G6 and identifying its values for S and B.

EXAMPLE 3: Find the slope and y-intercept of the line whose equation is $3x - 2y - 14 = 0$.

SOLUTION:

$3x - 2y - 14 = 0,$ (The given equation)

$-2y = -3x + 14,$ (Transposing $3x - 14$)

$y = \tfrac{3}{2}x - 7,$ (Dividing by $-2 = -2$)

Here $S = \frac{3}{2}$ and $B = -7$. Hence the slope of the given line must be $\frac{3}{2} = 1.5$, and its y-intercept must be -7.

Practice Exercise No. 51

A. Write the equations of the lines passing through the following points with the following slopes.

	Points	Slopes
(1)	(1,2)	2
(2)	(−4,6)	−1
(3)	(5,−4)	3
(4)	(0,2)	2
(5)	(0,−7)	−2
(6)	(0,0)	1
(7)	(5,0)	−3
(8)	(7,0)	$\frac{1}{2}$
(9)	(4,5)	0
(10)	(−1,−2)	$\frac{1}{4}$

B. Find the slopes and y-intercepts of the lines with the following equations:

(1) $y = 3$ (4) $-2y = 3x + 4$

(2) $2x - 3y + 7 = 0$ (5) $x = 2y - 9$

(3) $x = 5$ (6) $y - 2 = 3(x - 4)$

Two-Point Line Formulas

It is also a postulate of elementary geometry that: *Two points determine a straight line*, or: *Through any two given points one, and only one, straight line can be drawn.*

The next question therefore is: How can we express analytically — in terms of equations — the fact that a point $P(x,y)$ is on the line L which passes through two given points, $P_1(x_1,y_1)$ and $P(x_1,y_2)$?

This time the answer is quicker and easier. Having already found the point-slope formula for a line, we may now recall that the two points P_1 and P_2 determine the slope of the line L which passes through them as

$$S = \frac{y_2 - y_1}{x_2 - x_1}. \qquad \text{(Formula G4)}$$

Simply by substitution in Formula G6, therefore, we have as the two-point formula of a straight line,

G7: $\dfrac{y - y_1}{x - x_1} = \dfrac{y_2 - y_1}{x_2 - x_1}$

Although this formula may be re-written in non-fractional form, it is most conveniently cleared of fractions after specific values have been assigned for x_1, y_1 etc.

By cross-multiplying its denominators, transposing terms, and applying *Formula M3*

(Chapter VII), however, we may re-write this two-point line formula in the more convenient determinant form:

G7m: $\begin{vmatrix} x & y & 1 \\ x_1 & y_1 & 1 \\ x_2 & y_2 & 1 \end{vmatrix} = 0.$

EXAMPLE 4: Write the equation of the straight line through the points (3,9) and (11,13).

SOLUTION: Since $x_1 = 3$, $y_1 = 9$, $x_2 = 11$, $y_2 = 13$, we have

$\dfrac{y - 9}{x - 3} = \dfrac{13 - 9}{11 - 3}$ (Substitution in G7)

$\dfrac{y - 9}{x - 3} = \dfrac{4}{8} = \dfrac{1}{2},$ (Simplifying numerically)

$2y - 18 = x - 3,$ (Cross multiplying denominators)

$x - 2y + 15 = 0,$ Answer.   (Transposing terms)

Or, using the determinant form:

$\begin{vmatrix} x & y & 1 \\ 3 & 9 & 1 \\ 11 & 13 & 1 \end{vmatrix} = 0,$ (Substitution in G7m)

$-4x + 8y - 60 = 0,$ (Expanding by minors of row 1)

$x - 2y + 15 = 0.$ Answer.   (Dividing by $-4 = -4$)

An important special case of these formulas is that in which the fixed points P_1 and P_2 are $(A,0)$ and $(0,B)$ respectively. The abscissa A of P_1 is then called the **x-intercept** of the line L, and Formula G7 becomes,

$\dfrac{y - 0}{x - A} = \dfrac{B - 0}{0 - A},$ (Substitution in G7)

$-Ay = Bx - AB,$ (Cross-multiplying denominators)

$-Bx - Ay = -AB,$ (Transposing Bx)

or:

G7': $\dfrac{x}{A} + \dfrac{y}{B} = 1.$ (Dividing by $-AB = -AB$)

The corresponding **determinant formula** follows more directly as

G7'm: $\begin{vmatrix} x & y & 1 \\ A & 0 & 1 \\ 0 & B & 1 \end{vmatrix} = 0.$ (Substitution in G6m)

EXAMPLE 5: Write the equation of the line with x-intercept $= -5$ and y-intercept $= -1$.

SOLUTION: Since $A = -5$, $B = -1$, we have

$$\frac{x}{-5} + \frac{y}{-1} = 1,$$ (Substitution in G6)

$$x + 5y = -5,$$ (Multiplying by $-5 = -5$)

$$x + 5y + 5 = 0,$$ Answer. (Transposing -5)

And the determinant solution follows as before.

Practice Exercise No. 52

A. Write the equations of the lines through the following pairs of points:

(1) $(0,1)$ and $(3,0)$.
(2) $(0,-2)$ and $(-4,0)$.
(3) $(-5,0)$ and $(0,-5)$.
(4) $(6,3)$ and $(4,7)$.
(5) $(-6,3)$ and $(-4,7)$.
(6) $(6,-3)$ and $(4,-7)$.
(7) $(6,-3)$ and $(-7,4)$.
(8) $(-45,-17)$ and $(-18,-5)$.

B. Write the equations of the lines through the following pairs of points in the truss diagram of Figure 31.

(9) P_1P_2　　　　(10) P_3P_4

Points on Lines

A problem frequently met in geometry is that of **determining whether a given point lies on a given line.** For instance, we may be required to find whether a point P_3 lies on the straight line determined by two other points P_1 and P_2.

To solve this problem analytically we need only to find the equation of the line through P_1 and P_2, and then to test by substitution whether the coordinates of P_3 satisfy the equation.

When Formula G7m is used for this purpose, the problem reduces to one of finding whether the resulting determinant is 0. For, when the coordinates of P_1, P_2, and P_3, are substituted in Formula G7m, we find **the analytic condition for three points being on the same straight line** to be given by the related formula,

G8:　　$$\begin{vmatrix} x_1 & y_1 & 1 \\ x_2 & y_2 & 1 \\ x_3 & y_3 & 1 \end{vmatrix} = 0.$$

EXAMPLE 6: Determine whether the point $(10,12)$, or the point $(12,13)$, lies on the straight line through the points $(-8,-3)$ and $(4,7)$.

SOLUTION: The equation of the line through $(-8,-3)$ and $(4,7)$ is

$$\frac{y - (-3)}{x - (-8)} = \frac{7 - (-3)}{4 - (-8)},$$ (Substitution in G7)

or,

$$\frac{y + 3}{x + 8} = \frac{10}{12} = \frac{5}{6}.$$ (Removing parentheses, etc.)

Substituting $x = 10$, $y = 12$, in this equation, we get

$$\frac{12 + 3}{10 + 8} = \frac{15}{18} = \frac{5}{6}. \checkmark$$ (The values check)

Hence $(10,12)$ must be on the given line.
However, substituting $x = 12$, $y = 13$, we get

$$\frac{13 + 3}{12 + 8} = \frac{16}{20} = \frac{4}{5} \neq \frac{5}{6}.$$ (The values do *not* check)

Hence $(12,13)$ cannot be on the given line.

EXAMPLE 7: A parallelogram-shaped structure, $ABCD$, is to have a strut added from its vertex D to the midpoint M of its opposite side AB. Verify that this strut will intersect the diagonal AC at the latter's trisection point T closest to A.

SOLUTION: *Step One:* To simplify the equations and make use of the fact that lines through points with the same ordinate are parallel to the x-axis, set up the diagram as in Figure 45.

Step Two: The specified trisection point of AC is now

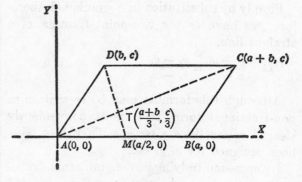

Fig. 45

$$T = \left(\frac{a+b}{3}, \frac{c}{3}\right).$$ (Formula G2)

The midpoint of AB is

$$M = \left(\frac{a}{2}, 0\right).$$ (Formula G2')

And the equation of the line through points D and M is

$$\frac{y-0}{x-\frac{a}{2}} = \frac{c-0}{b-\frac{a}{2}},$$ (Formula G7)

$$\left(b - \frac{a}{2}\right)y = cx - \frac{ac}{2},$$ (Cross-multiplying denominators)

or,

$$-cx + \left(b - \frac{a}{2}\right)y + \frac{ac}{2} = 0.$$ (Transposing terms)

Step Three: All we now need to show is that the coordinates of the point T satisfy this equation. The required substitution and the rest of *step four* is left to the student.

Observe that if we apply Formula G8 to this problem, instead of Formula G7 as above, then all we need to do is to show that the coordinates of D, M, and T satisfy the formula as in the following determinant equation,

$$\begin{vmatrix} b & c & 1 \\ \frac{a}{2} & 0 & 1 \\ \frac{a+b}{3} & \frac{c}{3} & 1 \end{vmatrix} = 0.$$ (Formula G8)

But this we have already done in a single step in Example 7 of Chapter VII (page 85).

A similar quick solution of the preceding Example by the use of formula G7 is as follows:

$$\begin{vmatrix} 10 & 12 & 1 \\ -8 & -3 & 1 \\ 4 & 7 & 1 \end{vmatrix} = \begin{vmatrix} 6 & 5 & 0 \\ -12 & -10 & 0 \\ 4 & 7 & 1 \end{vmatrix}$$ (First substituting in G8. Then subtracting row 3 from row 1 and from row 2)

$$= -60 + 60 = 0.$$ (Expanding by minors of col. 3)

Hence $(10,12)$ must be on the line through $(-8,-3)$ and $(4,7)$.

$$\begin{vmatrix} 12 & 13 & 1 \\ -8 & -3 & 1 \\ 4 & 7 & 1 \end{vmatrix} = \begin{vmatrix} 8 & 6 & 0 \\ -12 & -10 & 0 \\ 4 & 7 & 1 \end{vmatrix}$$ (As above)

$$= -80 + 72 \neq 0.$$ (As above)

Hence $(12,13)$ is not on this line.

Practice Exercise No. 53

Determine which of the following points lie on the following lines.

(a) $(2, 0)$ (a') $3x - 5y = 6$
(b) $(-1, 4)$ (b') $2x + 3y = 7$
(c) $(3, -1)$ (c') $x + y - 3 = 0$
(d) $(53, -33)$ (d') $2x - y + 6 = 0$
(e) $(6, -3)$ (e') $5x + 3z = 7$

Intersection Points of Lines

In Chapter III we interpreted the solutions of simultaneous pairs of first degree equations by locating the intersection points of their straight-line graphs (pages 33 to 36). Reversing that process, we may now solve geometric problems concerning the intersection of straight lines by solving the equations of these lines simultaneously.

EXAMPLE 8: A trapezoidal-shaped optical device, $ABCD$, has lenses mounted at the centers, E and F, of its two parallel sides, AB and DC, respectively. Lenses are also mounted at the ends of these sides. Determine analytically the points at which sightings through each pair of lenses, A and D, B and C, and E and F, coincide.

SOLUTION: *Step One:* To simplify the equations, place the diagram as in Figure 46:

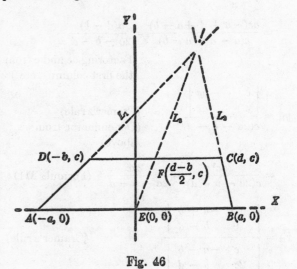

Fig. 46

Step Two: The lines of sighting are those labelled L_1, L_2, and L_3 in the diagram. Their equations are:

$$L_1: \begin{vmatrix} x & y & 1 \\ -b & c & 1 \\ -a & 0 & 1 \end{vmatrix} = 0, \qquad \text{(Formula G7m)}$$

$$L_2: \begin{vmatrix} x & y & 1 \\ d & c & 1 \\ a & 0 & 1 \end{vmatrix} = 0, \qquad \text{(Formula G7m)}$$

$$L_3: \begin{vmatrix} x & y & 1 \\ \dfrac{d-b}{2} & c & 1 \\ 0 & 0 & 1 \end{vmatrix} = \tfrac{1}{2} \begin{vmatrix} 2x & y & 1 \\ d-b & c & 1 \\ 0 & 0 & 1 \end{vmatrix} = 0.$$

(Formula G7m simplified by multiplying column 1 by 2, Formula M5)

Or:

$$L_1: \ cx - (a-b)y = -ac,$$
$$L_2: \ cx + (a-d)y = ac,$$
$$L_3: \ 2cx + (b-d)y = 0.$$

(Expanding the above by minors of the first rows)

Step Three: To find the required intersection points (x_{12}, y_{12}), etc., we need to solve each pair of the above equations simultaneously.

Step Four:

$$x_{12} = \frac{\begin{vmatrix} -ac & -a+b \\ ac & a-d \end{vmatrix}}{\begin{vmatrix} c & -a+b \\ c & a-d \end{vmatrix}} \qquad \text{(Cramer's rule)}$$

$$= \frac{ac(-a+d+a-b)}{c(a-d+a-b)} = \frac{a(d-b)}{2a-b-d}$$

(Factoring ac and c from the first columns, etc.)

$$y_{12} = \frac{\begin{vmatrix} c & -ac \\ c & ac \end{vmatrix}}{c(2a-b-d)} \qquad \begin{array}{l}\text{(Cramer's rule)} \\ \text{(Denominator from} \\ \text{above)}\end{array}$$

$$= \frac{2ac^2}{c(2a-b-d)} = \frac{2ac}{2a-b-d} \qquad \text{(Formula M1)}$$

$$x_{13} = \frac{\begin{vmatrix} -ac & -a+b \\ 0 & b-d \end{vmatrix}}{\begin{vmatrix} c & -a+b \\ 2c & b-d \end{vmatrix}} \qquad \text{(Cramer's rule)}$$

$$= \frac{ac(d-b)}{c(b-d+2a-2b)} = \frac{a(d-b)}{2a-b-d}. \qquad \begin{array}{l}\text{(Factoring and} \\ \text{expanding as} \\ \text{above)}\end{array}$$

$$y_{13} = \frac{\begin{vmatrix} c & -ac \\ 2c & 0 \end{vmatrix}}{c(2a-b-d)} = \frac{2ac}{2a-b-d}. \qquad \begin{array}{l}\text{(As for } y_{12} \\ \text{above)}\end{array}$$

Hence all three lines must intersect in the common point.

$$P = \left(\frac{a(d-b)}{2a-b-d}, \ \frac{2ac}{2a-b-d} \right). \qquad \text{(Formula G1)}$$

Readers who have not yet studied Chapter VII may verify the above steps by using Formula G6 to re-write the equation of L_1 as,

$$\frac{y-0}{x-(-a)} = \frac{c-0}{-b-(-a)}, \text{ etc.}$$

The greater length of the algebraic steps which are then required in Step Four illustrates the value of the determinant methods explained in Chapter VII.

The same readers may also temporarily skip the material immediately following and come back to it when they later finish Chapter VII.

Some geometric problems require us **to determine whether three lines are concurrent** — intersect in a common point — but do not require us to find the precise coordinates of the point in which they intersect. It is then convenient to have a single formula which will enable us to do this quickly.

To that end, let the equations of any three straight lines be:

$$L_1: \ a_1x + b_1y = c_1,$$
$$L_2: \ a_2x + b_2y = c_2,$$
$$L_3: \ a_3x + b_3y = c_3.$$

If L_1 and L_2 intersect in one, and only one, common point, then the coordinates of this point must be the simultaneous solution of their equations,

$$x = \frac{\begin{vmatrix} c_1 & b_1 \\ c_2 & b_2 \end{vmatrix}}{\begin{vmatrix} a_1 & b_1 \\ a_2 & b_2 \end{vmatrix}}, \ y = \frac{\begin{vmatrix} a_1 & c_1 \\ a_2 & c_2 \end{vmatrix}}{\begin{vmatrix} a_1 & b_1 \\ a_2 & b_2 \end{vmatrix}}, \qquad \text{(Cramer's rule)}$$

with the common denominator determinant $\neq 0$. For, if this determinant $= 0$, then the equations must be indeterminate and L_1 and L_2 either coincide or are parallel (Chapter VII).

If, now, line L_3 also passes through the same point, the above values of x and y must also satisfy its equation as follows,

$$a_3 \frac{\begin{vmatrix} c_1 & b_1 \\ c_2 & b_2 \end{vmatrix}}{\begin{vmatrix} a_1 & b_1 \\ a_2 & b_2 \end{vmatrix}} + b_3 \frac{\begin{vmatrix} a_1 & c_1 \\ a_2 & c_2 \end{vmatrix}}{\begin{vmatrix} a_1 & b_1 \\ a_2 & b_2 \end{vmatrix}} = c_3. \text{ (Substitution in } L_3)$$

Cleared of fractions, multiplied by $-1 = -1$, and with a change of sign in the first term effected by interchanging the order of its columns (Formula M10), this becomes the determinant equation.

G9: $\begin{vmatrix} a_1 & b_1 & c_1 \\ a_2 & b_2 & c_2 \\ a_3 & b_3 & c_3 \end{vmatrix} = 0.$

This equation also happens to hold good if two or more of the lines, L_1, L_2, L_3, are parallel (page 132). In that case the above steps of reasoning cannot be retraced to prove that L_1, L_2, L_3, meet in a common point, for then required denominator determinants will be $= 0$, etc. (as mentioned just before). But otherwise these steps can be so retraced. Hence Formula G9 states **the analytic condition for three non-parallel lines to be concurrent.**

Suppose, for instance, that in the preceding Example 8 we were required only to prove that the three lines of sighting, L_1, L_2, L_3, are concurrent, and not to find the coordinates of their common point of intersection. Since the figure is specified to be a trapezoid, it is clear that these lines are not parallel. Hence to prove that they are concurrent we have merely to show that the determinant of the constant coefficients in their equations $= 0$, as follows:

$\begin{vmatrix} c & -a+b & -ac \\ c & a-d & ac \\ 2c & b-d & 0 \end{vmatrix}$ (Substitution in G9)

$= \begin{vmatrix} c & -a+b & ac \\ c & a-d & -ac \\ c & a-d & -ac \end{vmatrix} = 0.$ (Subtracting row 1 from row 3 to make rows 2 and 3 identical. Then equating to 0 by Formula M6, page 84.)

EXAMPLE 9: Struts are to be secured to a triangular-shaped structure connecting each vertex with the mid-point of the opposite side. Determine analytically that they will intersect in one common point.

SOLUTION: *Step One:* Set up the diagram as in Figure 47:

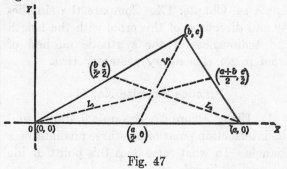

Fig. 47

Step Two: The equations of the three median lines are now:

L_1: $\begin{vmatrix} x & y & 1 \\ 0 & 0 & 1 \\ \frac{a+b}{2} & \frac{c}{2} & 1 \end{vmatrix} = \frac{1}{2}\begin{vmatrix} x & y & 1 \\ 0 & 0 & 1 \\ a+b & c & 2 \end{vmatrix} = 0$

L_2: $\begin{vmatrix} x & y & 1 \\ \frac{b}{2} & \frac{c}{2} & 1 \\ a & 0 & 1 \end{vmatrix} = \frac{1}{2}\begin{vmatrix} x & y & 1 \\ b & c & 2 \\ a & 0 & 1 \end{vmatrix} = 0$

L_3: $\begin{vmatrix} x & y & 1 \\ b & c & 1 \\ \frac{a}{2} & 0 & 1 \end{vmatrix} = \frac{1}{2}\begin{vmatrix} x & y & 1 \\ b & c & 1 \\ a & 0 & 2 \end{vmatrix} = 0$

(Formula G7m, simplified by Formula M5)

or:

L_1: $\quad -cx + (a+b)y + 0 = 0$
L_2: $\quad cx + (2a - b)y - ac = 0$
L_3: $\quad 2cx + (a - 2b)y - ac = 0$

(Multiplying by $2 = 2$ and expanding by M2)

Step Three: Since none of these medians can be parallel, no two of the above equations can be inconsistent. We need therefore show only that the determinant of the constant coefficients $= 0$.

Step Four:

$\begin{vmatrix} -c & a+b & 0 \\ c & 2a-b & -ac \\ 2c & a-2b & -ac \end{vmatrix} = \begin{vmatrix} -c & a+b & 0 \\ c & 2a-b & -ac \\ c & 2a-b & -ac \end{vmatrix} = 0.$

(Adding row 1 to row 3 by M8, and equating to 0 by M6)

Note that this demonstration proves the geometric theorem that *the medians of a triangle meet in a point* without assuming prior knowledge of where the meeting point is as in Example 14, Chapter IX. Compare the simplicity and directness of this proof with the length and tediousness of the synthetic method of proof in an elementary geometry text.

Practice Exercise No. 54

(1) Using Figure 47, Example 9 above, find the intersection point of the three medians of a triangle. In what ratio does this point divide the median lines?

(2) Using the same figure as for problem 5, Exercise 50, prove that the diagonals of a trapezoid intersect each other at their common intersection point with the line joining the midpoints of the parallel sides.

(3) Determine the intersection point of the preceding problem.

(4) A quadrilateral-shaped structure has struts joining the midpoints of its opposite sides. Find the intersection point of these struts in terms of the coordinates of the vertices.

Parallel Lines

Corresponding to the theory of parallels in *elementary* geometry (*MMS*, Chapter XIV), there is a series of formulas in analytic geometry concerning the conditions under which lines are parallel. All of these formulas depend upon the fact that, since a line's direction is expressed analytically by its slope (page 116), the analytic way of specifying that *two lines are parallel* is to state — in an equation — that *they have the same slope.*

For instance, let L_1 and L_2 be any two lines with the equations,

$$L_1: a_1x + b_1y + c_1 = 0,$$
$$L_2: a_2x + b_2y + c_2 = 0;$$

or with the equations,

$$L_1: y = S_1x + B_1, \qquad \text{(Formula G6')}$$
$$L_2: y = S_2x + B_2,$$

where

$$S_1 = -a_1/b_1, \quad B_1 = -c_1/b_1,$$
$$S_2 = -a_2/b_2, \quad B_2 = -c_2/b_2.$$

The analytic condition that lines L_1 and L_2 be parallel is then stated by any of the formulas,

G10: $S_1 = S_2$,   (As explained above)

or,

G10': $a_1/b_1 = a_2/b_2$,   (Substitution in G10, multiplied by $-1 = -1$)

or,

G10'm: $\begin{vmatrix} a_1 & b_1 \\ a_2 & b_2 \end{vmatrix} = 0.$ (G10' in determinant form)

Now we have only to apply these formulas to determine whether any two given lines are parallel, or to derive further formulas concerning parallel lines.

EXAMPLE 10: Determine whether each of the following pairs of lines are parallel:

(a) $y = 3x + 7$, and $y = 3x - 4$.

(b) $3x - 4y + 5 = 0$, and $3x - 7y + 9 = 0$.

SOLUTION (a): Here

$S_1 = 3$, and $S_2 = 3$.   (Formula G6')

Hence,

$S_1 = S_2$.   (Substitution)

Since their equations thus fulfill the requirement of Formula G10, the lines must be parallel.

SOLUTION (b): Here

$a_1/b_1 = 3/-4 = -3/4$,   (Substituting
$a_2/b_2 = 3/-7 = -3/7 \neq -3/4$. $a_1 = 3$, etc.)

Hence,

$a_1/b_1 \neq a_2/b_2$ (Substitution)

Since their equations thus do *not* fulfill the requirement of Formula G10', the lines cannot be parallel.

Or, by the determinant method,

$\begin{vmatrix} 3 & -4 \\ 3 & -7 \end{vmatrix} = -21 - (-12) \neq 0.$ (Substitution in G10'm)

Hence the same conclusion from the failure of the equations to fulfill the requirement of Formula G10'm.

This chapter began with the axiom of elementary geometry that only one straight line can be drawn through a given point parallel to a given line (page 125). Now we shall derive the equation of that line.

Let the given point be $P_1(x_1, y_1)$, let the equation of the given line L be $y = Sx + B$, and let the required line through P_1 parallel to L be designated L_1, as in Figure 48.

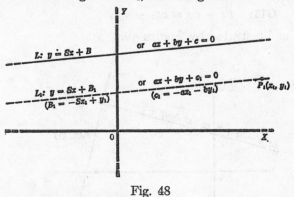

Fig. 48

Since L_1 is specified to be parallel to L, we may let its equation tentatively be

$$y = Sx + B_1, \qquad \text{(Formulas G6' and G10)}$$

where S is the same as in the equation of L, but B_1 is as yet an unknown quantity.

However, since L_1 is also specified to pass through the point P_1, the coordinates of the latter must satisfy this tentative equation by the equality:

$$y_1 = Sx_1 + B_1. \qquad \begin{array}{l}\text{(Substituting}\\ x = x_1, y = y_1)\end{array}$$

But since B_1 is the only unknown in this equality, we may solve for it as

$$B_1 = -Sx_1 + y_1. \qquad \text{(Transposing, etc.)}$$

Hence **the equation of the line** L_1 **which passes through the point** P_1 **and is parallel to the line** L **is**

G11': $y = Sx - Sx_1 + y_1.$ (Substitution for B_1)

Or, if the equation of L is originally given in the form $ax + by + c = 0$, then by making the substitution $S = -a/b$ in equation G11', we find its alternative form to be

$$y = -\frac{a}{b}x + \frac{a}{b}x_1 + y_1, \qquad \begin{array}{l}\text{(The stated sub-}\\ \text{stitution)}\end{array}$$

$$by = -ax + ax_1 + by_1. \qquad \begin{array}{l}\text{(Multiplying by}\\ b = b)\end{array}$$

or,

G11: $ax + by = ax_1 + by_1.$ (Transposing $-ax$)

EXAMPLE 11: Find the equation of the line which passes through the point $(4, -13)$ and is parallel to the line $2x + y - 7 = 0$.

SOLUTION: Here $a = 2$, $b = 1$, $x_1 = 4$, $y_1 = -13$. Hence the required equation is

$$2x + y = 2(4) + 1(-13), \qquad \begin{array}{l}\text{(Substitution in}\\ \text{G11)}\end{array}$$

$$= 8 - 13 = -5, \qquad \begin{array}{l}\text{(Removing paren-}\\ \text{theses)}\end{array}$$

or,

$$2x + y + 5 = 0. \quad \text{Answer} \quad \text{(Transposing } -5)$$

EXAMPLE 12: Find the equation of the line which is parallel to the same line and passes through the point $(11, -15)$.

SOLUTION: Here $x_1 = 11$, $y_1 = -15$. Hence the required equation is

$$2x + y = 2(11) + 1(-15)$$
$$= 22 - 15 = 7, \qquad \text{(Steps as above)}$$

or,

$$2x + y - 7 = 0.$$

But this is the equation of the line we were given! *Explanation:* The given point must be on the given line, and through this point only one line can be drawn with the slope of the given line. *Check:*

$$2(11) + (-15) - 7 = 22 - 15 - 7 = 0. \checkmark$$
$$\text{(Substitution)}$$

Practice Exercise No. 55

A. Formula G10'm above is also the condition under which we cannot solve the equations of L_1 and L_2 simultaneously by Cramer's Rule (pages 88–9). Why?

B. Why are the following *analytic conditions that two lines* L_1 *and* L_2 *coincide:*

G10a : $S_1 = S_2$, and $B_1 = B_2$?
G10'a: $a_1/a_2 = b_1/b_2 = c_1/c_2$?

C. Determine whether each of the following pairs of lines are parallel:

(1) $2x - 3y + 4 = 0$ (3) $4x - 3y = 7$
     $2x - 3y + 5 = 0$ $9y = 12x + 4$

(2) $3x + 6y - 3 = 0$ (4) $3x - 2y + 5 = 0$
     $2y = -x + 1$ $2x + 3y + 5 = 0$

D. Write the equations of the lines passing through the following given points and parallel to the following given lines:

(1) $(-2,3)$; $2x - 3y + 4 = 0$
(2) $(-1,1)$; $3x - 2y + 5 = 0$
(3) $(8,-3)$; $4x - 7y + 8 = 0$
(4) $(-1,3)$; $y = 2$
(5) $(-2,-2)$; $y = x$

Explain any unusual results you may obtain.

Perpendicular Lines

Problems concerning perpendicular lines always have to be treated *indirectly* in *elementary* geometry. To prove that L_1 is perpendicular to L_2 you must first show that L_1 is parallel to some other line given as perpendicular to L_2, or else you must show that the supplementary angles at which L_1 and L_2 intersect are also equal to each other and therefore to 90° (*MMS*, Chapter XIV).

In analytic geometry, however, problems concerning perpendiculars may be handled just as *directly* as problems concerning parallels. For, just as the analytic condition that two lines be parallel is that their slopes be equal, **the analytic condition that two lines be perpendicular is that the product of their slopes be $= -1$.** This is shown as follows:

Two lines L_1 and L_2 are perpendicular, by definition, if the angle θ between them is 90°. Then, however, their angles of incidence, θ_1 and θ_2, differ by 90°; for

$$\theta = 90° = \theta_2 - \theta_1. \qquad \text{(Formula G5)}$$

We know from trigonometry, moreover, that then (provided θ_1 and $\theta_2 \neq 0°$ or 180°),

$$(\tan \theta_1)(\tan \theta_2) = -1. \qquad \text{(Chapter VIII)}$$

But

$$\tan \theta_1 = S_1, \text{ and } \tan \theta_2 = S_2. \qquad \text{(Formula G4')}$$

Corresponding to formulas G10 and G10' above, therefore, we have the following formulas which state the analytic condition that lines L_1 and L_2 be perpendicular:

G12: $S_1 S_2 = -1.$ (Substitution above)

G12': $a_1/b_1 = -b_2/a_2.$ (Substituting $S_1 = -a_1/b_1$, etc.)

And corresponding to equations G11 and G11' above, we have the following **equations for the line L_1** which passes through the point P_1 and is perpendicular to the line L:

G13': $y = -\dfrac{1}{S} x + \dfrac{1}{S} x_1 + y_1,$

Or,

G13: $bx - ay = bx_1 - ay_1,$

as illustrated by Figure 49.

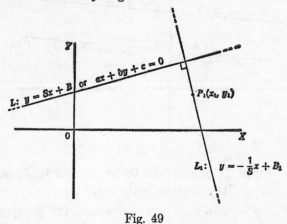

Fig. 49

Except for the difference in diagram illustrated by Figure 49, the algebraic steps required to derive this G12,13 series of formulas is exactly the same as for the preceding G10, 11 series of formulas (pages 132–3). The details are left to the student in Practice Exercise No. 56 below.

The method of applying these formulas is also exactly the same.

EXAMPLE 13: Determine whether each of the following pairs of lines are perpendicular:
(a) $y = 2x + 9$, and $y = -\frac{1}{2}x + 13$.
(b) $3x - 4y + 5 = 0$, and $3x - 7y + 9 = 0$.

SOLUTION (a): Here

$S_1 = 2$, and $S_2 = -\frac{1}{2}$. (Formula G6')

Hence,

$S_1 S_2 = 2(-\frac{1}{2}) = -1.$ (Substitution)

Since their equations thus fulfill the condition of Formula G12, the lines must be perpendicular. Answer.

SOLUTION (b): Here

$a_1/b_1 = 3/-4 = -3/4$, (Substituting $a_1 = 3$, etc.)
$-b_2/a_2 = -(-7)/3 = 7/3 \neq -3/4.$

Hence,

$$a_1/b_1 \neq -b_2/a_2.$$ (Substitution)

Since their equations thus do *not* fulfill the condition of Formula G12′, the lines cannot be perpendicular. Answer.

EXAMPLE 14: Find the equation of the line which passes through the point $(4, -13)$ and is perpendicular to the line $2x + y - 7 = 0$. (Compare with Example 11, page 133 above).

SOLUTION: Here again, $a = 2$, etc., as in the referenced example. But this time the required equation is

$$(1)x - 2y = (1)4 - 2(-13),$$ (Substitution in G13)
$$x - 2y = 4 + 26 = 30,$$ (Removing parentheses)

or,

$$x - 2y - 30 = 0, \text{ Answer.}$$ (Transposing 30)

EXAMPLE 15: Find the equation of the line which passes through the origin and is perpendicular to the line $y = Sx + B$.

SOLUTION: Since $P_1(x_1, y_1) = (0,0)$, $x_1 = y_1 = 0$, and the required equation is

$$y = -\frac{1}{S}x + \frac{1}{S}(0) + 0,$$ (Substitution in G13′)

or,

$$y = -\frac{1}{S}x, \text{ Answer.}$$ (Removing parentheses)

Practice Exercise No. 56

A. Derive Formula G13′ above, using Figure 49 and the method by which Formula G11′ is derived in the preceding text (page 133).

B. Do the same for Formula G13, referring to the steps by which Formula G11 is derived in the preceding text (page 133).

C. Determine whether each of the pairs of lines in Problem C, Practice Exercise No. 55, are perpendicular.

D. Write the equations of the lines passing through each of the points listed in the Practice Exercise No. 55D, and perpendicular to the corresponding lines.

Perpendicular Distances

The distance from a point to a line is, by definition, *the length of the perpendicular from the given point to the given line.*

For instance, the distance from the origin to the line L in Figure 50 is d — the length of the segment OQ on the line l which passes through O and is perpendicular to L at the point of intersection Q.

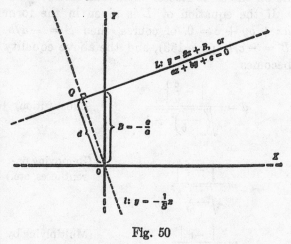

Fig. 50

To find a formula for d, let the equation of the given line L be

$$y = Sx + B.$$ (Formula G6′)

Then the equation of the line l, through O and perpendicular to L, is

$$y_1 = -\frac{1}{S}x.$$ (Example 15)

By solving these two equations simultaneously (as in Chapter II), we find the coordinates of the intersection point Q of L and l to be

$$(x,y) = \left(\frac{-BS}{S^2 + 1}, \frac{B}{S^2 + 1}\right)$$

Hence the required length of OQ is

$$d = \sqrt{\frac{B^2 S^2}{(S^2 + 1)^2} + \frac{B^2}{(S^2 + 1)^2}}$$ (Formula G3)
$$= \sqrt{\frac{B^2(S^2 + 1)}{(S^2 + 1)^2}} = \sqrt{\frac{B^2}{S^2 + 1}}$$
$$= \frac{|B|}{\sqrt{S^2 + 1}}.$$ (Adding fractions, etc.)

The absolute value sign is placed around B in this last expression because d, being the length of an undirected line, is always positive. But B will have a positive value only when the point Q is above the x-axis, and a negative value whenever Q is below the x-axis. For, B is the y-intercept of the given line L (Formula G6′), and it is clear from the diagram that Q must lie on the same side of the x-axis as B.

If the equation of L is given in the form $ax + by + c = 0$, of course, then $S = -a/b$, $B = -c/b$ (page 133), and the above equality becomes

$$d = \frac{\left| -\dfrac{c}{b} \right|}{\sqrt{\left(-\dfrac{a}{b}\right)^2 + 1}} \qquad \text{(Substitution)}$$

$$= \frac{\left| -\dfrac{c}{b} \right|}{\sqrt{\dfrac{a^2 + b^2}{b^2}}} \qquad \begin{array}{l}\text{(Removing pa-}\\\text{rentheses, etc.)}\end{array}$$

$$= \frac{\left| -c \right|}{\sqrt{a^2 + b^2}} ; \qquad \begin{array}{l}\text{(Multiplying by}\\ b/b = 1)\end{array}$$

Or, since $|-c| = |c|$, **the formula for the distance from the origin to the given line L is**

G14: $\quad d = \dfrac{\left| B \right|}{\sqrt{S^2 + 1}} = \dfrac{\left| c \right|}{\sqrt{a^2 + b^2}}.$ (Substitution)

EXAMPLE 16: Find the distance from the origin to the line $y = -\frac{4}{3}x + 15$.

SOLUTION: Here $S = -\frac{4}{3}$, $B = 15$. Hence

$$d = \frac{\left| 15 \right|}{\sqrt{\left(-\dfrac{4}{3}\right)^2 + 1}} \qquad \text{(Substitution in G14)}$$

$$= \frac{15}{\sqrt{\dfrac{16 + 9}{9}}} = \frac{15}{\dfrac{5}{3}} = 9. \qquad \begin{array}{l}\text{(Removing paren-}\\\text{theses, etc.)}\end{array}$$

EXAMPLE 17: Verify the result above by finding d when the equation of the line is given in the equivalent form $4x + 3y - 45 = 0$.

SOLUTION: Here $a = 4$, $b = 3$, $c = -45$. Hence

$$d = \frac{\left| -45 \right|}{\sqrt{4^2 + 3^2}} = \frac{45}{5} = 9. \quad \text{(Substitution in G14)}$$

Let L and L' be any two given parallel lines with the equations, $y = Sx + B$ and $y = Sx$ + B', respectively (Formulas G6′ and G10). Also, let D be the distance between L and L', and let d and d' be the distances from origin to L and L', respectively.

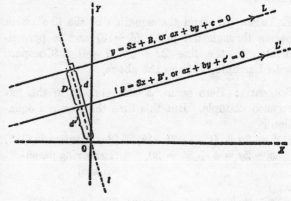

Fig. 51

If the relative positions of L and L' with respect to the origin are as in Figure 51, then

$$D = d - d' = \frac{\left| B \right| - \left| B' \right|}{\sqrt{S^2 + 1}}. \quad \text{(Formula G14)}$$

But if the relative positions of L and L' are interchanged in the diagram, then

$$D = d' - d = \frac{\left| B' \right| - \left| B \right|}{\sqrt{S^2 + 1}}. \quad \text{(Formula G14)}$$

Or, if L and L' lie on opposite sides of the origin, then

$$D = d + d' = \frac{\left| B \right| + \left| B' \right|}{\sqrt{S^2 + 1}}. \quad \text{(Formula G14)}$$

However, since B and B' are the y-intercepts of L and L' (Formula G6′), they will always have the same sign when L and L' are on the same side of the origin, and opposite signs when L and L' are on opposite sides of the origin. Hence their algebraic difference, $B-B'$, will always have the same *absolute value*. And, notwithstanding the various possible relationships between D, d, and d', noted above, we can always express the distance D between the parallel lines L and L' by the single equality,

$$D = \frac{\left| B - B' \right|}{\sqrt{S^2 + 1}}.$$

Or, if the equation of the given parallel lines are $ax + by + c = 0$ and $ax + by + c' = 0$ (Formula G10′), then $B = -c/b$, $B' = -c'/b$, $S = -a/b$ (page 133), and

$$D = \frac{\left| -\dfrac{c}{b} - \left(-\dfrac{c'}{b} \right) \right|}{\sqrt{\left(-\dfrac{a}{b} \right)^2 + 1}}. \quad \text{(Substitution)}$$

$$= \frac{\left| \dfrac{c' - c}{b} \right|}{\sqrt{\dfrac{a^2 + b^2}{b^2}}} \quad \begin{array}{l}\text{(Removing parentheses,}\\ \text{etc.)}\end{array}$$

Hence we have as **the formula for the distance D between parallel lines L and L′:**

G14′: $D = \dfrac{|B - B'|}{\sqrt{S^2 + 1}} = \dfrac{|c' - c|}{\sqrt{a^2 + b^2}}.$ $\begin{array}{l}\text{(Multiplying}\\ \text{by } b/b = 1)\end{array}$

EXAMPLE 18: Find the distance between the parallel lines, $y = -\frac{4}{3}x + 15$ and $y = -\frac{4}{3}x - 20$.

SOLUTION: Here $S = -\frac{4}{3}$ $B = 15$, $B' = -20$.

$$D = \frac{\left| 15 - (-20) \right|}{\sqrt{\left(-\dfrac{4}{3} \right)^2 + 1}} = \frac{35}{\dfrac{5}{3}} = 21. \quad \begin{array}{l}\text{(Substitution}\\ \text{in G14′)}\end{array}$$

EXAMPLE 19: Find the distance between the same pair of parallel lines when their equations are written in the equivalent forms, $4x + 3y - 45 = 0$ and $4x + 3y + 60 = 0$.

SOLUTION: Here $a = 4$, $b = 3$, $c = -45$, $c' = 60$. Hence

$$D = \frac{|60 - -45|}{\sqrt{4^2 + 3^2}} = \frac{105}{5} = 21.$$

(Substitution in G14′)

Finally, to find the formula for the distance D from any given point $P_1(x_1, y_1)$ to any given line L, let L' be the line drawn through P_1 parallel to L as in Figure 52.

Since D is now also the distance between the parallel lines L' and L, we can find it by means of Formula G14′ as follows: If we let the equation of the given line L be $ax + by + c = 0$, then the equation of the parallel line L' through P_1 is

$$ax + by + (-ax_1 - by_1) = 0. \quad \begin{array}{l}\text{(Formula G11,}\\ \text{transposed)}\end{array}$$

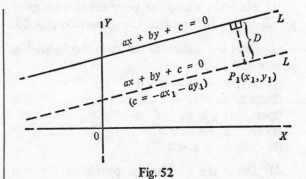

Fig. 52

Hence $c' = -ax_1 - by_1$, and

$$D = \frac{\left| -ax_1 - by_1 - c \right|}{\sqrt{a^2 + b^2}}. \quad \text{(Substitution in G 14′)}$$

Thus we have for **the distance D from the given point $P_1(x_1, y_1)$ to the given line L, the formula**

G15: $D = \dfrac{|ax_1 + by_1 + c|}{\sqrt{a^2 + b^2}}.$ $\begin{array}{l}\text{(Since } |-n|\\ = |n|)\end{array}$

EXAMPLE 20: Find the distance from the point $(-3, 9)$ to the line $3x - 4y + 15 = 0$.

SOLUTION: Here $a = 3$, $b = -4$, $c = 15$, $x_1 = -3$, $y_1 = 9$. Hence

$$D = \frac{\left| 3(-3) - 4(9) + 15 \right|}{\sqrt{3^2 + 4^2}} = \frac{|-30|}{5} = \frac{30}{5} = 6.$$

(Substitution in G15)

EXAMPLE 21: Find the distance from the point $(11, 12)$ to the same line.

SOLUTION: Here $x_1 = 11$, $y_1 = 12$, etc. Hence

$$D = \frac{\left| 3(11) - 4(12) + 15 \right|}{\sqrt{3^2 + 4^2}} = \frac{0}{5} = 0.$$

(Substitution in G15)

Explanation: The given point $(11, 12)$ is actually on the given line, and hence is at a zero distance from it. The numerator in the above fraction is itself a substitution *check* on this.

Practice Exercise No. 57

A. Find the distances of the following lines from the origin:

(1) $3y - 2x + 5 = 0$ (4) $y = 2x - 3$

(2) $x = 5$ (5) $2x = y - 7$

(3) $y = 7$ (6) $(y - 3) = 2(x - 4)$

B. Find the distances between the pairs of parallel lines in Problem C — Exercise No. 55.

C. Find the distances between the following points and lines:

(1) $(2,3)$ $3x - 2y + 5 = 0$

(2) $(2,2)$ $4x - 5y - 2 = 0$

(3) $(4,-1)$ $y = 3x - 4$

(4) $(2,3)$ $x = 7$

(5) $(0,0)$ $2y + x = 3$

D. Draw the graphs for problems (1) and (2) in C and graphically verify your results.

Problem Solving Technique

The formulas developed in this chapter add considerably to the power of the general problem-solving technique introduced in the preceding chapter.

EXAMPLE 22: Struts are to be secured to the vertices of a triangular-shaped structure, perpendicular to the opposite sides. Show analytically that they will intersect in a common point.

SOLUTION: *Step One:* Set up the diagram as in Figure 53.

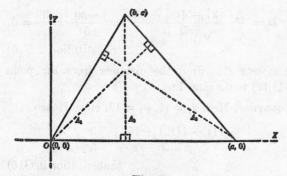

Fig. 53

Step Two: The slope of the side to which strut A_1 is perpendicular is

$$\frac{c - 0}{b - a} = \frac{c}{b - a}. \qquad \text{(Formula G4)}$$

Hence the equation of A_1 is

$$A_1: \quad \frac{y - 0}{x - 0} = \frac{-1}{\dfrac{c}{b - a}} = \frac{a - b}{c},$$

$$\text{(Formulas G6 and G12)}$$

or,

$$(a - b)x - cy = 0. \qquad \text{(Cross multiplying, etc.)}$$

Similarly, the equation of A_2 is

$$A_2: \quad bx + cy - ab = 0. \qquad \text{(Steps as above)}$$

And the equation of A_3 is

$$A_3: \quad x = b, \text{ or } x - b = 0. \qquad \text{(The same)}$$

Step Three: Since none of the lines A_1, A_2, and A_3 can be parallel, no pair of equations in the above redundant system can be inconsistent. Hence it is sufficient to show that the determinant D of their constant coefficient vanishes.

Step Four:

$$D = \begin{vmatrix} a - b & -c & 0 \\ b & c & -ab \\ 1 & 0 & -b \end{vmatrix} \qquad \begin{array}{l}\text{(Substitution,} \\ \text{formula G9)}\end{array}$$

$$= \begin{vmatrix} 0 & 0 & 0 \\ b & c & -ab \\ 1 & 0 & -b \end{vmatrix} = 0, \text{ as required.}$$

(Adding row 2 and $-a$ times row 3 to row 1, Formula M8; and equating to 0 by Formula M3)

Practice Exercise No. 58

(1) Determine analytically the intersection point of the preceding Example 22.

(2) Verify that braces erected perpendicular to the midpoints of the sides of a triangular shaped structure intersect in one common point. Place the vertices of the triangle as in Figure 53.

(3) Find the intersection point of the preceding example.

(4) Demonstrate that the intersection points of problems 1 and 3 above, and of problem 1, Practice Exercise No. 54, all lie on the same straight line. *Suggestion:* applying formula G8, show that

$$\begin{vmatrix} \dfrac{a + b}{3} & \dfrac{c}{3} & 1 \\[2mm] b & \dfrac{ab - b^2}{c} & 1 \\[2mm] \dfrac{a}{2} & \dfrac{b^2 + c^2 - ab}{2c} & 1 \end{vmatrix} = 0$$

Note: This amounts to proving that the intersection points of the medians of a triangle, of the altitudes of a triangle, and of the perpendicular bisectors of the sides of a triangle, all lie on the same straight line. See if you can demonstrate the same theorem by synthetic methods!

5. Show that the area A of the triangle with vertices $P_1(x_1,y_1)$, $P_2(x_2,y_2)$, and $P_3(x_3,y_3)$, is given by the formula:

G16: $A = \pm\frac{1}{2} \begin{vmatrix} x_1 & y_1 & 1 \\ x_2 & y_2 & 1 \\ x_3 & y_3 & 1 \end{vmatrix}$.

Suggestion: First find the base P_1P_2 by Formula G3. Next find the altitude as the distance from P_3 to the line through P_1 and P_2 by Formula G15. Then apply the Review Formula 39 and write the result in determinant form.

6. Using Formula G16 (above) compute the area of each of the triangles in the construction truss diagram, Figure 31.

Summary

As **further analytic tools** we now have the following additional **formulas** —

G6: $\dfrac{y - y_1}{x - x_1} = S.$

G6m: $\begin{vmatrix} x & y & 1 \\ x_1 & y_1 & 1 \\ 1 & S & 0 \end{vmatrix} = 0.$

The straight line passing through point P_1 with slope $= S$.

G6′: $y = Sx + B.$

G6′m: $\begin{vmatrix} x & y & 1 \\ 0 & B & 1 \\ 1 & S & 0 \end{vmatrix} = 0.$

The straight line with y-intercept $= B$ and slope $= S$.

G7: $\dfrac{y - y_1}{x - x_1} = \dfrac{y_2 - y_1}{x_2 - x_1}.$

G7m: $\begin{vmatrix} x & y & 1 \\ x_1 & y_1 & 1 \\ x_2 & y_2 & 1 \end{vmatrix} = 0.$

The straight line passing through points P_1 and P_2.

G7′: $\dfrac{x}{A} + \dfrac{y}{B} = 1.$

G7′m: $\begin{vmatrix} x & y & 1 \\ A & 0 & 1 \\ 0 & B & 1 \end{vmatrix} = 0.$

The straight line with x-intercept $= A$ and y-intercept $= B$.

G8: $\begin{vmatrix} x_1 & y_1 & 1 \\ x_2 & y_2 & 1 \\ x_3 & y_3 & 1 \end{vmatrix} = 0.$

Three points, P_1, P_2, and P_3, on the same straight line.

G9: $\begin{vmatrix} a_1 & b_1 & c_1 \\ a_2 & b_2 & c_2 \\ a_3 & b_3 & c_3 \end{vmatrix} = 0.$

Three non-parallel lines, L_1, L_2, and L_3, passing through the same point.

G10: $S_1 = S_2.$

G10′: $a_1/a_2 = b_1/b_2.$

G10′m: $\begin{vmatrix} a_1 & b_1 \\ a_2 & b_2 \end{vmatrix} = 0.$

Two lines, L_1 and L_2, parallel.

G11: $ax + by = ax_1 + by_1.$

G11′: $y = Sx - Sx_1 + y_1.$

The line through the point P_1, parallel to the line L.

G12: $S_1S_2 = -1.$

G12′: $a_1/b_1 = -b_2/a_2.$

Lines L_1 and L_2 mutually perpendicular.

G13: $bx - ay = bx_1 - ay_1.$

G13′: $y = -\frac{1}{S}x + \frac{1}{S}x_1 + y_1$

The line through P_1 perpendicular to line L.

G14: $d = \dfrac{|B|}{\sqrt{S^2 + 1}} = \dfrac{|c|}{\sqrt{a^2 + b^2}}.$

The distance d from the origin to line L.

G14′: $D = \dfrac{|B - B'|}{\sqrt{S^2 + 1}} = \dfrac{|c' - c|}{\sqrt{a^2 + b^2}},$

The distance D between parallel lines L and L'.

G15: $\quad D = \dfrac{|\,ax_1 + by_1 + c\,|}{\sqrt{a^2 + b^2}}$.

The distance D from point P_1 to line L.

G16: $\quad A = \pm\frac{1}{2} \begin{vmatrix} x_1 & y_1 & 1 \\ x_2 & y_2 & 1 \\ x_3 & y_3 & 1 \end{vmatrix}$.

Area of the triangle with vertices P_1, P_2, P_3.

CHAPTER XI

CONIC SECTIONS — PARABOLAS

Conic Sections as Loci

Parabolas, ellipses, circles, and hyperbolas are called **conic sections** — the lines along which a flat plane intersects a right circular cone.

By definition, a **right circular cone** is the surface generated by rotating one straight line about another straight line, intersected at an oblique angle. The fixed line is the cone's **axis**. The possible positions of the generating line in its rotation — for instance, any of the straight-line silhouettes in the following diagrams — are the cone's **elements**. The common intersection points of all the elements is the cone's **vertex**. And the two symmetrical parts of the generated surface on each side of the vertex are the **nappes** of the cone.

When the intersecting plane cuts through both nappes of the cone the curve has two

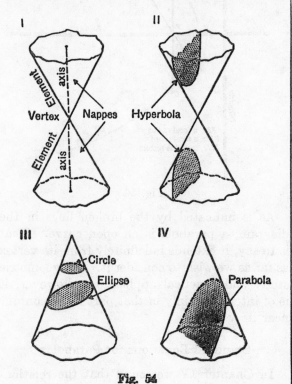

Fig. 54

parts and is called a **hyperbola** (Diagram II, Figure 54). When the intersecting plane is at right angles to the axis the curve is a circle (Diagram III, Figure 54). When the intersecting plane cuts completely across one nappe at an oblique angle to the axis the curve is an ellipse (Diagram III, Figure 54). When the intersecting plane is parallel to an element the curve is a **parabola** (Diagram IV, Figure 54). All these curves are called **non-degenerate conics**.

Examples of **degenerate conics** are the cases in which the intersecting plane cuts the cone only in the single point of its vertex, only in a single element, or only in a single pair of elements.

Although the ancients studied conic sections by synthetic methods for their fascinating aesthetic properties and theoretical interest, *these curves have been found to be of great practical value for modern science and technology because of their many mechanical, optical, and acoustical properties.* Again and again the same contours turn up in the design of such widely different structures as reflectors, suspension bridges, and machine gears. Indeed, the very orbits of the comets and planets, including that of our own earth, have been found to be "conics."

It has also been found more convenient to study these curves, by modern analytic methods, as loci. The word **locus** (plural: *loci*) means *place* in Latin. But in geometry it has the special meaning of "the place of those points which satisfy certain specified geometric conditions." For, by defining curves as loci in this way, we can more readily derive the equations which are their analytic equivalents.

Parabolas

Definitions and Construction

A **parabola** may be defined as the locus of

points in a plane equidistant from a given point, called its **focus**, and from a given line, called its **directrix**.

Draw a straight vertical line on a blank sheet of paper and mark a point to one side of this line near its middle. Set a pencil-compass to an opening a little wider than half the distance from the point to the line and strike two short arcs, one above the point and one below it. Keeping the compass to the same opening, and using it as a gauge-scriber with the metallic tip along the original vertical line, draw two short vertical line-segments intersecting the previously drawn arcs. Open the compass a bit wider and repeat the same operations several times with successively wider openings of the compass. Then connect all intersections of arcs and vertical line-segments with a single smooth curve as in Figure 55:

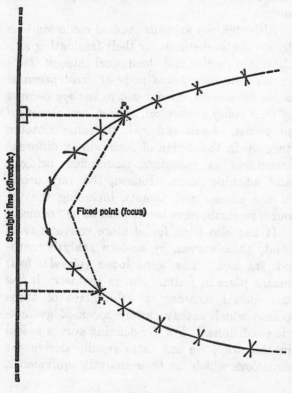

Fig. 55

The result is an approximate parabola with the original vertical line as directrix and the original fixed point as focus. For, by the con-

struction, every intersection point of an arc and vertical line-segment, such as P_1 or P_2 in the diagram, is equidistant from the given directrix line and focal point.

The auxiliary line which may be drawn through the focus, perpendicular to the directrix of the parabola, is its **axis** (Figure 56).

The point in which the axis intersects the parabola is its **vertex** (point V in the diagram). Any straight line from the focus to the parabola is called a **focal radius**. Any straight line joining two points on the parabola and passing through the focus is called a **focal chord**. And that focal chord which is perpendicular to the axis is called the **latus rectum** of the parabola.

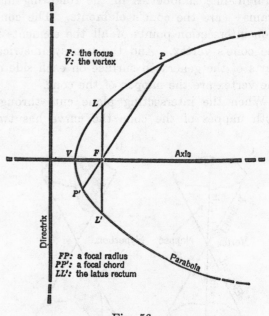

Fig. 56

As is indicated by the broken lines in the diagram, a parabola is an **open curve**. That is to say, it extends indefinitely from its vertex as far as we wish to consider it. For geometric study and practical applications, however, it is of interest mainly in that part of its contour near its vertex.

Standard Equations for Parabolas

In Chapter IV we found that the relationship between the variables of equations in the typical form $y = ax^2 + bx + c$ or $x = ay^2 + by + c$ can be represented by parabolic graphs. Reversing our procedure there, we shall now see that we can express the geometric properties of parabolas by means of such quadratic equations.

Let the line $x = -p$ be the directrix of a parabola whose focus is at the point $F(p,0)$, and let $P(x,y)$ be any point on the parabola as in Figure 57:

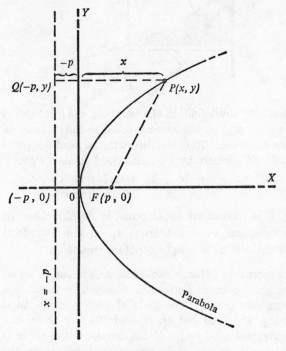

Fig. 57

The point $Q(-p,y)$ will then be the foot of the perpendicular from P to the directrix. And, by the locus-definition of a parabola,

$$|PQ| = |PF|.$$

But

$$|PQ| = x + p, \text{ and} \qquad \text{(Formula G3)}$$
$$|PF| = \sqrt{(x-p)^2 + (y-0)^2}. \quad \text{(Formula G3)}$$

Hence

$$x + p = \sqrt{(x-p)^2 + y^2}, \quad \text{(Substitution)}$$
$$x^2 + 2px + p^2 = x^2 - 2px + p^2 + y^2.$$
$$\text{(Squaring both sides)}$$

And

$$y^2 = 4px. \qquad \text{(Transposing terms)}$$

This means that every point which satisfies the locus-definition of a parabola with focus $(p,0)$ and directrix $x = -p$ has coordinates which satisfy the above equation. Since the above steps may be reversed we can just as readily show that all values of x and y which satisfy this equation are also coordinates of points which lie on the same parabola.

Hence, **the analytic equivalent of the parabola with focus $(p,0)$ and directrix $x = -p$** is given by the formula,

G17: $y^2 = 4px.$

If we let $(-p,0)$ be the focus and $x = p$ be the directrix, the parabola opens to the left instead of to the right and the formula is

G17': $y^2 = -4px.$

Interchanging variables simply makes the parabolas symmetric with the y-axis instead of with the x-axis and produces the formulas,

G17'': $x^2 = 4py.$
G17''' $x^2 = -4py.$

Formulas of this G17-group are called **standard equations** of parabolas because they place the vertex conveniently at the origin for setting up problems and provide a simple means of determining the curve's focus and directrix.

EXAMPLE 1: What are the focus and directrix of the parabola whose equation is $y^2 = 12x$?

SOLUTION: Since the standard equation of such a parabola is

$$y^2 = 4px, \qquad \text{(Formula G17)}$$

it follows that

$$4p = 12, \text{ and} \qquad \text{(Substitution)}$$
$$p = 3. \qquad \text{(Dividing by } 4 = 4)$$

Hence the focus is the point,

$$F = (3,0) \qquad \text{(Formula G17)}$$

and the directrix is the line,

$$x = -3. \qquad \text{(The same)}$$

EXAMPLE 2: Write the equation of the parabola whose focus is $(0,2)$ and whose directrix is $y = -2$.

SOLUTION: Since the parabola is symmetric with the y-axis and has its focus above the x-axis, the standard equation is

$$x^2 = 4py. \qquad \text{(Formula G17'')}$$

But

$$p = 2. \qquad \text{(The given condition)}$$

Hence the required equation is

$$x^2 = 4(2)y = 8y. \qquad \text{(Substitution)}$$

Practice Exercise No. 59

A. What are the focus and directrix of each of the following parabolas?

(1) $y^2 = 7x$ (2) $x^2 = 9y$

(3) $y^2 = -21x$ (4) $x^2 = -73y$

B. Write the equations of the parabolas with the following

Focus	Directrix
(1) (5,0)	$x = -5$
(2) (0,−3)	$y = 3$
(3) (2,0)	$x = 1$
(4) (1,11)	$x = y$

Applications of Parabola Formulas

When a particular focal point is specified in a practical problem involving parabolas we must first assign the required value to p in an appropriate standard formula.

EXAMPLE 3: A parabolic reflector is to be designed with a light (or heat) source at its focus $2\frac{1}{4}$ inches from its vertex. If the reflector is to be ten inches deep, how broad must it be, and how far will the outer rim be from the source?

SOLUTION: *Step One:* Set up the diagram with the vertex of the reflector's parabola-shaped cross-section at the origin and the focus at $(2\frac{1}{4},0)$ as in Figure 58:

Step Two: The standard equation of the cross-section is now

$$y^2 = 4px \qquad \text{(Formula G17)}$$
$$= 4(9/4)x = 9x. \qquad \text{(Substituting } p = 2\frac{1}{4} = 9/4\text{)}$$

Step Three: Since the reflector is to be 10″ deep we may let a point on its outer rim be $P(10,k)$. We now need to compute the breadth $= 2k$ and focal radius FP.

Step Four: Since $P(10,k)$ is a point on the parabola,

$$k^2 = 9(10) = 90, \qquad \text{(Substitution in G17)}$$
$$k = 9.486. \qquad (MMS, \text{Chapter IX})$$

And the breadth of the reflector is

$$2k = 18.972 \text{ inches.} \qquad \text{(Multiplying by 2 = 2)}$$

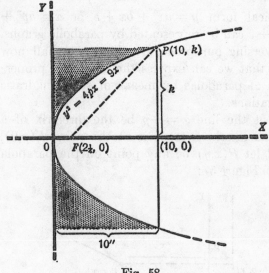

Fig. 58

Since, by the definition of a parabola, the focal radius to any point on the curve is equal to the distance of the same point from the directrix, we need not compute FP directly by Formula G3, but may write

$$FP = x + p = 10 + 2\frac{1}{4} = 12.25 \text{ inches.}$$
$$\text{(Substitution)}$$

If a particular focal point is *not* specified in a problem we must treat $4p$ in the standard equations *as a single constant quantity*.

EXAMPLE 4: When a cable suspends a load of equal weight for equal horizontal distances it assumes a parabolic curve. The ends of such a cable on a bridge are 1000 feet apart, and 100 feet above the horizontal road bed. If the center of the cable is level with the road bed, compute its height above the road bed at a distance of 300 feet from either end.

SOLUTION: *Step One:* Set up the diagram with the vertex of the parabola at the origin, the road bed along the x-axis, and the cable-ends at the points (500,100) and (−500,100) as in Figure 59:

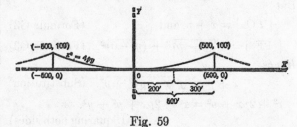

Fig. 59

Step Two: The standard equation of the parabola is now

$$x^2 = 4py \qquad \text{(Formula G17'')}$$

Step Three: What we need to compute is the value of y when $x = 500 - 300 = 200$. But we may do this by means of the above standard equation only after we have first found the value of $4p$.

Step Four: Since the parabola passes through the point (500,100),

$$(500)^2 = 4p(100), \qquad \text{(Substitution in G17'')}$$

and

$$4p = \frac{250{,}000}{100} = 2{,}500. \qquad \text{(Transposing terms, etc.)}$$

Hence the equation is

$$x^2 = 2{,}500y \qquad \text{(Substitution)}$$

or

$$y = .0004x^2. \qquad \text{(Dividing by 2,500)}$$

And when $x = 200$,

$$
\begin{aligned}
y &= .0004(200)^2 && \text{(Substitution)} \\
&= .0004(40{,}000) && \text{(Squaring 200)} \\
&= 16 \text{ feet.} && \text{(Removing parentheses)}
\end{aligned}
$$

EXAMPLE 5: When a liquid is rotated in a cylindrical container its hollow upper surface is parabolic. A centrifuge is 50 inches across and the rotated fluid in it is 10 inches higher at the circumference than at the center. Compute the distance from the center at which the height of the fluid is .4 inches higher.

SOLUTION: *Step One:* Set up a cross-section diagram with the vertex of the parabola at the origin and the outer extremities of the surface segment at the points (25,10) and (−25,10) as in Figure 60:

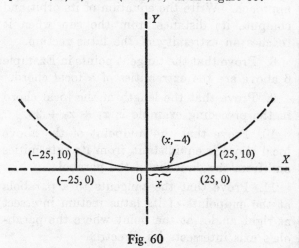

Fig. 60

Step Two: The standard equation of the parabola is now

$$x^2 = 4py \qquad \text{(Formula G17'')}$$

Step Three: What we need to compute is the value of x when $y = .4$. But first we must find the value of $4p$ in the above standard equation.

Step Four: Since the parabola passes through the point (25,10),

$$25^2 = 4p(10) \qquad \text{(Substitution)}$$

and

$$4p = \frac{625}{10} = 62.5 \qquad \text{(Transposing, etc.)}$$

Hence the equation is

$$x^2 = 62.5y \qquad \text{(Substitution)}$$

And when $y = .4$

$$x^2 = 62.5(.4) = 25, \qquad \text{(Substitution)}$$

or

$$x = \pm 5 \text{ inches.} \qquad \text{(Taking the square roots)}$$

The standard equations of a parabola may also be used in connection with the formulas of the two preceding chapters as part of our general analytic technique.

EXAMPLE 6: It is easily shown in calculus that the slope of the tangent to a parabola $y^2 = 4px$ at any point $P(x,y)$ is $2p/y$. Prove that if two tangents to a parabola are perpendicular to each other they intersect on the parabola's directrix.

SOLUTION: *Step One:* Let the two tangent lines be L_1 and L_2, touching the parabola $y^2 = 4px$ at the points $P_1(x_1, y_1)$ and $P_2(x_2, -y_2)$ as in Figure 61:

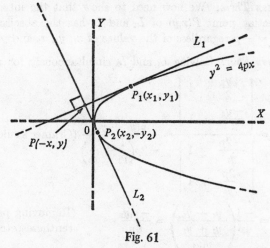

Fig. 61

Step Two: Since P_1 and P_2 are points on the parabola,

$$y_1{}^2 = 4px_1 \text{ and } y_2{}^2 = 4px_2. \qquad \text{(Substitution)}$$

Hence

$$x_1 = \frac{y_1{}^2}{4p}, \qquad x_2 = \frac{y_2{}^2}{4p}, \qquad \text{(Dividing by } 4p = 4p)$$

And

$$P_1 = \left(\frac{y_1^2}{4p}, y_1\right), \ P_2 = \left(\frac{y_2^2}{4p}, -y_2\right). \quad \text{(Formula G1)}$$

The point-slope equations of the tangents are therefore:

$$L_1: \quad \begin{vmatrix} x & y & 1 \\ \dfrac{y_1^2}{4p} & y_1 & 1 \\ 1 & \dfrac{2p}{y_1} & 0 \end{vmatrix} = 0,$$

$$\text{(Formula G6m)}$$

$$L_2: \quad \begin{vmatrix} x & y & 1 \\ \dfrac{y_2^2}{4p} & -y_2 & 1 \\ 1 & \dfrac{2p}{-y_2} & 0 \end{vmatrix} = 0.$$

Or:

$$L_1: \quad -\frac{2p}{y_1}x + y = y_1 - \frac{y_1}{2} = \frac{y_1}{2},$$

$$\text{(Formula M2')}$$

$$L_2: \quad \frac{2p}{y_2}x + y = -y_2 + \frac{y_2}{2} = -\frac{y_2}{2}.$$

Moreover, since L_1 and L_2 are stated to be perpendicular,

$$\frac{2p}{y_1} = \frac{-1}{\dfrac{2p}{-y_2}} = \frac{y_2}{2p}, \quad \text{(Formula G12)}$$

and,

$$y_1 y_2 = 4p^2. \quad \begin{array}{l}\text{(Multiplying} \\ \text{by } 2py_1 = 2py_1)\end{array}$$

Step Three: We now need to show that the intersection point $P(x,y)$ of L_1 and L_2 has the abscissa $x = -p$ regardless of the values of x_1, y_1, x_2 and y_2.

Step Four: Solving L_1 and L_2 simultaneously for x,

$$x = \frac{\begin{vmatrix} \dfrac{y_1}{2} & 1 \\ \dfrac{-y_2}{2} & 1 \end{vmatrix}}{\begin{vmatrix} \dfrac{-2p}{y_1} & 1 \\ \dfrac{2p}{y_2} & 1 \end{vmatrix}} = \frac{\frac{1}{2}(y_1 + y_2)}{-2p\left(\dfrac{1}{y_1} + \dfrac{1}{y_2}\right)} \quad \text{(Cramer's rule)}$$

$$= \frac{-1}{4p}\left(\frac{y_1 + y_2}{\dfrac{y_1 + y_2}{y_1 y_2}}\right) = \frac{-y_1 y_2}{4p} \quad \begin{array}{l}\text{(Removing pa-} \\ \text{rentheses, etc.)}\end{array}$$

$$= \frac{-4p^2}{4p} = -p. \quad \begin{array}{l}\text{(Substituting} \\ y_1 y_2 = 4p^2)\end{array}$$

And since $x = -p$ is the equation of the directrix, the theorem is proven.

Practice Exercise No. 60

1. Compute the breadth of the parabolic reflector in Example 3 above if it is designed to be 5 inches deep. Or 15 inches deep. What is the length of the focal radius to the rim of the reflector in each case?

2. Compute the depth of the same reflector if it is designed to be 20 inches broad. Or 15 inches broad. What is the length of the focal radius to the rim of the reflector in each case?

3. Find the height of the cable described in Example 4 above at distances of 10 feet, 300 feet, and 400 feet from the center.

4. Compute the distances from the center at which the cable will be 10 feet, 50 feet, and 90 feet above the road bed.

5. Compute the height above the center of the liquid in the centrifuge described in Example 5 at points 10 inches, 15 inches, and 24 inches from the side.

6. Compute the distances from the center of the centrifuge at which the height of the liquid will be .1, inches, 5 inches, and 9 inches higher than at the center.

7. Certain comets (the non-returning type) have parabolic orbits about the sun as focus. Such a comet is observed to come within 16 million miles of the sun at its point of closest approach. Write the equation of its orbit and compute its distance from the sun when it reaches an extremity of the latus rectum.

8. Prove that the tangent points in Example 6 above are the extremities of a focal chord.

9. Prove that the length of the focal chord in the preceding example is $x_1 + x_2 + 2p$.

10. Prove that the midpoint of the above focal chord is equidistant from its extremities and from the directrix of the parabola.

11. Prove that the tangents to a parabola at the endpoints of its latus rectum intersect at right angles at the point where the parabola's axis intersects its directrix.

The Summary of this chapter is included with that of Chapter XIII on pages 160–1.

CHAPTER XII

ELLIPSES AND CIRCLES

Ellipses

Definitions and Construction of Ellipses

An **ellipse** is the locus of points in a plane, the sum of whose distances from two given points, called its **foci**, is constant.

Tie the ends of a slack piece of string to two thumb tacks in a sheet of paper. With the point of the pencil always pulling the string taut, draw the complete curve which can thus be traced by the pencil point as in Figure 62:

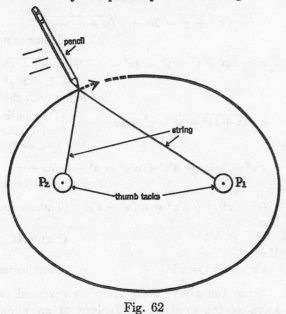

Fig. 62

The result is an ellipse with the thumb tack points as foci. This is because, by the construction, every point on the curve has the same total distance from the two thumb tack points — namely, the length of the piece of string.

The line of indefinite length through the foci of an ellipse (F_1 and F_2 in Figure 63) is the **transverse axis**.

That part of the transverse axis which is contained within the ellipse (A_1A_2 in the diagram) is the **major axis**.

The midpoint (0) of the line (F_1F_2) joining the foci of the ellipse is its **center**.

The line of indefinite length perpendicular to the transverse axis at the center is the **conjugate axis**.

That part (B_1B_2) of the conjugate axis which is contained within the ellipse is the **minor axis**.

The extremities of the major axis (A_1 and A_2) are the **vertices**.

Both **focal radii** and **focal chords** are defined for an ellipse as for a parabola (page 142).

Also, each of the two focal chords (G_1G_1' and G_2G_2') perpendicular to the major axis is called a **latus rectum** (plural: *latera rectae*).

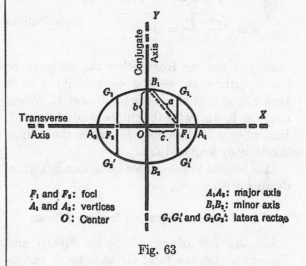

F_1 and F_2: foci	A_1A_2: major axis
A_1 and A_2: vertices	B_1B_2: minor axis
O: Center	G_1G_1' and G_2G_2': latera rectae

Fig. 63

Eccentricity of Ellipses

Although all parabolas have the same basic shape and differ only as to the scale on which they are drawn, the same is not true of ellipses. If in the above construction you were to lengthen the string between the two thumb tacks, or place the thumb tacks closer together, you would get "rounder" ellipses. With a very long string, or with the thumb tacks very close together, you could hardly tell the ellipse from a circle merely by looking at it. But if you were to shorten the string, or place the thumb tacks farther apart, you would get "flatter" ellipses. In the special case where the length of the string is exactly equal to the distance between the two foci, the ellipse "degenerates" into the straight line F_1F_2.

Hence we have as another feature of an ellipse its **eccentricity**. This is its degree of "non-roundness" or "flatness" and is expressed numerically as the ratio

$$e = \frac{c}{a}$$

where c and a are the indicated arm and hypotenuse of the right triangle in the first quadrant of the above ellipse diagram. Or, since the other arm of the same right triangle $= b$, and $c = \sqrt{a^2 - b^2}$,

$$e = \frac{\sqrt{a^2 - b^2}}{a} = \sqrt{1 - \frac{b^2}{a^2}} \qquad \text{(Substitution)}$$

As you can see from either the diagram or the equations above, the eccentricity e of a true ellipse is always between 0 and 1. When e is close to 0 in value the ellipse is nearly round. But when e close to 1 in value the ellipse is relatively long and flat.

It is helpful to keep these facts in mind when dealing with —

The Standard Equation for an Ellipse

Let the foci of an ellipse be $F_1(c,0)$ and $F_2(-c,0)$. Let the total distance to F_1 and to F_2 from any point $P(x,y)$ on the ellipse be designated $2a$. And let half the minor axis be designated b as in Figure 64:

Now

$$| \, F_1P \, | + | \, PF_2 \, | = 2a. \qquad \text{(Definition of an ellipse)}$$

But

$$| \, F_1P \, | = \sqrt{(x - c)^2 + y^2}$$
$$| \, PF_2 \, | = \sqrt{(x + c)^2 + y^2} \qquad \text{(Formula G3)}$$

So that

$$\sqrt{(x - c)^2 + y^2} + \sqrt{(x + c)^2 + y^2} = 2a \qquad \text{(Substitution)}$$

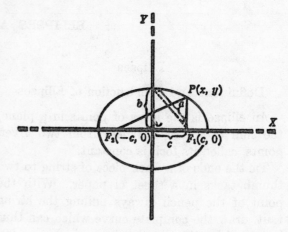

Fig. 64

$$\sqrt{(x - c)^2 + y^2} = 2a - \sqrt{(x + c)^2 + y^2} \qquad \text{(Transposing)}$$

$$x^2 - 2cx + c^2 + y^2 = 4a^2 - 4a\sqrt{(x + c)^2 + y^2} + x^2 + 2cx + c^2 + y^2 \qquad \text{(Squaring both sides)}$$

$$a\sqrt{(x + c)^2 + y^2} = a^2 + cx \qquad \text{(Transposing and dividing by } 4 = 4)$$

$$a^2x^2 + 2a^2cx + a^2c^2 + a^2y^2 = a^4 + 2a^2cx + c^2x^2 \qquad \text{(Squaring both sides)}$$

$$(a^2 - c^2)x^2 + a^2y^2 = a^2(a^2 - c^2). \qquad \text{(Transposing)}$$

But

$$a^2 - c^2 = b^2. \qquad \text{(Figure 64)}$$

Hence

$$b^2x^2 + a^2y^2 = a^2b^2. \qquad \text{(Substitution)}$$

Since the above steps may be reversed, we have as **the analytic equivalent of an ellipse** with foci $F_1(c,0)$ and $F_2(-c,0)$, major axis $= 2a$, and minor axis $= 2b$, the formula,

$$\text{G18:} \quad \frac{x^2}{a^2} + \frac{y^2}{b^2} = 1, \qquad \text{(Dividing by } a^2b^2 = a^2b^2)$$

or,

$$y = \pm b\sqrt{1 - \frac{x^2}{a^2}}, \qquad \text{(Solving for } y)$$

or,

$$x = \pm a\sqrt{1 - \frac{y^2}{b^2}}. \qquad \text{(Solving for } x)$$

The $+$ and $-$ signs, of course, correspond to the two possible values of x and y as two-valued functions of each other (Chapter V). And the above formulas are called **standard equations** because they enable us to place the diagram of an ellipse conveniently for solving problems with the center at the origin.

When the transverse axis of the ellipse is the x-axis of coordinates, as in the above derivation, the value of a will always be larger than the value of b (See Figure 64). But when the value of b is larger than that of a the same formula gives the equation of a corresponding ellipse with its transverse axis as the y-axis of coordinates and with the roles of the constants a and b reversed (See Example 2 below).

EXAMPLE 1: Write the equation in standard form of the ellipse whose major and minor axes are 10 and 8 respectively. What are the foci and eccentricity?

SOLUTION: The standard equation is

$$\frac{x^2}{a^2} + \frac{y^2}{b^2} = 1. \qquad \text{(Formula G18)}$$

But

$2a = 10,$	(Since $2a$ = the major axis)
$a = 5.$	(Dividing by $2 = 2$)

And

$2b = 8,$	(Since $2b$ = the minor axis)
$b = 4.$	(Dividing by $2 = 2$)

Hence the equation is

$$\frac{x^2}{25} + \frac{y^2}{16} = 1. \qquad \text{(Substitution)}$$

Also

$$c = \sqrt{a^2 - b^2} \qquad \text{(Figure 64)}$$

$$= \sqrt{25 - 16} = \sqrt{9} = 3. \qquad \text{(Substitution)}$$

Hence the foci are

$$F_1 = (3,0), \ F_2 = (-3,0) \qquad \text{(Formula G1)}$$

And the eccentricity is

$$c = \frac{c}{a} = \frac{3}{5} = .6 \qquad \text{(Substitution)}$$

EXAMPLE 2: Identify the quadratic curve whose equation is

$$x^2/576 + y^2/625 = 1.$$

What are its foci and eccentricity?

SOLUTION: Since the equation has the form of Formula G18 we recognize that the curve is an ellipse. However, since the denominator of y^2 is larger than that of x^2, the transverse axis must be the y-axis of coordinates and we shall write it as

$$\frac{x^2}{b^2} + \frac{y^2}{a^2} = 1. \qquad \text{(Formula G18 with } a \text{ and } b \text{ reversed)}$$

Now

$a^2 = 625$	(Substitution)
$a = 25$	(Taking the square roots)
$2a = 50,$ the major axis	(Multiplying by $2 = 2$)
$b^2 = 576$	(Substitution)
$b = 24$	(Taking the square roots)
$2b = 48,$ the minor axis	(Multiplying by $2 = 2$)

Also,

$$c = \sqrt{a^2 - b^2} = \sqrt{625 - 576} \qquad \text{(Figure 64)}$$

$$= \sqrt{49} = \pm 7. \qquad \text{(Extracting the square roots)}$$

Hence the foci are

$$F_1(0,7), \ F_2(0,-7) \qquad \text{(Formula G1)}$$

And the eccentricity is

$$e = \frac{c}{a} = \frac{7}{25} = .28 \qquad \text{(Substitution)}$$

Practice Exercise No. 61

A. From the information derived in the solutions of Examples 1 and 2 above, sketch the graphs of the ellipses.

B. Write the equations of the following ellipses and sketch their graphs:
   (1) Foci $(2,0)$, $(-2,0)$, major axis of length 10.
   (2) Foci $(0,2)$, $(0,-2)$, major axis of length 8.
   (3) Foci $(1,2)$, $(1,3)$, eccentricity = 2/5.
   (4) Minor axis 6 along y-axis, and distance between ends of axes = 5.
   (5) A curve such that the distance of each point to the line $x = 6$, divided by its distance to $(3,0)$ is $\sqrt{2}$.

Applications of Ellipse Formulas

Standard ellipse formulas may be applied in much the same manner as standard parabola formulas (pages 142–3 above).

EXAMPLE 3: The central span of London Bridge is the upper half of an ellipse 152 feet wide and 38 feet high.   Compute its height 19 feet from the center and the distance from the center at which it is 19 feet high.

SOLUTION: *Step One:* Set up the diagram with the center of the ellipse at the origin, the vertices at the points (76,0) and (−76,0), and the *y*-intercept at the point (0,38), as in Figure 65:

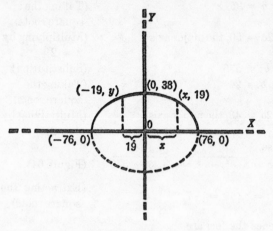

Fig. 65

Step Two: The equation of the ellipse is now

$$\frac{x^2}{76^2} + \frac{y^2}{38^2} = 1$$

or

$$y = 38\sqrt{1 - \left(\frac{x}{76}\right)^2},$$ (Formula G18)

and

$$x = 76\sqrt{1 - \left(\frac{y}{38}\right)^2}.$$

Step Three: We now need to substitute the appropriate values of the variables *x* and *y* in these equations.

Step Four: When $x = 19$,

$$y = 38\sqrt{1 - \left(\frac{19}{76}\right)^2}$$ (Substitution)

$$= 38\sqrt{1 - \left(\frac{1}{4}\right)^2} = 38\sqrt{\frac{15}{16}}$$

(Squaring $\frac{1}{4}$ and adding fractions)

$$= \frac{38}{4}\sqrt{15} = 9.5(3.873)$$

(Extracting the square roots)

$$= 36.8 \text{ ft., approx.}$$ (Removing parentheses)

When $y = 19$,

$$x = 76\sqrt{1 - \left(\frac{19}{38}\right)^2}$$ (Substitution)

$$= 76\sqrt{1 - \left(\frac{1}{2}\right)^2} = 76\sqrt{\frac{3}{4}}$$

(Squaring $\frac{1}{2}$ and adding fractions)

$$= \frac{76}{2}\sqrt{3} = 38(1.732)$$

(Extracting the square roots)

$$= 65.8 \text{ ft., approx.}$$ (Removing parentheses)

EXAMPLE 4: The orbit of the earth is an ellipse with a major axis of about 186,000,000 miles and foci about 3,000,000 miles apart.   Compute its eccentricity.

SOLUTION: *Steps One and Two:* Since the distance between the foci is $3(10)^6$ miles,

$$2c = 3(10)^6$$ (Substitution)
$$c = 1.5(10)^6$$ (Dividing by 2 = 2)

And since the major axis is $186(10)^6$ miles,

$$2a = 186(10)^6$$ (Substitution)
$$a = 93(10)^6$$ (Dividing by 2 = 2)

Steps Three and Four: We may therefore compute the eccentricity as

$$e = \frac{c}{a} = \frac{1.5(10)^6}{93(10)^6} = .016 \text{ approx.}$$

(Substitution)

Note that the small value thus found for *e* means that the earth's orbit is very nearly circular.   This is to be expected from the fact that the distance between the foci is short as compared to the length of the major axis.   It also illustrates the general point that the eccentricity of an ellipse is a function of its *shape* and *not* of its *size*.

In solving less routine problems involving ellipses it is sometimes necessary, in order to simplify the algebraic expressions, to make one or more substitutions from the derived equation,

$$b^2x^2 + a^2y^2 = a^2b^2.$$ (Formula G18 multiplied by $a^2b^2 = a^2b^2$)

EXAMPLE 5: A straight-edged machine-part rocks upon an elliptical cam surface so that its straight edge is always tangent to the surface.   It is readily shown in calculus that the slope of the tangent to the ellipse $x^2/a^2 + y^2/b^2 = 1$ is $-b^2x/a^2y$ at any point $P(x,y)$ on the ellipse.   Prove that the product of the distances from the foci of the ellipse to the tangent straight edge is a constant.

SOLUTION: *Step One:* Locating the ellipse in the standard position, let $P(k,l)$ be the point of tangency, let $F_1(c,0)$ and $F_2(-c,0)$ be the foci, and let P_1 and P_2 be the feet of the perpendiculars from F_1 and F_2 to the tangent line, as in Figure 66:

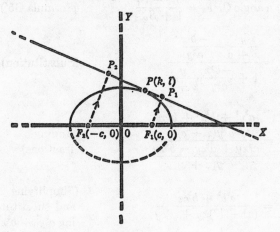

Fig. 66

Step Two: The equation of the ellipse is now

$$\frac{x^2}{a^2} + \frac{y^2}{b^2} = 1. \qquad \text{(Formula G18)}$$

Since $P(k,l)$ is a point on the ellipse,

$$b^2k^2 + a^2l^2 = a^2b^2. \qquad \text{(Substitution, as above)}$$

And the equation of the tangent line is

$$\begin{vmatrix} x & y & 1 \\ k & l & 1 \\ 1 & \dfrac{-b^2k}{a^2l} & 0 \end{vmatrix} = \frac{1}{a^2l} \begin{vmatrix} x & y & 1 \\ k & l & 1 \\ a^2l & b^2k & 0 \end{vmatrix} = 0,$$

$$\left(\text{Formula G6m, slope} = \frac{-b^2k}{a^2l} \right)$$

$$b^2kx + a^2ly - (b^2k^2 + a^2l^2) = 0,$$

$$\text{(Formula M2')}$$

or,

$$b^2kx + a^2ly - a^2b^2 = 0. \qquad \begin{array}{l}\text{(Substituting}\\ b^2k^2 + a^2l^2 = a^2b^2)\end{array}$$

Step Three: We now need to show that $(F_1P_1)(F_2P_2)$ = a constant.

Step Four:

$$F_1P_1 = \frac{\left| b^2kc - a^2b^2 \right|}{\sqrt{b^4k^2 + a^4l^2}}$$

$$\text{(Formula G15)}$$

$$F_2P_2 = \frac{\left| b^2k(-c) - a^2b^2 \right|}{\sqrt{b^4k^2 + a^4l^2}}$$

$$F_1P_1 \cdot F_2P_2 = \frac{\left| b^2kc - a^2b^2 \right|}{\sqrt{b^4k^2 + a^4l^2}} \cdot \frac{\left| b^2k(-c) - a^2b^2 \right|}{\sqrt{b^4k^2 + a^4l^2}}$$

$$\text{(Substitution)}$$

$$= \frac{\left| b^4k^2(-c^2) + a^2b^2(a^2b^2) \right|}{b^4k^2 + a^4l^2}$$

$$\text{(Multiplying fractions)}$$

$$= \frac{b^2 \left| b^2k^2(b^2 - a^2) + a^2(b^2k^2 + a^2l^2) \right|}{b^4k^2 + a^4l^2}$$

$$\begin{array}{l}\text{(Substituting } -c^2 = b^2 - a^2\\ \text{and } a^2b^2 = b^2k^2 + a^2l^2)\end{array}$$

$$= \frac{b^2 \left| b^4k^2 + a^4l^2 \right|}{b^4k^2 + a^4l^2} = b^2, \text{ a constant.}$$

$$\text{(Removing parentheses, etc.)}$$

Note that the demonstration is essentially complete in the fourth last line above, since all the terms in the righthand side of the equation are constants. For practical purposes of computation, however, the result remains algebraically cumbersome until simplified by the substitutions in the second last line. *Note* also that the quantities removed from the absolute value signs are "squared" and therefore known to be positive.

EXAMPLE 6: A reflector is to be designed so as to concentrate all radiation (light, heat, or sound) from a point source on to another point. The physical principle to be applied is that when radiation is reflected from a curved surface the lines of incidence and the lines of reflection make equal angles with the plane tangent to the surface at the point of incidence. Demonstrate by a cross-section analysis that an ellipsoidal surface — one with elliptical cross-sections through the major axis — has the required optical and acoustical properties in that the focal radii to any point on an ellipse make equal angles with the tangent to the curve at that point.

SOLUTION: *Step One:* Let the line QR be the cross section of the tangent plane, touching the elliptical cross section of the reflecting surface at $P(x,y)$ as in Figure 67:

Step Two: Since the equation of the ellipse is

$$\frac{x^2}{a^2} + \frac{y^2}{b^2} = 1, \qquad \text{(Formula G18)}$$

the slope S_t of the tangent at $P(x,y)$ is

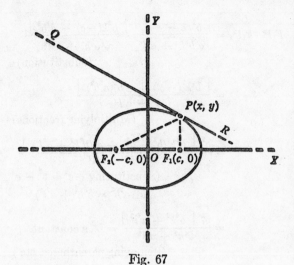

Fig. 67

$$S_t = \frac{-b^2x}{a^2y}.$$ (Example 5, above)

Also, the slope S_1 of the focal radius F_1P is

$$S_1 = \frac{y}{x - c},$$ (Formula G4)

and the slope S_2 of the focal radius F_2P is

$$S_2 = \frac{y}{x + c}.$$ (Formula G4)

Step Three: We now need to prove that angle F_1PR = angle QPF_2. Since both are *first* quadrant angles, this will follow if we show that their *tangents* are equal.

Step Four:

Tan angle $F_1PR = \dfrac{S_t - S_1}{1 + S_1S_t}$ (Formula G5′)

$$= \frac{\dfrac{-b^2x}{a^2y} - \dfrac{y}{x - c}}{1 + \dfrac{y}{x - c} \cdot \dfrac{-b^2x}{a^2y}}$$ (Substitution)

$$= \frac{\dfrac{-b^2x^2 + b^2cx - a^2y^2}{a^2y(x - c)}}{\dfrac{a^2xy - a^2cy - b^2xy}{a^2y(x - c)}}$$ (Adding fractions)

$$= \frac{b^2cx - a^2b^2}{(a^2 - b^2)xy - a^2cy}$$ (Simplifying and substituting $-b^2x^2 - a^2y^2 = -a^2b^2$)

$$= \frac{b^2(cx - a^2)}{c^2xy - a^2cy}.$$ (Factoring and substituting $a^2 - b^2 = c^2$)

$$= \frac{b^2(cx - a^2)}{cy(cx - a^2)} = \frac{b^2}{cy}$$ (Factoring and cancelling the common factor)

Tan angle $QPF_2 = \dfrac{S_2 - S_t}{1 + S_tS_2}$ (Formula G5′)

$$= \frac{\dfrac{y}{x + c} - \dfrac{-b^2x}{a^2y}}{1 + \dfrac{-b^2x}{a^2y} \cdot \dfrac{y}{x + c}}$$ (Substitution)

$$= \frac{\dfrac{a^2y^2 + b^2x^2 + b^2cx}{a^2y(x + c)}}{\dfrac{a^2xy + a^2cy - b^2xy}{a^2y(x + c)}}$$ (Adding fractions)

$$= \frac{a^2b^2 + b^2cx}{(a^2 - b^2)xy + a^2cy}$$ (Simplifying and substituting $a^2y^2 + b^2x^2 = a^2b^2$)

$$= \frac{b^2(a^2 + cx)}{c^2xy + a^2cy}$$ (Factoring and substituting $a^2 - b^2 = c^2$)

$$= \frac{b^2(a^2 + cx)}{cy(a^2 + cx)} = \frac{b^2}{cy}.$$ (Factoring and cancelling the common factor)

Hence angle F_1PR = angle QPF_2.

(Substitution, etc.)

This means that all radiation emanating from either focus will be so reflected that all the rays will be concentrated on the other focus.

Practice Exercise No. 62

1. Compute the height of the bridge in Example 3 above at a distance of 38 feet from the center.

2. Compute the distance from the center of the same bridge at which the height is $9\frac{1}{2}$ feet.

3. The longest and shortest focal radii of an ellipse are those which lie along the major axis. Assuming the sun to be at the focus of the earth's orbit, what are the maximum and minimum distances of the earth from the sun according to the data in Example 4 above?

4. Applying the data of Example 5 above, show that for all ellipses with a common major axis = $2a$, but with different minor axes = b_1 b_2, etc., the lines tangent at the extremities of the latera rectae on the same side of the major axis pass through a common point.

5. Find the common point of the preceding problem.

6. Applying the data of Example 6, Chapter XI, and the method of Example 6 this chapter, show that a parabolic reflector has the optical property of reflecting all radiation emanating from its focus in a single beam of rays parallel to its axis.

Circles

The standard equation of a **circle** may be derived like that of any other conic from its elementary definition as the locus of the points in a plane equidistant from a given point called its **center**.

From an analytic point of view, however, we may just as conveniently define a circle as an ellipse whose foci coincide with its center. Then, since $a = b = r$, the **radius** of the circle, we have

$$\frac{x^2}{r^2} + \frac{y^2}{r^2} = 1.$$ (G18, substituting $a = b = r$)

Hence, the **standard equation of a circle**, is

G19: $x^2 + y^2 = r^2$,   (Multiplying by $r^2 = r^2$)

or,

$$x = \pm\sqrt{r^2 - y^2},$$

or,   (Solving for x and y)

$$y = \pm\sqrt{r^2 - x^2}.$$

Since these equations have the same basic form as those of an ellipse, they are applied in the same way.

EXAMPLE 7: The underside of a viaduct has a semicircular contour 50 feet in diameter. Compute its clearance at a distance of 15 feet from the center, and compute the distance from the center at which it will have a clearance of 7 feet.

SOLUTION: *Step One:* Let the center of the semicircle be at the origin as in Figure 68:

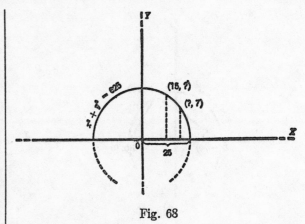

Fig. 68

Step Two: Since $r = \frac{1}{2}(50) = 25$, the equation is

$$x^2 + y^2 = 25^2 = 625,$$ (Formula G19)

or

$$y = \sqrt{625 - x^2},$$ (Solving for y)

and

$$x = \sqrt{625 - y^2}.$$ (Solving for x)

Step Three: We now need to find y when $x = 15$ and x when $y = 7$.

Step Four:

$y = \sqrt{625 - 15^2} = \sqrt{625 - 225}$ (Substitution)

$= \sqrt{400} = 20$ feet clearance.   (Extracting the square root)

$x = \sqrt{625 - 7^2} = \sqrt{625 - 49}$ (Substitution)

$= \sqrt{576} = 24$ ft from the center.   (Extracting the square root)

EXAMPLE 8: Under a series of semicircular arches of different diameters pairs of girders are to be erected from the outer extremities to meet at points on the circumferences. Prove that each pair of girders may be bored so as to be riveted at right angles at their point of meeting on the circumference.

SOLUTION: *Step One:* Let the center of a typical semicircular arch be at the origin, let $Q(-r,0)$ and $R(r,0)$ be its extremities, and let $P(x,y)$ be the point on the circumference where the typical pair of girders, PQ and PR, meet as in Figure 69 (page 154).

Step Two: The equation of the contour is now

$$x^2 + y^2 = r^2,$$

and   (Formula G19)

$$y = \sqrt{r^2 - x^2}.$$

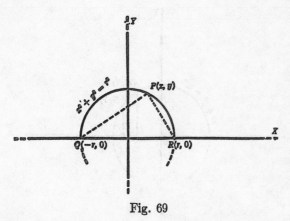

Fig. 69

Step Three: Letting S_q = the slope of PQ and S_r = the slope of PR, we may now prove that PQ is perpendicular to PR by showing that $S_qS_r = -1$ in accordance with formula G12.

Step Four:

$$S_q = \frac{y-0}{x-(-r)} = \frac{y}{x+r} \qquad \text{(Formula G4)}$$

$$= \frac{\sqrt{r^2-x^2}}{x+r}. \qquad \begin{array}{l}\text{(Substituting}\\ y = \sqrt{r^2-x^2})\end{array}$$

$$S_r = \frac{y-0}{x-r} = \frac{\sqrt{r^2-x^2}}{x-r}; \qquad \begin{array}{l}\text{(As for } S_q\\ \text{above)}\end{array}$$

Hence

$$S_qS_r = \frac{\sqrt{r^2-x^2}}{x+r}\;\frac{\sqrt{r^2-x^2}}{x-r} \qquad \text{(Substitution)}$$

$$= \frac{r^2-x^2}{x^2-r^2} = \frac{-(x^2-r^2)}{x^2-r^2} = -1, \qquad \begin{array}{l}\text{(Multiplying}\\ \text{fractions, etc.)}\end{array}$$

as was to be shown.

Practice Exercise No. 63

1. Compute the clearance of the viaduct arch in Example 7 at a distance of 7 feet from the center.

2. Compute the distance from the center of the same arch at which the clearance will be 15 feet.

3. Prove that the clearance at any point under a semicircular arch equals the square root of the product of its horizontal distances from the sides of the arch. *Suggestion:* let the point be $P(x,0)$ and let the equation of the arch be $x^2 + y^2 = r^2$.

4. A beam has one end, Q, on the ground and rests upon a hemispherical surface so that it is tangent to the hemisphere at the point P. It is easily shown in calculus that the slope of the tangent to the circle $x^2 + y^2 = r^2$ at any point (x,y) is $-x/y$. Using a cross-section analysis, show that $PQ = d - r$, where r is the radius of the sphere and d is the distance of Q from the center of the sphere.

5. After an arch is designed to have the formula $x^2/k^2 + y^2/4k^2 = 1$ it is decided to double the clearance at all points. Find the equation of the new arch and identify its curve.

The Summary of this chapter is included with that of Chapter XIII on pages 160–1.

CHAPTER XIII

HYPERBOLAS

Definitions and Construction

A **hyperbola** is the locus of points in a plane the *difference* of whose distances from two given points, called its foci, is constant.

To construct a hyperbola, begin by marking two points, F_1 and F_2, about an inch and a half apart near the center of a sheet of paper. Well out of the way — near the bottom of the sheet, for instance — draw a straight line and mark on it a series of points, Q_1, Q_2, R, S, T, U, V, W, etc., so that Q_1 and Q_2 are about one inch apart and all the others are about a half inch apart.

Using the series of auxiliary points as a gauge, set a compass to an opening $= Q_2R$ and strike four arcs: one above F_1 with F_1 as center, one below F_1 with F_1 as center, one above F_2 with F_2 as center, and one below F_2 with F_2 as center.

Resetting the compass to an opening $= Q_1R$, strike four similar arcs: the new arc above F_1 with F_1 as center intersecting the first arc above F_2 with F_2 as center, the new arc below F_1 with F_1 as center intersecting the first arc be-

low F_2 with F_2 as center, and so on for the others. If the radii of these two sets of arcs are too short to intersect, erase them and try the wider compass settings Q_2S and Q_1S. Once you get a first set to intersect, repeat the process several times with successively wider pairs of compass settings. Finally, draw two smooth curves through the intersections as in Figure 70.

Each of the resulting curves is a **branch** of a hyperbola which always consists of two such branches. For, by the construction, the difference of the distances of any point on either branch from the two focal points F_1 and F_2 is the same constant quantity — namely, the length of the line Q_1Q_2.

Note: An alternative *thread-and-thumbtack method* of constructing a hyperbola is as follows. Set two thumbtacks at the intended foci, F_1 and F_2, as in the corresponding construction of an ellipse (page 147 above). Roll the two ends of a piece of thread around a spool until a small loop is left, and tie the point of a pencil slightly to one side of the middle of this loop. Taking one turn about each thumb tack with each segment thus formed in the

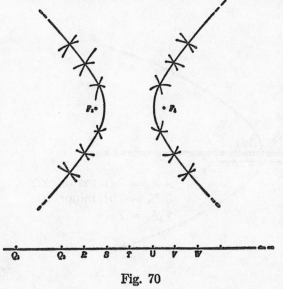

Fig. 70

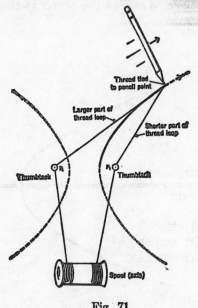

Fig. 71

loop, and holding the spool so that its axis is parallel to F_1F_2, draw the thread taut with the spool so that the pencil point is held at a point between F_1 and F_2. (If the pencil point does not come between F_1 and F_2, spread the thumb tacks wider apart or retie the pencil point closer to the center of the loop so that it does.) Then, allowing the thread to unwind slowly from the spool, and always keeping it taut with the pencil point, trace out a curve as in Figure 71. The result will be a segment of one branch of a hyperbola. The opposite half of the other branch may be similarly drawn by beginning with the spool on the other side of F_1F_2. The remaining halves of both branches may be completed in the same way merely by turning the spool over lengthwise so that the sides of the shorter and longer lengths of thread-loop are reversed. For, by this construction, the difference of the distances of the pencil point from the two foci is a constant — namely, twice the distance it is tied from the center of the loop.

The line of indefinite length through the foci of a hyperbola (F_1 and F_2 in Figure 72) is the **transverse axis.**

The points (A_1 and A_2) in which the branches of the hyperbola intercept the transverse axis are the **vertices.**

That segment of the transverse axis between the vertices ($A_1A_2 = 2a$ in the diagram) is the **major axis.**

The midpoint (0) of the major axis is the **center.**

The line of indefinite length perpendicular to the transverse axis at the center is the **conjugate axis.**

That segment of the conjugate axis (B_1B_2) which has its midpoint at the center and the length $2b$ defined by the equation $b = \sqrt{c^2 - a^2}$ is the **minor axis.**

Latera rectae are defined as for a parabola (page 142). The same is true of **focal radii** and **focal chords** except that the former are defined to either branch from either focus.

Eccentricity

From the above construction and diagrams it is clear that hyperbolas, like parabolas, are open curves. Each branch extends indefinitely from its vertex however far we wish to consider it. But unlike parabolas, and like ellipses, *hyperbolas have different shapes.*

If in the preceding construction (Figure 70) the foci F_1 and F_2 are taken further apart or the auxiliary line segment Q_1Q_2 is taken shorter, each branch of the resulting hyperbola tends to "open out" wider from the transverse axis and both branches tend to come relatively close to each other, as shown by Figure 73, diagram A. In the special case where Q_1Q_2 degenerates into a single point the hyperbola itself degenerates into a single vertical straight line, the conjugate axis, as shown by Figure 73, diagram B.

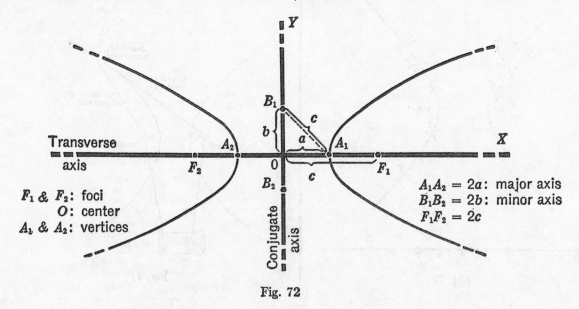

F_1 & F_2: foci
O: center
A_1 & A_2: vertices

$A_1A_2 = 2a$: major axis
$B_1B_2 = 2b$: minor axis
$F_1F_2 = 2c$

Fig. 72

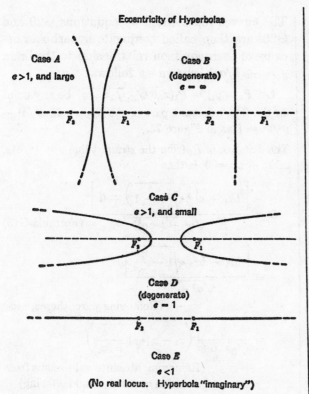

Eccentricity of Hyperbolas

Case *A*
e > 1, and large

Case *B*
(degenerate)
e = ∞

Case *C*
e > 1, and small

Case *D*
(degenerate)
e = 1

Case *E*
e < 1
(No real locus. Hyperbola "imaginary")

Fig. 73

If, on the other hand, the foci F_1 and F_2 are taken closer together or the auxiliary line segment Q_1Q_2 is taken longer, each branch of the resulting hyperbola tends to narrow down closer to the transverse axis and both branches tend to spread relatively far from each other, as shown by Figure 73, diagram C. The auxiliary line segment Q_1Q_2 can never be longer than the distance between the foci F_1F_2, for then it would be impossible to carry out the construction. This you can verify by trial. But in the special case where $Q_1Q_2 = F_1F_2$, the hyperbola again degenerates, this time into those two segments of the transverse axis on either side of F_1F_2 (which then becomes the major axis) as shown by Figure 73, diagram D.

The eccentricity of a hyperbola is therefore its degree of "open-ness" or "non-flatness" which may be expressed numerically by the same ratio as for an ellipse:

$$e = \frac{c}{a} = \frac{\sqrt{a^2 + b^2}}{a} = \sqrt{1 + \frac{b^2}{a^2}}$$

Note, however, that whereas the eccentricity of a real ellipse is always between 0 and 1, the eccentricity of a real hyperbola is always greater than 1. As you can verify by varying the values of $c = \frac{1}{2}F_1F_2$ and $a = \frac{1}{2}Q_1Q_2$ in the above construction, larger values of *e* correspond to more "open" hyperbolas as in Figure 73, diagram A, smaller values of *e* correspond to "flatter" hyperbolas as in Figure 73, diagram C. But when *e* = 1 the hyperbola degenerates into segments of the transverse axis as in Figure 73, diagram D, and when *e* < 1, the construction is impossible as noted in Case E of Figure 73.

Standard Equation for a Hyperbola

Let the foci of a hyperbola be $F_1(c,0)$ and $F_2(-c,0)$, let its vertices be $A_1(a,0)$ and $A_2(-a,0)$, and let $P(x,y)$ be any point on either branch as in Figure 74:

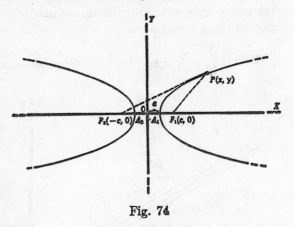

Fig. 74

Since

$$|PF_2| - |PF_1| = 2a, \qquad \text{(Definition of a hyperbola)}$$

we arrive, by steps similar to those for the derivation of the ellipse formula, at the relationship:

$$(c^2 - a^2)x^2 - a^2y^2 = a^2(c^2 - a^2). \quad \text{(Page 148)}$$

Or, letting $\qquad b^2 = c^2 - a^2,$

$$b^2x^2 - a^2y^2 = a^2b^2. \qquad \text{(Substitution)}$$

And since all these steps are retraceable in reverse order, we have as the standard equation of a hyperbola with major axis $= 2a$ and minor axis $= 2b$, the formula:

G20: $\dfrac{x^2}{a^2} - \dfrac{y^2}{b^2} = 1$.   (Dividing by $a^2b^2 = a^2b^2$)

or,

$$y = \pm b\sqrt{\dfrac{x^2}{a^2} - 1}\,,$$ (Solving for y)

or,

$$x = \pm a\sqrt{\dfrac{y^2}{b^2} + 1}\,.$$ (Solving for x)

If the foci are taken on the y-axis at the points $F_1'(0,c)$ and $F_2'(0,-c)$, the corresponding equation of a hyperbola with the y-axis of coordinates as its transverse axis is

$$\dfrac{y^2}{a^2} - \dfrac{x^2}{b^2} = 1.$$ (Steps as above)

If in addition to this interchange of transverse and conjugate axes, the lengths of the major and minor axes are interchanged so that the major axis $= 2b$ and the minor axis $= 2a$, the resulting equation is

G20': $\dfrac{y^2}{b^2} - \dfrac{x^2}{a^2} = 1.$ (Steps as above)

The curves represented by equations G20 and G20' are then called **conjugate hyperbolas** because of their common relationship to the lines, $y = \pm(b/a)x$, shown as follows:

Let $P_1(x_1,y_1) = P_1(x_1, b\sqrt{x_1^2/a^2 - 1})$ be any point on the first quadrant part of the hyperbola $x^2/a^2 - y^2/b^2 = 1$ as in Figure 75.

The distance of P_1 from the straight line $y = (b/a)x$, or $bx - ay = 0$, is then

$$D = \dfrac{\left| bx_1 - a\left(b\sqrt{\dfrac{x_1^2}{a^2} - 1}\right) + 0 \right|}{\sqrt{b^2 + (-a)^2}}$$ (Formula G15)

$$= \dfrac{\left| bx_1 - bx_1\sqrt{1 - \dfrac{a^2}{x_1^2}} \right|}{\sqrt{a^2 + b^2}}$$

(Removing parentheses, etc.)

$$= \dfrac{b}{\sqrt{a^2 + b^2}}\left(x_1 - x_1\sqrt{1 - \dfrac{a^2}{x_1^2}} \right).$$

(Removing absolute value signs from positive quantities and factoring)

The quantity outside the parenthesis is now a constant. Moreover, as x_1 becomes larger,

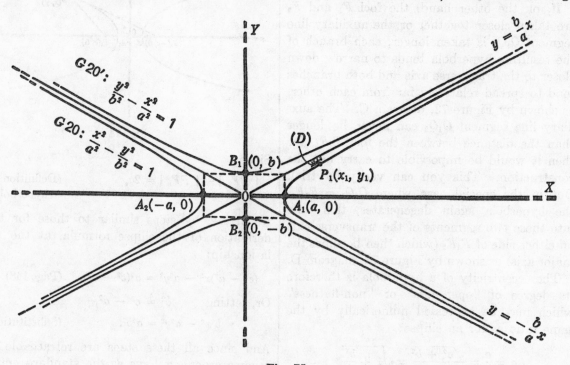

Fig. 75

$a^2/x_1{}^2$ becomes smaller, $1 - a^2/x_1{}^2$ becomes more nearly equal to 1, and the total expression within parentheses becomes more nearly equal to 0. As we shall be able to show more rigorously later in the calculus, if we take x_1 sufficiently large we shall find D to be smaller in absolute value than any pre-assignable quantity we may choose. For any given definite value of x_1, however, D will also have some definite absolute value. This means that as we follow the hyperbola further and further from its vertex A_1 we shall find it approaching closer and closer to the straight line $y = (b/a)x$ without ever actually touching it. Hence we call the line $y = (b/a)x$ an **asymptote** of the curve $y = b\sqrt{x^2/a^2 - 1}$ which we say **approaches the line asymptotically.**

From the symmetry of the equations G20, G20′, and $y = \pm(b/a)x$, moreover, it is clear that both branches of both hyperbolas have the same pair of intersecting straight lines as asymptotes. That is what is meant by calling them **conjugate hyperbolas.**

In the special case when the major and minor axes of a hyperbola are equal, $a = b$, and equation G20 becomes

G20″: $\dfrac{x^2}{a^2} - \dfrac{y^2}{a^2} = 1.$ (Substituting $b = a$)

The resulting curve is called a **rectangular hyperbola** because its asymptotes,

$$y = \pm \frac{b}{a}x = \pm \frac{a}{a}x = \pm x,$$ (Substitution)

intersect at right angles. As you can verify by repeating the above derivation with foci at $F_1(k,k)$ and $F_2(-k,-k)$, the equation of the rectangular hyperbola with major and minor axes $= 2k$, with $c = \sqrt{2}k$, and with the x and y axes of coordinates as asymptotes, is

G20‴: $xy = k^2/2.$ (Steps as suggested)

Applications of Hyperbola Formulas

The algebraic routine of applying hyperbola formulas is much like that for the other conics already discussed. In some cases, however, we may be required to find the values of the constants from a specification of asymptotes, which other conics do not possess.

EXAMPLE 1: The surface design of a rectangular ceiling panel, 20′ by 25′, calls for tracing out a pair of conjugate hyperbolas with the diagonals of the panel as asymptotes and with the vertices closest to the center 2 feet apart. Write the equations of the required curves and compute how broad the least eccentric hyperbola will be at a distance on its transverse axis 5 feet from the center.

SOLUTION: *Step One:* Set up the diagram with the center of the rectangle at the origin and the vertices at the points $P_1(12\frac{1}{2},10)$, etc., as in Figure 76:

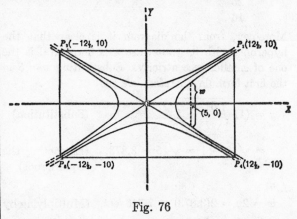

Fig. 76

Step Two: The standard equations of the parabolas are now

$$\frac{x^2}{a^2} - \frac{y^2}{b^2} = 1, \text{ and}$$ (Formula G20)

$$\frac{y^2}{b^2} - \frac{x^2}{a^2} = 1,$$ (Formula G20′)

and the standard equations of the asymptotes are

$$y = \pm \frac{b}{a}x.$$ (Above)

Step Three: We now need to determine the values of the constants in these equations, and the value of $2y$ in the first when $x = 5$.

Step Four: From the diagram, the equations of the asymptotes are

$$y = \pm \frac{20}{25}x = \pm \frac{4}{5}x.$$ (Figure 76)

Hence

$$\frac{b}{a} = \frac{4}{5};$$ (Substitution)

and

$$a = \frac{5}{4} b. \qquad \text{(Solving for } a\text{)}$$

But since it is given that the shortest major axis $= 2b = 2$, it follows that $b = 1$ and

$$a = \frac{5}{4}(1) = \frac{5}{4}. \qquad \text{(Substitution)}$$

Hence the equations of the hyperbolas are

$$\frac{x^2}{25} - \frac{y^2}{1} = 1, \text{ and}$$

$$\qquad \qquad \text{(Substitution)}$$

$$\frac{y^2}{1} - \frac{x^2}{\frac{25}{16}} = 1.$$

Moreover, from the diagram it is clear that the hyperbola with its transverse axis horizontal is the one of smallest eccentricity. Substituting $x = 5$ in the first formula, therefore, we get

$$y = (1)\sqrt{\frac{5^2}{\left(\frac{5}{4}\right)^2} - 1} \qquad \text{(Substitution)}$$

$$= \sqrt{16 - 1} = \sqrt{15} = 3.873. \qquad \begin{array}{l}\text{(Extracting the} \\ \text{square root)}\end{array}$$

And

$$w = 2y = 2(3.873) = 7.746 \text{ feet}, \quad \begin{array}{l}\text{(Multiplying by} \\ 2 = 2\text{)}\end{array}$$

the width of the hyperbola for $x = 5$.

EXAMPLE 2: Show that the product of the distances from any point on a hyperbola to the asymptotes is a constant.

SOLUTION: *Steps One and Two:* From Figure 75 and the steps there followed to explain the relationship of the asymptotes to the curve of a hyperbola, we have already seen that the distance D from any point $P(x,y)$ on the hyperbola $x^2/a^2 - y^2/b^2 = 1$ to the asymptote $y = (b/a)x$ is

$$D = \frac{b}{\sqrt{a^2 + b^2}} (x - \sqrt{x^2 - a^2}). \quad \text{(Page 158)}$$

By the same steps, changing only one sign, the distance D' from the same point to the other asymptote $y = -(b/a)x$ is

$$D' = \frac{b}{\sqrt{a^2 + b^2}} (x + \sqrt{x^2 - a^2}). \quad \text{(As above)}$$

Steps Three and Four: Hence the product of these distances

$$D \cdot D' = \left(\frac{b}{\sqrt{a^2 + b^2}}\right)^2 (x - \sqrt{x^2 - a^2})(x + \sqrt{x^2 - a^2})$$
$$\qquad \qquad \text{(Substitution)}$$

$$= \frac{b^2}{a^2 + b^2} (x^2 - x^2 + a^2) \qquad \text{(Multiplying)}$$

$$= \frac{a^2 b^2}{a^2 + b^2}, \text{ a constant.} \qquad \begin{array}{l}\text{(Removing} \\ \text{parentheses,} \\ \text{etc.)}\end{array}$$

Practice Exercise No. 64

1. Compute the foci and eccentricities of the conjugate hyperbolas in Example 1 above.

2. What are the dimensions of the largest rectangular area which may be laid out between the vertices of these hyperbolas and with the same lines as diagonals? How does this rectangle compare in shape with the larger rectangle of the entire panel?

3. How wide will the more eccentric of the above hyperbolas be at a point on its transverse axis 6 feet from the center?

4. It is shown in calculus that the slope of the tangent to a standard hyperbola at any point $P(x,y)$ is b^2x/a^2y. Prove that the product of the distances from the foci of a hyperbola to a tangent is constant regardless of the position of the tangent. You will find that the constant product is the same as in the corresponding theorem for an ellipse, Example 5, Chapter XII.

5. Prove that the product of the distances from the foci to any point on an equilateral hyperbola is equal to the square of the distance of that point from the center.

Summary

Defining **conic sections as loci**, we have arrived at the following **standard equations** for treating the geometric properties of these curves analytically:

G17: $\qquad y^2 = 4px.$

The parabola with focus $(p,0)$ and directrix $x = -p.$

G18: $\dfrac{x^2}{a^2} + \dfrac{y^2}{b^2} = 1.$

The ellipse with major axis = $2a$, minor axis = $2b$, and foci $(c,0)$ and $(-c,0)$ where $c = \sqrt{a^2 - b^2}$.

G19: $x^2 + y^2 = r^2.$

The circle with center at the origin and radius = r.

G20: $\dfrac{x^2}{a^2} - \dfrac{y^2}{b^2} = 1.$

The hyperbola with major axis = $2a$, minor axis = $2b$, and foci $(c,0)$ and $(-c,0)$ where $c = \sqrt{a^2 + b^2}$.

G20′: $\dfrac{y^2}{b^2} - \dfrac{x^2}{a^2} = 1.$

The hyperbola conjugate to G20.

G20″: $\dfrac{x^2}{a^2} - \dfrac{y^2}{a^2} = 1.$

The rectangular hyperbola with major axis = minor axis = $2a$, and with the lines $y = \pm x$ as asymptotes.

G20‴: $xy = \dfrac{k^2}{2}.$

The rectangular hyperbola with foci (k,k) and $(-k,-k)$, with major axis = minor axis = $2k$ with the coordinate axes as asymptotes.

CHAPTER XIV

POINTS AND DIRECTIONS IN SPACE

Solid analytic geometry extends the methods of the five preceding chapters to the study of points, lines, surfaces, and volumes in three-dimensional space.

Three-Dimensional Rectangular Coordinates

At the origin of a pair of *x-y* rectangular coordinate axes add a third — termed the **z-coordinate axis** — at right angles to the other two. The result is a **system of three-dimensional rectangular coordinates.**

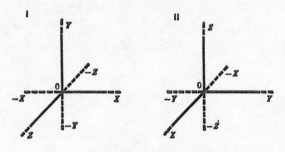

Fig. 77

Diagram I of Figure 77 shows the *x* and *y*-axes of such a system in their familiar positions so that the positive direction of the *z*-axis appears in perspective as a line projecting out toward the reader, below and to the left of his line of vision. Diagram II of the Figure shows the same system from a different perspective angle so that the positive direction of the *x*-axis appears to be projecting toward the reader.

The set of rectangular coordinate axes shown from different perspectives by the two diagrams of Figure 77 is called a **right-handed system.** For, if you were to turn a faucet, screw, or any ordinary threaded device with a *right-handed twist* from the positive direction on the *x*-axis to the positive direction on the

y-axis, the result would be a thrust along the positive direction of the *z*-axis.

Interchanging any two of the letters, *x*, *y*, and *z*, in either of the above diagrams, produces a **left-handed system.** For then a corresponding *left-handed twist* would be necessary in order to effect a thrust along the positive direction of the *z*-axis.

Since right-handed and left-handed systems of coordinate axes are interchangeable in ordinary analytic geometry and calculus, the diagrams of many texts on these subjects use either or both. But since certain of the basic definitions of vector analysis normally specify a right-handed system, we shall follow the practice of using right-handed axis diagrams here from the beginning.

The three planes determined by the three possible sub-pairs of coordinate axes are called the **(x,y)**, the **(x,z)** and the **(y,z) coordinate planes.**

Just as the axes of a two-dimensional set of coordinates divide the entire reference plane into four *quadrants*, the three coordinate planes of a three-dimensional set divide all space into eight regions called **octants.**

The octant in which *x*, *y*, and *z*, are all positive is known as the **first octant.** The other seven are not usually numbered but are designated according to the sign of their corresponding values of *x*, *y*, and *z*, as the $(+,+,-)$ octant, the $(+,-,-)$ octant, the $(-,-,+)$ octant, etc.

Given any three values x_1, y_1, and z_1 of the variables x, y, and z, a corresponding point in space P_1 is uniquely determined with respect to a three-dimensional system of rectangular coordinates.

Let us assume for the moment that x_1, y_1,

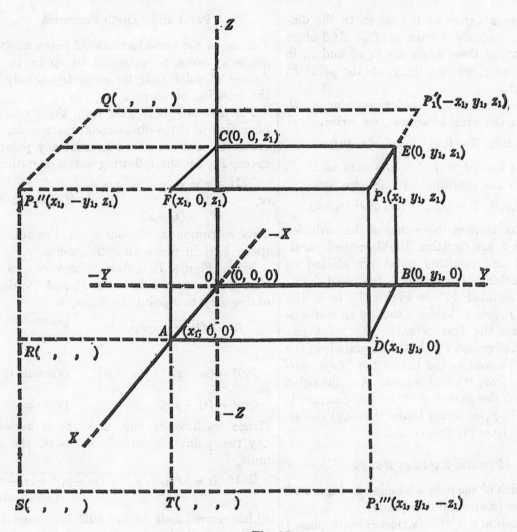

Fig. 78

and z_1, are all positive. By measuring off x_1 units from the origin along the positive direction on the x-axis, we arrive at the point designated A in Figure 78. Likewise, by measuring off y_1 and z_1 units from the origin along the positive directions on the y- and z-axes respectively, we arrive at the points designated B and C in the diagram.

If from the point A on the x-axis we measure off y_1 units along a line parallel to the y-axis in the positive direction of that axis, we arrive at the point D in the x-y coordinate plane. Or, if from the point B on the y-axis we measure off x_1 units along a line parallel to the x-axis in the positive direction of that axis, we

again arrive at the point D. But having arrived at the latter point by either route, if we then measure off z_1 units along a line parallel to the z-axis in the positive direction on that axis, we arrive at the point P_1 in the first octant of the diagram.

We arrive at precisely the same point P_1, moreover, if we take corresponding steps via the alternative routes:

O to B to E to P_1, or $\quad O$ to C to E to P_1, or
O to C to F to P_1, or $\quad O$ to A to F to P_1.

In each of the above cases we simply follow the edges of a rectangular parallelopiped (called a "rectangular solid" in *MMS*, Chapter

XIV) from a vertex at the origin to the diametrically opposite vertex at P_1. And since the lengths of these edges are x_1, y_1, and z_1, in every instance, we may designate the point P_1 as (x_1, y_1, z_1).

In the special case of points such as A, B, and C on the axes, of course, we write:

$A = (x_1, 0, 0)$, $B = (0, y_1, 0)$, and $C = (0, 0, z_1)$.

Or, in the special case of points such as D, E, and F, in the coordinate planes, we write:

$D = (x_1, y_1, 0)$, $E = (0, y_1, z_1)$, and $F = (x_1, 0, z_1)$.

If one or more of the values of the variables x, y, and z, are negative, the described parallelopiped and resulting point are shifted to another octant. For instance, the point $(-x_1, y_1, z_1)$ designated P_1' in Figure 78 is in the $(-, +, +)$ *octant* behind the (y, z) *coordinate plane* from the first octant. The point $(x_1, -y_1, z_1)$ designated P_1'' in the diagram is in the $(+, -, +)$ *octant* to the left of the (x, z) *coordinate plane* from the first octant. And the point $(x_1, y_1, -z_1)$ designated P_1''' in the diagram is in the $(+, +, -)$ *octant* below the (x, y) *coordinate plane* from the first octant.

Practice Exercise No. 65

1. Which of the points labelled in Figure 78 are in the (x, y) coordinate plane?
2. Which are in the (y, z) coordinate plane?
3. In the (x, z) coordinate plane?
4. What are the appropriate parenthetical designations of the points Q, R, S, and T in Figure 78?
5. Locate and label by additions to the diagram of Figure 78 the points $(0, -y_1, 0)$, $(0, 0, -z_1)$, $(0, y, -z_1)$, $(-x_1, y_1, 0)$, $(-x_1, 0, 0)$, $\left(\frac{x_1}{2}, 0, 0\right)$, $\left(0, \frac{y_1}{2}, 0\right)$, $\left(0, 0, \frac{z_1}{2}\right)$, $\left(\frac{x_1}{2}, \frac{y_1}{2}, 0\right)$, $\left(\frac{x_1}{2}, \frac{y_1}{2}, \frac{-z_1}{2}\right)$.

6. If you were to extend the line P_1O a distance equal to its own length back through the origin, what would be the end-point of its extension?

Point and Length Formulas

Some of the basic formulas of plane analytic geometry may be extended to apply to the figures of solid analytic geometry simply by the addition of z-terms.

For instance, it is clear from the above explanation of three-dimensional rectangular coordinates that we may represent **any point in space**, P_1, by the following set of equations:

G21: $x = x_1$, $y = y_1$, $z = z_1$,

or,   (Pages 162–3)

$P_1 = (x_1, y_1, z_1)$.

This corresponds, of course, to Formula G1 (page 112) in plane analytic geometry.

From Figure 79 below, moreover, we see that the distance D between P_1 and P_2, designating any two points in space, is

$$D = \sqrt{P_1Q^2 + QP_2{}^2}.$$ (Formula R40, page 16)

But

$P_1Q^2 = (x_2 - x_1)^2 + (y_2 - y_1)^2$,   (Formula G3)

and

$QP_2{}^2 = (z_2 - z_1)^2$.   (The same)

Hence we have for the **distance D between any two points P_1 and P_2 in space**, the formula,

G23: $D = \sqrt{(x_2 - x_1)^2 + (y_2 - y_1)^2 + (z_2 - z_1)^2}$,

(Substitution)

which corresponds to Formula G3 (page 114) in plane analytic geometry.

The plane analytic geometry formulas G2 and G2′ may be extended to three dimensions in the same way. The point P which divides the line P_1P_2 in the ratio m/n is given by the formula,

G22: $P = \left(\dfrac{nx_1 + mx_2}{m + n}, \dfrac{ny_1 + my_2}{m + n}, \dfrac{nz_1 + mz_2}{m + n}\right).$

In the special case where $m = n = 1$, this becomes the formula for the midpoint M of the line P_1P_2,

G22′: $M = \left(\dfrac{x_1 + x_2}{2}, \dfrac{y_1 + y_2}{2}, \dfrac{z_1 + z_2}{2}\right).$

As in several other cases throughout the rest of this chapter, however, we shall omit the details of the ordinary analytic proof because

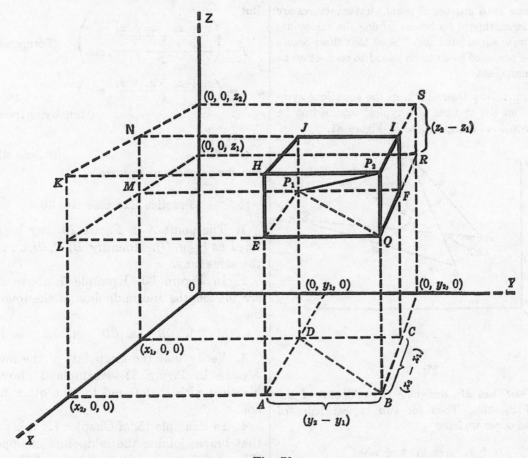

Fig. 79

these may be given more simply by the methods of vector analysis explained in the next volume.

EXAMPLE 1: The main girders of one quadrant of a tower are shown in Figure 80. What are the length and midpoint of the cross-member *AE?*

SOLUTION:

$$AE = \sqrt{(0 - 9)^2 + (1 - 0)^2 + (15 - 0)^2}$$

(Formula G23)

$$= \sqrt{81 + 1 + 225} = \sqrt{307}$$

(Removing parentheses)

$$= 17.52.$$

(Extracting the square root)

$$M = \left(\frac{9 + 0}{2}, \frac{0 + 1}{2}, \frac{0 + 15}{2}\right)$$

(Formula G22′)

$$= (4.5, .5, 7.5).$$

(Simplifying fractions)

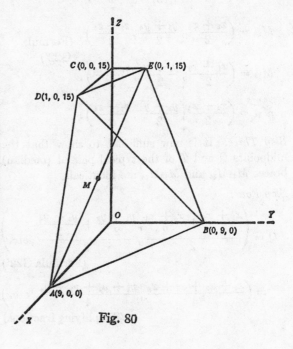

Fig. 80

EXAMPLE 2: A number of tetrahedral structures are to be strengthened by braces joining the midpoints of opposite edges (medians). Show that these braces may be prepared so as to be joined to each other at their midpoints.

SOLUTION: *Step One:* To keep the equations symmetric, let the vertices of a typical tetrahedron be $P_1(x_1,y_1,z_1),\ldots P_4(x_4,y_4,z_4)$ as in Figure 81.

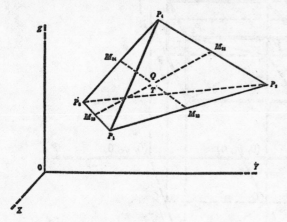

Fig. 81

Step Two: Let M_{12} designate the midpoint of the edge P_1P_2, etc. Then for two typical pairs of opposite edges we have

$$M_{12} = \left(\frac{x_1 + x_2}{2}, \frac{y_1 + y_2}{2}, \frac{z_1 + z_2}{2}\right),$$

$$M_{34} = \left(\frac{x_3 + x_4}{2}, \frac{y_3 + y_4}{2}, \frac{z_3 + z_4}{2}\right),$$ (Formula G22′)

$$M_{13} = \left(\frac{x_1 + x_3}{2}, \frac{y_1 + y_3}{2}, \frac{z_1 + z_3}{2}\right),$$

$$M_{24} = \left(\frac{x_2 + x_4}{2}, \frac{y_2 + y_4}{2}, \frac{z_2 + z_4}{2}\right).$$

Step Three: It is now sufficient to show that the midpoints Q and T of the typical pair of (median) braces $M_{12}M_{34}$ and $M_{13}M_{24}$ are identical.

Step Four:

$$Q = \left(\frac{\frac{x_1 + x_2}{2} + \frac{x_3 + x_4}{2}}{2}, \frac{\frac{y_1 + y_2}{2} + \frac{y_3 + y_4}{2}}{2}, \text{etc.}\right)$$

(Formula G22′)

$$= \left(\frac{x_1 + x_2 + x_3 + x_4}{4}, \frac{y_1 + y_2 + y_3 + y_4}{4}, \text{etc.}\right)$$

(Simplifying fractions)

But

$$T = \left(\frac{\frac{x_1 + x_3}{2} + \frac{x_2 + x_4}{2}}{2}, \text{etc.}\right)$$ (Formula G22′)

$$= \left(\frac{x_1 + x_2 + x_3 + x_4}{4}, \text{etc.}\right)$$

(Simplifying fractions)

Hence

$$Q = T,$$ (Substitution)

and the proposition is proven.

Practice Exercise No. 66

1. The point A in Figure 79 may be identified as (x_2,y_1,O). Identify $B, C, D, \ldots S$, in the same way.

2. In Figure 80, Example 1 above, what are the lengths and midpoints of the following lines:

   a. *OA*   b. *OB*   c. *CD*   d. *CE*   e. *BD*

3. Verify that the midpoints of the median $M_{14}M_{23}$ in Figure 81, Example 2 above, is identical with that found for the other medians.

4. In Example 13, of Chapter IX it is shown that braces joining the midpoints of opposite sides of a plane quadrilateral structure intersect at their midpoints. Verify that this is still true even when the structure is bent along a diagonal line so as to become a three-dimensional quadrilateral with a triangular part in each of two intersecting planes.

5. Prove that the diagonals of a rectangular parallelopiped intersect at their midpoints.

6. Prove that the sum of the squares of the diagonals of any (two- or *three*-dimensional) quadrilateral equals twice the sum of the squares of the lines joining the midpoints of the opposite sides.

7. The center of gravity of a triangle is the point where its medians intersect (Problem 1, Practice Exercise No. 54). Show that for the triangle with vertices at $P_1(x_1,y_1,z_1)$, $P_2(x_2,y_2,z_2)$, $P_3(x_3,y_3,z_3)$, this point is

$$G = \left(\frac{x_1 + x_2 + x_3}{3}, \frac{y_1 + y_2 + y_3}{3}, \frac{z_1 + z_2 + z_3}{2}\right)$$

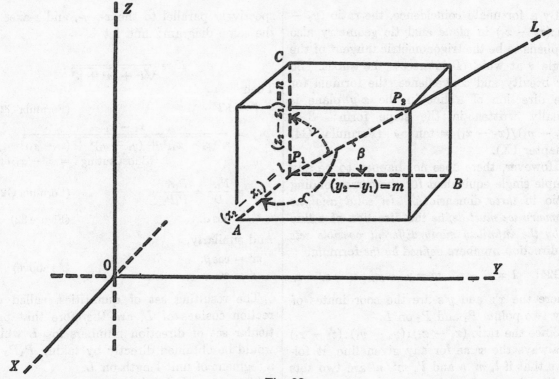

Fig. 82

8. Show that the lines joining the vertices of a tetrahedron with the centers of gravity of the opposite faces intersect in a common point which divides them in the ratio 3 to 1. *Suggestion:* Find the indicated point on each of two typical lines and verify that it is the same in both cases.

Direction Numbers and Cosines

Although the remaining formulas of solid analytic geometry are also essentially similar to those of plane analytic geometry, some require more than one equation, and some are usually written in a form which makes them appear unrelated to the corresponding two-dimensional formulas. This is especially true of those for specifying directions.

Just as the direction of a line L in the x-y plane is defined in terms of the coordinates of any two of its points by the unchanging ratio $(y_2 - y_1):(x_2 - x_1)$, the direction of a line L in space is defined by the corresponding ratio $(x_2 - x_1):(y_2 - y_1):(z_2 - z_1)$.

In the two-dimensional case the quantities $(x_2 - x_1)$ and $(y_2 - y_1)$ are always the lengths of sides of similar right triangles formed upon segments P_1P_2 of L as hypotenuse (Figures 30 and 32, Chapter IX). In the three-dimensional case the quantities $(x_2 - x_1)$, $(y_2 - y_1)$, $(z_2 - z_1)$ — hereafter designated l, m, and n, for short — are always the lengths of the edges of similar rectangular parallelopipeds formed about segments P_1P_2 of L as diagonal (Figure 82).

When P_1 and P_2 are taken close together, the quantities l, m, n are relatively small. When P_1 and P_2 are taken farther apart, the quantities l, m, n are relatively large. When P_1 and P_2 are taken in the opposite order the quantities l, m, n have opposite signs, since $x_1 - x_2 = -(x_2 - x_1)$, etc. But whether large or small, positive or negative, **the infinitely many different possible sets of such numbers for a given line L always have the same fixed ratio $l:m:n$ which is different from the corresponding ratio for any other line not parallel to L.**

By a fortunate coincidence, the ratio $(y_2 - y_1):(x_2 - x_1)$ in *plane* analytic geometry also happens to be the trigonometric tangent of the angle θ at which L intersects the x-axis. So for brevity and convenience, the formula for the direction of a line in the x-y plane is usually written in the slope form: $S = (y_2 - y_1)/(x_2 - x_1) = \tan \theta$ (Formula G4, Chapter IX).

However, there does not happen to be any simple single equivalent for the corresponding ratio in *three* dimensions. *In solid analytic geometry we must define the direction of a line L by the infinitely many different possible sets of direction numbers defined by the* formula.

G24: $l = x_2 - x_1,$ $m = y_2 - y_1,$ $n = z_2 - z_1,$

where the x's and y's are the coordinates of any two points P_1 and P_2 on L.

Since the ratio $(x_2 - x_1):(y_2 - y_1):(z_2 - z_1)$ is always the same for any given line, it follows that if l, m, n and l', m', n' are two sets of direction numbers for the same line L, they are related by the formula:

G24′: $l' = Kl,$ $m' = Km,$ $n' = Kn,$

or

$$\frac{l'}{l} = \frac{m'}{m} = \frac{n'}{n} = K,$$

where K is some constant which may be positive or negative.

For instance, the line which passes through the points $P_1(2,1,0)$ and $P_2(4,7,-8)$ has as one possible set of direction numbers:

$$\begin{aligned} l &= 4 - 2 = 2, \\ m &= 7 - 1 = 6, \\ n &= -8 - 0 = -8. \end{aligned} \qquad \text{(Formula G24)}$$

But, by Formula G24′, the proportional sets of quantities $1,3,-4$, or $4,12,-16$, or $-2,-6,8$, have the same ratio and will therefore serve just as well as direction numbers for the given line.

For an important application of Formula G24′, let l, m, n be the set of direction numbers derived for L from the points P_1 and P_2 in Figure 82; let α, β, γ be the angles between P_1P_2 and the lines P_1A, P_1B, and P_1C, re-

spectively parallel to the x-, y-, and z-axes in the same diagram; and let

$$K = \frac{1}{\sqrt{l^2 + m^2 + n^2}}.$$

Then

$$\begin{aligned} l' &= Kl \qquad\qquad\qquad\qquad \text{(Formula G24′)} \\ &= \frac{x_2 - x_1}{\sqrt{(x_2 - x_1)^2 + (y_2 - y_1)^2 + (z_2 - z_1)^2}} \\ &\qquad\qquad \text{(Substituting } l = x_2 - x_1, \text{ etc.)} \\ &= \frac{P_1A}{D} = \frac{P_1A}{P_1P_2} \qquad\quad \text{(Formula G23)} \\ &= \cos \alpha. \qquad\qquad\qquad \text{(Figure 82)} \end{aligned}$$

And similarly,

$$\begin{aligned} m' &= \cos \beta. \\ n' &= \cos \gamma. \end{aligned} \qquad\qquad \text{(As above)}$$

The resulting set of quantities, called direction cosines of L, are therefore that particular set of direction numbers for L which would be obtained directly by taking P_1P_2 as a segment of unit length on L.

The above derivation, however, is based on the relative positions of P_1 and P_2 in Figure 82. If we were to interchange the positions of P_1 and P_2, we would be regarding the line L as directed in the opposite direction. The direction angles of L with respect to the positive directions of the coordinate axes would then be the supplements of α, β, γ:

$$\alpha' = 180° - \alpha, \quad \beta' = 180° - \beta, \quad \gamma' = 180° - \gamma.$$

And the direction cosines of L would be the negatives of those derived above:

$$\begin{aligned} \cos \alpha' &= \cos(180° - \alpha) = -\cos \alpha, \\ \cos \beta' &= \cos(180° - \beta) = -\cos \beta, \quad \text{(Chapter VIII)} \\ \cos \gamma' &= \cos(180° - \gamma) = -\cos \gamma. \end{aligned}$$

The two possible sets of direction cosines, which correspond to the two possible directions in which a line may be regarded as directed, are therefore given by the general formula.

G25: $\cos \alpha = \dfrac{\pm(x_2 - x_1)}{D} = \dfrac{\pm l}{\sqrt{l^2 + m^2 + n^2}},$

$$\cos \beta = \frac{\pm(y_2 - y_1)}{D} = \frac{\pm m}{\sqrt{l^2 + m^2 + n^2}},$$

$$\cos \gamma = \frac{\pm(z_2 - z_1)}{D} = \frac{\pm n}{\sqrt{l^2 + m^2 + n^2}}.$$

By squaring and adding the terms in the columns of these equations, we find that the **direction cosines of a line are related by the formula,**

G25': $\cos^2 \alpha + \cos^2 \beta + \cos^2 \gamma = 1.$

EXAMPLE 3: At what angles does the girder BE in Figure 80, page 165, intersect the positive direction of the y-axis and directed lines parallel to the positive directions of the x and y axes?

SOLUTION: Let $P_1 = B(0,9,0)$ and $P_2 = E(0,1,15)$. Then the corresponding direction numbers for BE are:

$$l = 0 - 0 = 0,$$
$$m = 1 - 9 = -8, \qquad \text{(Formula G24)}$$
$$n = 15 - 0 = 15.$$

And the direction cosines of BE are:

$$\cos \alpha = \frac{0}{\sqrt{0^2 + (-8)^2 + 15^2}} = 0,$$

$$\cos \beta = \frac{-8}{\sqrt{289}} = \frac{-8}{17} = -.4706, \quad \text{(Formula G25)}$$

$$\cos \gamma = \frac{15}{17} = .8824.$$

Hence the angles α, β, γ are:

$$\alpha = \cos^{-1} 0 = 90°,$$
$$\beta = \cos^{-1} - .4706 = 180° - 61°56' = 118°4',$$
$$\gamma = \cos^{-1} .8824 = 28°4'.$$

(Substitution)

Note that angle $\alpha = 90°$ because BE is in the y-z plane perpendicular to the x-axis; cosine β is negative because BE makes an obtuse angle with the positive direction of the y-axis; and cosine γ is positive because BE makes an acute angle with the positive direction of the z-axis.

EXAMPLE 4: Line L makes an obtuse angle with the z-axis and has the direction cosines, $\cos \alpha = \frac{1}{3}$ and $\cos \beta = -\frac{2}{3}$. What is its third direction cosine?

SOLUTION: Since

$$\cos^2 \alpha + \cos^2 \beta + \cos^2 \gamma = 1, \quad \text{(Formula G25')}$$

$$\left(\frac{1}{3}\right)^2 + \left(\frac{-2}{3}\right)^2 + \cos^2 \gamma = 1. \quad \text{(Substitution)}$$

Hence

$$\cos^2 \gamma = 1 - \tfrac{5}{9} = \tfrac{4}{9}, \qquad \begin{array}{l}\text{(Solving for} \\ \cos^2 \gamma)\end{array}$$

and

$$\cos \gamma = \pm \tfrac{2}{3}. \qquad \begin{array}{l}\text{(Extracting} \\ \text{square roots)}\end{array}$$

But since it is given that L makes an obtuse angle with the z-axis, we reject the positive value and conclude that

$$\cos \gamma = -\tfrac{2}{3}. \qquad \text{(Chapter VIII)}$$

Parallel and Perpendicular Directions

Two line segments L_1 and L_2 are **parallel** if their *direction numbers* satisfy the **formula,**

G36: $\dfrac{l_1}{l_2} = \dfrac{m_1}{m_2} = \dfrac{n_1}{n_2} = K.$

Or, two line segments L_1 and L_2 are parallel if their *direction cosines* satisfy the **formula,**

G36': $\dfrac{\cos \alpha_1}{\cos \alpha_2} = \dfrac{\cos \beta_1}{\cos \beta_2} = \dfrac{\cos \gamma_1}{\cos \gamma_2} = \pm 1,$

L_1 and L_2 being like directed if the "+" sign obtains, and oppositely directed if the "−" sign obtains. These observations follow at once from the definitions of direction numbers and direction cosines.

Two straight lines will, in general, not meet in three dimensional space even when they are not parallel. We therefore define the **angle between any two non-intersecting directed lines** as the angle between the positive directions of two intersecting lines having the same directions as the given lines.

As can most simply be shown by vector methods, the **angle θ between any two directed lines L_1 and L_2** is given by the *direction-cosine* formula:

G26: $\cos \theta = \cos \alpha_1 \cos \alpha_2 + \cos \beta_1 \cos \beta_2$
$\qquad\qquad\qquad + \cos \gamma_1 \cos \gamma_2.$

By substitution from Formula G25, this becomes the *direction-number* formula,

G26': $\cos \theta = \dfrac{l_1 l_2 + m_1 m_2 + n_1 n_2}{\sqrt{l_1^2 + m_1^2 + n_1^2}\sqrt{l_2^2 + m_2^2 + n_2^2}}.$

Since $\cos 90° = 0$, it follows that two line segments L_1 and L_2 are mutually perpendicular when, and only when, their direction numbers or cosines satisfy the formulas,

> G38: $l_1 l_2 + m_1 m_2 + n_1 n_2 = 0,$

or

> G38′: $\cos\alpha_1 \cos\alpha_2 + \cos\beta_1 \cos\beta_2$
> $+ \cos\gamma_1 \cos\gamma_2 = 0.$

EXAMPLE 5: Compute the angle B between the girders BE and BA in the tower diagram of Figure 80.

SOLUTION: For $BE = L_1$ we have already found:

$l_1 = 0, \; m_1 = -8, \; n_1 = 15,$ (Example 3 above)

$\sqrt{l_1{}^2 + m_1{}^2 + n_1{}^2} = 17.$

Letting $BA = L_2$, we now find:

$$l_2 = 9 - 0 = 9,$$
$$m_2 = 0 - 9 = -9,$$ (Formula G24)
$$n_2 = 0 - 0 = 0.$$

However, the proportional set of numbers $1, -1, 0$ obtained by dividing the above by 9, are numerically simpler, so we shall use them to compute:

$$\cos B = \frac{0(1) - 8(-1) + 15(0)}{17\sqrt{1^2 + 1^2 + 0^2}}$$ (Formula G26′)

$$= \frac{8}{17\sqrt{2}} = \frac{8\sqrt{2}}{17(2)} = \frac{4}{17}\sqrt{2}$$ (Simplifying)

$$\frac{4}{17}(1.414) = \frac{5.565}{17} = .333$$

(Substituting $\sqrt{2} = 1.414$)

Hence

$$B = \cos^{-1} .333 = 70°33'.$$ (By tables)

EXAMPLE 6: Show that the direction numbers of the line L perpendicular to each of two intersecting lines L_1 and L_2 is given by the formula,

G38″: $l = \begin{vmatrix} m_1 & n_1 \\ m_2 & n_2 \end{vmatrix}, \quad m = \begin{vmatrix} n_1 & l_1 \\ n_2 & l_2 \end{vmatrix}, \quad n = \begin{vmatrix} l_1 & m_1 \\ l_2 & m_2 \end{vmatrix}.$

SOLUTION: Since L is perpendicular to L_1 and L_2,

$$l_1 l + m_1 m + n_1 n = 0,$$ (Formula C38)
$$l_2 l + m_2 m + n_2 n = 0.$$

These equations are a defective simultaneous system in l, m, n as three unknown variables, and will therefore have infinitely many solutions for l, m, n when consistent (Chapter VII). This is to be expected, since there are always infinitely many possible sets of direction numbers for any one line (page 168). But for

the same reason, we may always assign an arbitrary value to any one direction number of a set. Setting $n = 1$, we may solve the above equations for l and m to find

$$l = \frac{\begin{vmatrix} m_1 & n_1 \\ m_2 & n_2 \end{vmatrix}}{\begin{vmatrix} l_1 & m_1 \\ l_2 & m_2 \end{vmatrix}},$$

(Cramer's Rule and M10)

$$m = \frac{\begin{vmatrix} n_1 & l_1 \\ n_2 & l_2 \end{vmatrix}}{\begin{vmatrix} l_1 & m_1 \\ l_2 & m_2 \end{vmatrix}},$$

when $n = 1$. Having found this one set of direction numbers for L, however, we may now multiply the set by the determinant in the common denominator of the two above fractions to derive the proportional set:

$$l = \begin{vmatrix} m_1 & n_1 \\ m_2 & n_2 \end{vmatrix}, \quad m = \begin{vmatrix} n_1 & l_1 \\ n_2 & l_2 \end{vmatrix}, \quad n = \begin{vmatrix} l_1 & m_1 \\ l_2 & m_2 \end{vmatrix},$$

as required by the problem.

Practice Exercise No. 67

A. Derive the above formulas, G25′ and G26′, by the steps indicated in the text.

B. What are the angles between the coordinate axes and each of the following girders in the tower construction diagram, Figure 80:

(1) BD (2) AE (3) AD (4) AB (5) DE
(6) OA (7) OB (8) CD

C. What are the angles between each of the following pairs of girders in the same diagram?

(1) BE and BD.   (3) AD and AB
(2) AD and AE (4) AE and BD

D. Lines parallel to the coordinate planes always have at least one direction cosine = 0. Why?

E. When do lines have two direction cosines = 0?

F. Is it possible for a line to have three direction cosines = 0? Explain.

G. The direction cosines have the same signs as the variables in the octant toward which the line points. Why?

H. What angles does a line make with the axes if all its direction cosines are equal?

The Summary of this chapter is included with that of Chapter XV on pages 178–80.

SURFACES AND LINES IN SPACE

Normal Plane-Formulas

The "slant" or "tilt" of a surface at each point is specified by the direction of its normal — *the* line perpendicular to the surface at that point.

All normals to a (flat) **plane** are, of course, parallel to each other. And **parallel planes** are planes whose normals have the same direction.

For instance, the x-axis and every line parallel to the x-axis is at some point perpendicular, and therefore *normal, to* the (y,z) coordinate plane and to every plane parallel to the (y,z) coordinate plane. The common direction of the x-axis and of all lines parallel to the x-axis will therefore specify the "slant in space" of the (y,z) coordinate plane and of every plane parallel to the (y,z) coordinate plane.

Let S be the fixed plane which passes through the fixed point P_1, and for which all normals, such as P_1L in Figure 83, have the direction numbers l, m, n.

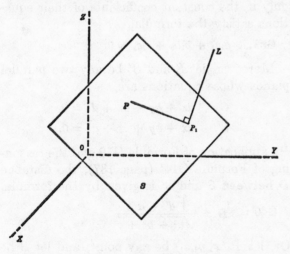

Fig. 83

Since we know from elementary geometry that a line perpendicular to a plane is perpendicular to all lines in the plane which intersect

it, the geometric condition that any other point P also lie in the plane S is that the line P_1P be perpendicular to the line P_1L. Analytically, this means that the direction numbers of P_1P and P_1L must satisfy formula G38 (page opposite).

The direction numbers of P_1L, however, are given as l, m, n; and the direction numbers of P_1P are

$$(x - x_1), \quad (y - y_1), \quad (z - z_1). \qquad \text{(Formula G24)}$$

Hence the analytic condition that P be a point in the plane S is that

$$l(x - x_1) + m(y - y_1) + n(z - z_1) = 0.$$
$$\text{(Formula G38)}$$

And **the equation of the plane S which passes through P_1 and to which the normals have the direction numbers l, m, n, is given by the formula:**

G28: $\quad lx + my + nz = lx_1 + my_1 + nz_1.$
(Removing parentheses, and transposing $-lx_1$, etc.)

If we multiply formula G38 through by $\pm 1/\sqrt{l^2 + m^2 + n^2}$ and take P_1 as the foot of the perpendicular from the origin to S so that l, m, $n = x_1$, y_1, z_1, we obtain as **the equation of the plane at distance D from the origin and with normals having the direction angles α, β, γ, the formula,**

G29: $\quad (cos\,\alpha)x + (cos\,\beta)y + (cos\,\gamma)z = \pm D.$
(Formula G25)

Two possible choices of sign must be prefixed here to the positive quantity D, representing an absolute distance, because the direction cosines on the left-hand side of the equation may have opposite signs depending upon the direction of the normals, *from* or *toward* the origin.

Since all the above-described steps are retraceable, Formula G28 may be applied to interpret any first-degree equation in three variables of the typical form,

$$S: \quad ax + by + cz + d = 0.$$

By comparing equation S with Formula G28, we gather that S must be the equation of a plane whose normals have the direction numbers a, b, c.

Transposing d and multiplying through by $1/\sqrt{a^2 + b^2 + c^2}$, we may rewrite S in the form,

$$S': \quad \frac{a}{\sqrt{a^2+b^2+c^2}}x + \frac{b}{\sqrt{a^2+b^2+c^2}}y$$
$$+ \frac{c}{\sqrt{a^2+b^2+c^2}}z = \frac{-d}{\sqrt{a^2+b^2+c^2}}$$

But by Formula G25, $a/\sqrt{a^2 + b^2 + c^2} = \cos \alpha$, etc. Hence we may again rewrite S in the form,

$$S'': \quad (\cos \alpha)x + (\cos \beta)y + (\cos \gamma)z = \frac{-d}{\sqrt{a^2+b^2+c^2}}.$$

And by comparing S'' with Formula G29 we gather that S must be the equation of the plane whose normals have direction cosines given by the formula,

$$G30: \quad \cos \alpha = \frac{a}{\sqrt{a^2+b^2+c^2}},$$
$$\cos \beta = \frac{b}{\sqrt{a^2+b^2+c^2}},$$
$$\cos \gamma = \frac{c}{\sqrt{a^2+b^2+c^2}},$$

and whose distance from the origin is given by the formula,

$$G40: \quad D = \frac{\mid -d \mid}{\sqrt{a^2+b^2+c^2}}.$$

Parallel and Perpendicular Planes

Several general observations now follow immediately for any two planes, S_1 and S_2, whose equations are:

$$S_1: \quad a_1x + b_1y + c_1z + d_1 = 0,$$
$$S_2: \quad a_2x + b_2y + c_2z + d_2 = 0.$$

By application of Formula G26', the angle θ between planes S_1 and S_2, defined as the angle between their normals, is given in terms of the constant coefficients of their equations by the formula,

$$G31: \quad \cos \theta = \frac{a_1a_2 + b_1b_2 + c_1c_2}{\sqrt{a_1^2 + b_1^2 + c_1^2}\sqrt{a_2^2 + b_2^2 + c_2^2}}.$$

By application of Formula G36, planes S_1 and S_2 are parallel if, and only if, the constant coefficients of their equations satisfy the formula,

$$G37: \quad \frac{a_1}{a_2} = \frac{b_1}{b_2} = \frac{c_1}{c_2} = K.$$

And planes S_1 and S_2 are identical if, and only if, the constant coefficients of their equations satisfy the formula,

$$G37': \quad \frac{a_1}{a_2} = \frac{b_1}{b_2} = \frac{c_1}{c_2} = \frac{d_1}{d_2} = K.$$

Under the conditions of Formula G37, of course, equations S_1 and S_2 are inconsistent (page 21). And under the conditions of Formula G37', equations S_1 and S_2 are equivalent (page 22).

Also, by application of Formula G38, planes S_1 and S_2 are mutually perpendicular if, and only if, the constant coefficients of their equations satisfy the formula,

$$G39: \quad a_1a_2 + b_1b_2 + c_1c_2 = 0.$$

Moreover, let S and S' be any two parallel planes whose equations are,

$$S: \quad ax + by + cz + d = 0,$$
$$S': \quad ax + by + cz + d' = 0.$$

By application of Formula G40 and the reasoning of Formula G14' (page 137), the distance D between S and S' is given by the formula,

$$G40': \quad D = \frac{\mid d' - d \mid}{\sqrt{a^2+b^2+c^2}}.$$

Or, let $P_1(x_1, y_1, z_1)$ be any point, and let S be the plane whose equation is,

$$S: \quad ax + by + cz + d = 0.$$

By a similar application of formula G40, the distance D between point P_1 and plane S is given by the formula,

G41: $D = \dfrac{|\, ax_1 + by_1 + cz_1 + d \,|}{\sqrt{a^2 + b^2 + c^2}}$.

EXAMPLE 1: What is the distance from the origin to the plane determined by the girders BE and BA in Figure 80?

SOLUTION: From Example 5, Chapter XIV above, we already know the direction numbers of these lines to be,

for BE: 0, −8, 15,

for BA: 1, −1, 0.   (Page 170)

Hence direction numbers of lines perpendicular to the plane of BE and BA must be,

$\begin{vmatrix} -8 & 15 \\ -1 & 0 \end{vmatrix}$, $\begin{vmatrix} 15 & 0 \\ 0 & 1 \end{vmatrix}$, $\begin{vmatrix} 0 & -8 \\ 1 & -1 \end{vmatrix}$ (Formula G38″)

= 15, 15, 8.   (Formula M2)

The equation of the plane of BE and BA must therefore be:

$15x + 15y + 8z = 15(0) + 15(9) + 8(0) = 135$

(Formula G28)

And the distance of this plane from the origin must be:

$D = \dfrac{|\, 135 \,|}{\sqrt{15^2 + 15^2 + 8^2}} = \dfrac{135}{\sqrt{514}}$ (Formula G40)

$= \dfrac{135}{22.67} = 5.955$ (Extracting the root, etc.)

When direction numbers are not specified or computed, most of the above equations are applied more directly in conjunction with —

Point Plane-Formulas

Many of the preceding solid analytic geometry formulas *for planes* are obviously analogous to the corresponding plane analytic geometry formulas *for lines*. Let us therefore examine the equation,

$\begin{vmatrix} x & y & z & 1 \\ x_1 & y_1 & z_1 & 1 \\ x_2 & y_2 & z_2 & 1 \\ x_3 & y_3 & z_3 & 1 \end{vmatrix} = 0,$

which would correspond in solid analytic geometry to the plane analytic geometry formula G7m, for the line through two fixed points (page 127).

If we expand the above determinant by minors of the first row (Formula M3, page 81), the resulting equation is of the first degree in x, y, z. By Formula G28, therefore, the equation must be that of a plane.

By Formula G21, however, the quantities x_1, x_2, x_3, y_1, etc., may be assigned as the coordinates of three points, P_1, P_2, P_3. Hence the equation of the plane through any three points, P_1, P_2, P_3, must actually be given by the formula,

G33: $\begin{vmatrix} x & y & z & 1 \\ x_1 & y_1 & z_1 & 1 \\ x_2 & y_2 & z_2 & 1 \\ x_3 & y_3 & z_3 & 1 \end{vmatrix} = 0.$

When $P_1 = (A,0,0)$, $P_2 = (0,B,0)$, $P_3 = (0,0,C)$, Formula G33 reduces to the intercept form of the equation of a plane,

G33′m: $\begin{vmatrix} x & y & z & 1 \\ A & 0 & 0 & 1 \\ 0 & B & 0 & 1 \\ 0 & 0 & C & 1 \end{vmatrix} = 0.$ (Substitution)

Or, by expanding the above determinant and factoring the resulting terms:

G33′: $\dfrac{x}{A} + \dfrac{y}{B} + \dfrac{z}{C} = 1.$

Since x, y, z in Formula G33 are the coordinates of any point P in the plane of P_1, P_2, and P_3, it also follows at once that **the analytic condition for any four points, P_1, P_2, P_3, and P_4, being in the same plane is that their coordinates satisfy the formula,**

G34: $\begin{vmatrix} x_1 & y_1 & z_1 & 1 \\ x_2 & y_2 & z_2 & 1 \\ x_3 & y_3 & z_3 & 1 \\ x_4 & y_4 & z_4 & 1 \end{vmatrix} = 0.$ (Substitution)

EXAMPLE 2: Solve the problem of Example 1, above, without computing direction numbers of BE or BA.

SOLUTION: Since the plane determined by the girders BE and BA passes through the points $E(0,1,15)$, $B(0,9,0)$, $A(9,0,0)$, its equation must be

$\begin{vmatrix} x & y & z & 1 \\ 0 & 1 & 15 & 1 \\ 0 & 9 & 0 & 1 \\ 9 & 0 & 0 & 1 \end{vmatrix} = 0.$ (Formula G33)

Or, expanding by minors of the first row,

$$\begin{vmatrix} 1 & 15 & 1 \\ 9 & 0 & 1 \\ 0 & 0 & 1 \end{vmatrix} x - \begin{vmatrix} 0 & 15 & 1 \\ 0 & 0 & 1 \\ 9 & 0 & 1 \end{vmatrix} y + \begin{vmatrix} 0 & 1 & 1 \\ 0 & 9 & 1 \\ 9 & 0 & 1 \end{vmatrix} z - \begin{vmatrix} 0 & 1 & 15 \\ 0 & 9 & 0 \\ 9 & 0 & 0 \end{vmatrix} = 0,$$

(Formula M3)

$$\begin{vmatrix} 1 & 15 \\ 9 & 0 \end{vmatrix} x + \begin{vmatrix} 0 & 15 \\ 9 & 0 \end{vmatrix} y + 9 \begin{vmatrix} 1 & 1 \\ 9 & 1 \end{vmatrix} z = 9 \begin{vmatrix} 1 & 15 \\ 9 & 0 \end{vmatrix},$$

(Formula M3)

$$-9(15)x - 9(15)y + 9(-8)z = 9(-135),$$

(Formula M2)

$$15x + 15y + 8z = 135. \quad \text{(Dividing by } -9 = -9)$$

From this point the steps are as before (page 173).

Practice Exercise No. 68

A. Write the equation of the plane through each of the following points, with the given direction numbers:

(1) $(2,1,-2)$; $3{:}-1{:}4$

(2) $(-3,1,0)$; $3{:}0{:}-1$

B. Write the equation of the plane:

(3) Through $(-1,2,-3)$ parallel to $2x-3y+z=8$

(4) Through $(2,-1,3)$ parallel to the (x,z) plane.

(5) Through $A(-1,-2,-1)$, $B(0,-2,-3)$, $C(1,4,1)$

C. Show that the volume V of the tetrahedron with vertices P_1, P_2, P_3, P_4, is given by the formula,

G42:
$$V = \pm \tfrac{1}{6} \begin{vmatrix} x_1 & y_1 & z_1 & 1 \\ x_2 & y_2 & z_2 & 1 \\ x_3 & y_3 & z_3 & 1 \\ x_4 & y_4 & z_4 & 1 \end{vmatrix}.$$

Lines as Intersections of Planes

By the same reasoning according to which we may write the formula for the line through two points P_1 and P_2 in *plane* analytic geometry as

G7:
$$\frac{y-y_1}{x-x_1} = \frac{y_2-y_1}{x_2-x_1}, \quad \text{(Page 127)}$$

or,
$$\frac{x-x_1}{x_2-x_1} = \frac{y-y_1}{y_2-y_1}, \quad \left(\text{Mult. by } \frac{x-x_1}{y_2-y_1}\right)$$

we may write the formula for the line through two points P_1 and P_2 in *solid* analytic geometry as,

G32:
$$\frac{x-x_1}{x_2-x_1} = \frac{y-y_1}{y_2-y_1} = \frac{z-z_1}{z_2-z_1}.$$

(Figure 84, below)

By substitution, therefore, the formula of the line through the point P_1, with direction numbers l,m,n, is:

G27:
$$\frac{x-x_1}{l} = \frac{y-y_1}{m} = \frac{z-z_1}{n}.$$

(Formula G24)

And the formula of the line through the point P_1 with direction angles α, β, γ, is:

G27′:
$$\frac{x-x_1}{\cos\alpha} = \frac{y-y_1}{\cos\beta} = \frac{z-z_1}{\cos\gamma}.$$

(Formula G25)

These formulas are, of course, not single equations but alternative choices among three possible combinations of two equations. For instance, formula G27 is equivalent to:

(1) S' : $\dfrac{x-x_1}{x_2-x_1} = \dfrac{y-y_1}{y_2-y_1}$, and S'' : $\dfrac{y-y_1}{y_2-y_1} = \dfrac{z-z_1}{z_2-z_1}$, or

(2) S' : $\dfrac{x-x_1}{x_2-x_1} = \dfrac{y-y_1}{y_2-y_1}$, and S''' : $\dfrac{x-x_1}{x_2-x_1} = \dfrac{z-z_1}{z_2-z_1}$, or

(3) S'' : $\dfrac{y-y_1}{y_2-y_1} = \dfrac{z-z_1}{z_2-z_1}$, and S''' : $\dfrac{x-x_1}{x_2-x_1} = \dfrac{z-z_1}{z_2-z_1}$.

This, however, is exactly what we should expect to find.

The familiar equation-like expression — $P_1 = (x_1,y_1)$ — by which we designate a point analytically in plane geometry is strictly speaking not a *single* equation either. Rather, it is a shorthand way of writing the two simultaneous equations,

$$x = x_1 \quad \text{and} \quad y = y_1,$$

for a pair of lines which *intersect* in the point (x_1,y_1). And two such equations are necessary to designate a point in plane geometry because a linear equation in x and/or y is in general the equation of a straight line with respect to two coordinate axes.

Likewise, since a linear equation in three variables is in general the equation of a *plane* with respect to three coordinate axes, two such equations are always required in solid geometry to designate a line analytically as the locus of their intersection.

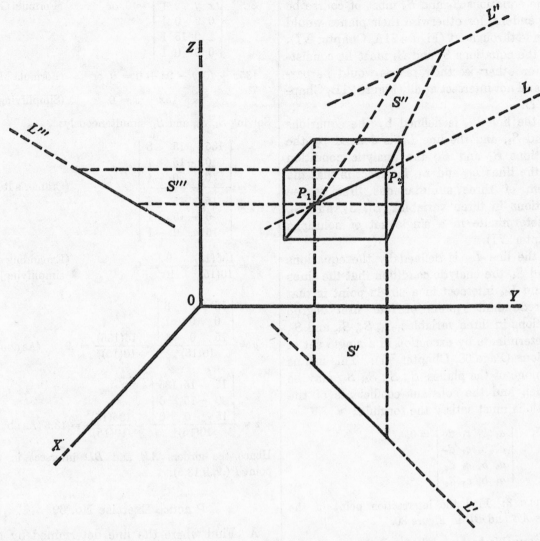

Fig. 84

The first equation in pair (1) above, for instance, is that of the plane S' which passes through L and is perpendicular to the (x,y) coordinate plane which it intersects in the line L' (Figure 84). The second equation in pair (1) is that of the plane S'' which also passes through L but is perpendicular to the (y,z) coordinate plane which it intersects in the line L''. The third equation which appears in pairs (2) and (3) is that of the plane S''' which also passes through L but is perpendicular to the (x,z) coordinate plane which it intersects in the line L'''.

Planes S', S'', and S''', are called the pro-jection planes of L. Lines L', L'', and L''', are called the **trace-lines** of the respective projection planes, or the **projection lines** of L, in the respective coordinate planes. And since any two of the projection planes serve to define the line L by their intersection, any two of the equations S', S'', and S''' will serve to specify the line L analytically.

It is not necessary, however, to have the equations of a line in *projection*-plane form. *If S_1 and S_2 are the equations of any two inter-secting planes, then S_1 and S_2 may be taken as the simultaneous equations of the line L of their intersection.*

The equations S_1 and S_2 must of course be **independent**, for otherwise their planes would coincide throughout (Figure 21A, Chapter VI). And the equations S_1 and S_2 must be **consistent**, for otherwise their planes would be parallel and not intersect at all (Figure 21B, Chapter VI).

If the line L_{12} is defined by the equations S_1 and S_2, and the line L_{13} is defined by the equations S_1 and S_3, **the analytic condition that the lines L_{12} and L_{13} intersect is that the system of three simultaneous first degree equations in three variables, S_1, S_2, and S_3, be determinate in a single set of solutions** (Chapter VI).

If the line L_{34} is defined by the equations S_3 and S_4, **the analytic condition that the lines L_{12} and L_{34} intersect in a single point is that the redundant system of four first degree equations in three variables, S_1, S_2, S_3, and S_4, be determinate by exception in a single set of solutions** (Page 68, Chapter VI). This means that none of the planes, S_1, S_2, S_3, S_4, may be parallel, and the constant coefficients of the equations must satisfy the formula:

G35:
$$\begin{vmatrix} a_1 & b_1 & c_1 & d_1 \\ a_2 & b_2 & c_2 & d_2 \\ a_3 & b_3 & c_3 & d_3 \\ a_4 & b_4 & c_4 & d_4 \end{vmatrix} = 0.$$

EXAMPLE 3: Find the intersection point of the girders AE and BD in Figure 80.

SOLUTION: We have already found the equation of the plane of AE and BD to be:

S_1: $15x + 15y + 8z = 135.$　　(Page 173)

The line AE, however, is also contained in the intersecting plane S_2 through the points A, E, and O. Equation:

S_2:
$$\begin{vmatrix} x & y & z & 1 \\ 9 & 0 & 0 & 1 \\ 0 & 1 & 15 & 1 \\ 0 & 0 & 0 & 1 \end{vmatrix} = 0, \text{ or}$$
　　　　(Formula G33)

$(0)x - 135y + 9z + 0 = 0$, or　　(Formula M3)

$- 15y + z = 0.$　　(Simplifying)

And the line BD is also contained in the intersecting plane S_3 through the points B, D, and O. Equation:

S_3:
$$\begin{vmatrix} x & y & z & 1 \\ 0 & 9 & 0 & 1 \\ 1 & 0 & 15 & 1 \\ 0 & 0 & 0 & 1 \end{vmatrix} = 0, \text{ or}$$
　　　　(Formula G33)

$135x + (0)y - 9z + 0 = 0$, or　　(Formula M3)

$15x - z = 0.$　　(Simplifying)

Solving S_1, S_2, and S_3, simultaneously:

$$x = \dfrac{\begin{vmatrix} 135 & 15 & 8 \\ 0 & -15 & 1 \\ 0 & 0 & -1 \end{vmatrix}}{\begin{vmatrix} 15 & 15 & 8 \\ 0 & -15 & 1 \\ 15 & 0 & -1 \end{vmatrix}}$$
　　(Cramer's Rule)

$$= \dfrac{135(15)}{10(15)^2} = \dfrac{9}{10} = .9$$
　　(Expanding and simplifying)

$$y = \dfrac{\begin{vmatrix} 15 & 135 & 8 \\ 0 & 0 & 1 \\ 15 & 0 & -1 \end{vmatrix}}{10(15)^2} = \dfrac{15(135)}{10(15)^2} = .9$$
　　(As above)

$$z = \dfrac{\begin{vmatrix} 15 & 15 & 135 \\ 0 & -15 & 0 \\ 15 & 0 & 0 \end{vmatrix}}{10(15)^2} = \dfrac{135(15)^2}{10(15)^2} = 13.5$$
　(As above)

Hence the girders AE and BD intersect in the point $P(.9, .9, 13.5)$.

Practice Exercise No. 69

A. Find where the line determined by the pair of equations, $x + y + 3z = 5$ and $x - 3y - 3z = -7$, intersects the coordinate planes.

B. Find the angles of the triangle formed by the intersections of the plane $2x + 3y - z = 2$ with each of the planes:

$$3x - 7y + 2z = 4,$$
$$4x - 2y + 3z = 7,$$
$$x - y + 9z = 5$$

Quadric Surfaces

A quadric surface is any surface defined by an equation of the second degree with respect to a system of three-dimensional coordinates.

A **cylinder** is a surface which may be generated by a straight line moving so as always to be parallel to a fixed line and always to intersect a fixed plane curve. The fixed plane curve is called the **directrix,** and all possible positions of the generating line are called **elements** of the cylinder. When the elements are perpendicular to the plane of the directrix, the generated surface is termed a **right cylinder.**

Just as the equations of straight lines in plane analytic geometry become the equations of planes when referred to a set of three rectangular coordinates, the equations of the conic sections in Chapters XI, XII, XIII above, become the equations of **right quadric cylinders** in solid analytic geometry.

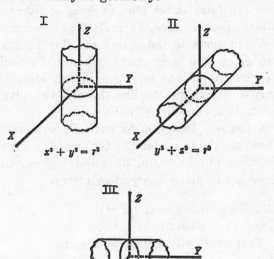

Fig. 85

For instance, the **right circular cylinder** in diagram I of Figure 85 is the three-dimensional graph of the equation $x^2 + y^2 = r^2$. Its directrix is the circle which has the same equation in the x-y coordinate plane. And all its elements are parallel to the axis of the variable z which does not appear in its equation.

Similarly, the right circular cylinder in diagram II of the same Figure is the three-dimensional graph of the equation $y^2 + z^2 = r^2$.

And the right circular cylinder in diagram III is the three-dimensional graph of the equation $x^2 + z^2 = r^2$.

Typical of the equations of non-cylindrical quadric surfaces is the **formula of the ellipsoid with axes 2a, 2b, 2c:**

G43: $\dfrac{x^2}{a^2} + \dfrac{y^2}{b^2} + \dfrac{z^2}{c^2} = 1.$ (Figure 86)

By setting each of the variables x,y,z, in turn equal to 0, you may easily verify that the traces of this surface in the y-z, x-z, and x-y coordinate planes are respectively the ellipses whose equations in those planes are $y^2/b^2 + z^2/c^2 = 1$, $x^2/a^2 + z^2/c^2 = 1$, and $x^2/a^2 + y^2/b^2 = 1$. The surface is called an **ellips-oid** because its real intersection with every plane parallel to a coordinate plane is also an ellipse.

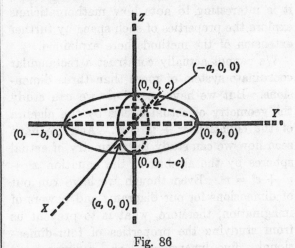

Fig. 86

The quantities 2a, 2b, 2c from the above equation are called the **major, mean,** and **minor axes** of the ellipsoid in the order of their magnitude (in Figure 86, for instance, 2b is the major axis because it is the longest).

When $a = b = c = r$, this equation becomes the **formula of the sphere of radius r with center at the origin:**

$$\frac{x}{r^2} + \frac{y}{r^2} + \frac{z}{r^2} = 1,$$ (Substitution)

or:

G45: $x^2 + y^2 + z^2 = r^2.$ (Mult. by $r^2 = r^2$)

In the above case, or even when only two

of the quantities a, b, c, are equal, the resulting figure is called a **surface of revolution**. The surface $x^2 + y^2 + z^2 = r^2$, for instance, may be generated by revolving the circle $x^2 + y^2 = r^2$ about either the x- or y-axis. And the surface,

G44: $\dfrac{x^2}{a^2} + \dfrac{y^2}{b^2} + \dfrac{z^2}{b^2} = 1,$

called an **ellipsoid of revolution**, may be generated by revolving the ellipse $x^2/a^2 + y^2/b^2 = 1$ about the x-axis. We shall study the properties of such surfaces further in integral calculus (*Advanced Algebra & Calculus Made Simple*).

Footnote on Hyper-Geometry

Meanwhile, although the geometry of **hyper-space** — space having more than three dimensions — is beyond the scope of this book, it is interesting to note how mathematicians explore the properties of such space by further extension of the methods here explained.

We cannot actually construct a rectangular coordinate system of more than three dimensions. But we have seen how we can study the geometry of actual circles by the algebra of the equation $x^2 + y^2 = r^2$. And we have seen how we can study the geometry of actual spheres by the algebra of the equation $x^2 + y^2 + z^2 = r^2$. Even though we have run out of dimensions for our diagrams and powers of imagination, therefore, what is to prevent us from studying the properties of four-dimensional, five-dimensional, or n-dimensional **hyper-spheres** by the algebra of such equations as

G46: $x^2 + y^2 + z^2 + u^2 + v^2 = r^2!$

Such is the power of the analytic method that it is not even limited by our inability to conceive of the infinitely many kinds of space to which it may be applied!

Summary

Extending the methods of the preceding chapters to geometric figures in three dimensions, we have added the formulas below to our geometric technique.

Some of these formulas differ from corresponding plane analytic geometry formulas only by the addition of z-terms. For instance, Formula G23 for the distance between two points in space, adds only z^2 to the expression under the radical sign in the denominator of Formula G3 for the distance between two points in a coordinate plane.

Other formulas differ also in their geometric interpretation, since they are applicable to planes instead of to lines, etc. For instance, Formula G33 is the equation of a plane through three given points, as contrasted with Formula G7m which is the equation of a line through two given points.

Still other formulas differ in the number of their equations rather than in form or interpretation. For instance, Formula G32 consists of two equations to determine the line through two given points in space, whereas Formula G7 consists of one equation of the identical type to determine the line through two given points in a coordinate plane.

A last set, finally, differ radically in content. These are the formulas like G25 which concern direction numbers and direction cosines, unnecessary in plane analytic geometry.

G21: $x = x_1, \quad y = y_1, \quad z = z_1,$
or $P_1 = (x_1, y_1, z_1).$
The point P_1 with coordinates x_1, y_1, z_1.

G22: $P = \left(\dfrac{nx_1 + mx_2}{m + n}, \dfrac{ny_1 + my_2}{m + n}, \dfrac{nz_1 + mz_2}{m + n} \right).$
The point P which divides P_1P_2 in the ratio m/n.

G22′: $M = \left(\dfrac{x_1 + x_2}{2}, \dfrac{y_1 + y_2}{2}, \dfrac{z_1 + z_2}{2} \right).$
The midpoint M of P_1P_2.

G23: $D = \sqrt{(x_2 - x_1)^2 + (y_2 - y_1)^2 + (z_2 - z_1)^2}.$
Distance D from P_1 to P_2.

G24: $l = x_2 - x_1, \quad m = y_2 - y_1, \quad n = z_2 - z_1.$
Direction numbers of the line through P_1 and P_2.

G24′: $\dfrac{l}{l'} = \dfrac{m}{m'} = \dfrac{n}{n'} = K.$

Relationship between $l, m, n,$ and $l', m', n',$ for the same line.

G25: $\cos \alpha = \dfrac{\pm(x_2 - x_1)}{D} = \dfrac{\pm l}{\sqrt{l^2 + m^2 + n^2}},$

$\cos \beta = \dfrac{\pm(y_2 - y_1)}{D} = \dfrac{\pm m}{\sqrt{l^2 + m^2 + n^2}},$

$\cos \gamma = \dfrac{\pm(z_2 - z_1)}{D} = \dfrac{\pm n}{\sqrt{l^2 + m^2 + n^2}}.$

Direction cosines of the line through P_1 and P_2.

G25': $\cos^2 \alpha + \cos^2 \beta + \cos^2 \gamma = 1.$

Relationship between the same.

G26: $\cos \theta = \cos \alpha_1 \cos \alpha_2 + \cos \beta_1, \cos \beta_2$
$+ \cos \gamma_1, \cos \gamma_2.$

G26': $\cos \theta = \dfrac{l_1 l_2 + m_1 m_2 + n_1 n_2}{\sqrt{l_1^2 + m_1^2 + n_1^2}\sqrt{l_2^2 + m_2^2 + n_2^2}}.$

The angle θ between lines L_1 and L_2.

G27: $\dfrac{x - x_1}{l} = \dfrac{y - y_1}{m} = \dfrac{z - z_1}{n}.$

The line through P_1 with direction numbers l, m, n.

G27': $\dfrac{x - x_1}{\cos \alpha} = \dfrac{y - y_1}{\cos \beta} = \dfrac{z - z_1}{\cos \gamma}.$

The line through P_1 with direction angles α, β, γ.

G28: $lx + my + nz = lx_1 + my_1 + nz_1.$

The plane through P_1 with normal direction numbers l, m, n.

G29: $(\cos \alpha)x + (\cos \beta)y + (\cos \gamma)z = \pm D.$

The plane S at distance D from the origin with normal direction angles, α, β, γ.

G30: $\cos \alpha = \dfrac{a}{\sqrt{a^2 + b^2 + c^2}},$

$\cos \beta = \dfrac{b}{\sqrt{a^2 + b^2 + c^2}},$

$\cos \gamma = \dfrac{c}{\sqrt{a^2 + b^2 + c^2}}.$

Direction cosines of normals to plane S.

G31: $\cos \theta = \dfrac{a_1 a_2 + b_1 b_2 + c_1 c_2}{\sqrt{a_1^2 + b_1^2 + c_1^2}\sqrt{a_2^2 + b_2^2 + c_2^2}}.$

The angle θ between planes S_1 and S_2.

G32: $\dfrac{x - x_1}{x_2 - x_1} = \dfrac{y - y_1}{y_2 - y_1} = \dfrac{z - z_1}{z_2 - z_1}.$

The line through P_1 and P_2.

G33: $\begin{vmatrix} x & y & z & 1 \\ x_1 & y_1 & z_1 & 1 \\ x_2 & y_2 & z_2 & 1 \\ x_3 & y_3 & z_3 & 1 \end{vmatrix} = 0.$

The plane through P_1, P_2, and P_3.

G33' : $\dfrac{x}{A} + \dfrac{y}{B} + \dfrac{z}{C} = 1.$

G33'm: $\begin{vmatrix} x & y & z & 1 \\ A & 0 & 0 & 1 \\ 0 & B & 0 & 1 \\ 0 & 0 & C & 1 \end{vmatrix} = 0.$

The plane with x,y,z-intercepts $= A, B, C$.

G34: $\begin{vmatrix} x_1 & y_1 & z_1 & 1 \\ x_2 & y_2 & z_2 & 1 \\ x_3 & y_3 & z_3 & 1 \\ x_4 & y_4 & z_4 & 1 \end{vmatrix} = 0.$

Four points P_1, P_2, P_3, and P_4, in the same plane.

G35: $\begin{vmatrix} a_1 & b_1 & c_1 & d_1 \\ a_2 & b_2 & c_2 & d_2 \\ a_3 & b_3 & c_3 & d_3 \\ a_4 & b_4 & c_4 & d_4 \end{vmatrix} = 0.$

Four non-parallel planes through one point.

G36: $\dfrac{l_1}{l_2} = \dfrac{m_1}{m_2} = \dfrac{n_1}{n_2} = K.$

G36': $\dfrac{\cos \alpha_1}{\cos \alpha_2} = \dfrac{\cos \beta_1}{\cos \beta_2} = \dfrac{\cos \gamma_1}{\cos \gamma_2} = \pm 1.$

Lines L_1 and L_2 parallel.

G37: $\dfrac{a_1}{a_2} = \dfrac{b_1}{b_2} = \dfrac{c_1}{c_2} = K.$

Planes S_1 and S_2 parallel.

G38: $l_1 l_2 + m_1 m_2 + n_1 n_2 = 0.$

G38': $\cos \alpha_1 \cos \alpha_2 + \cos \beta_1 \cos \beta_2 + \cos \gamma_1 \cos \gamma_2 = 0.$

Lines L_1 and L_2 mutually perpendicular.

G38'': $l = \begin{vmatrix} m_1 & n_1 \\ m_2 & n_2 \end{vmatrix}, \quad m = \begin{vmatrix} n_1 & l_1 \\ n_2 & l_2 \end{vmatrix}$

$n = \begin{vmatrix} l_1 & m_1 \\ l_2 & m_2 \end{vmatrix}.$

Line L perpendicular to lines L_1 and L_2.

G39: $a_1 a_2 + b_1 b_2 + c_1 c_2 = 0.$

Planes S_1 and S_2 mutually perpendicular.

G40: $D = \dfrac{\mid -d \mid}{\sqrt{a^2 + b^2 + c^2}}$.

Distance D of plane S from the origin.

G40′: $D = \dfrac{\mid d' - d \mid}{\sqrt{a^2 + b^2 + c^2}}$.

Distance D between parallel planes S and S'.

G41: $D = \dfrac{\mid ax_1 + by_1 + cz_1 + d \mid}{\sqrt{a^2 + b^2 + c^2}}$.

Distance D from P_1 to plane S.

G42: $V = \pm\frac{1}{6} \begin{vmatrix} x_1 & y_1 & z_1 & 1 \\ x_2 & y_2 & z_2 & 1 \\ x_3 & y_3 & z_3 & 1 \\ x_4 & y_4 & z_4 & 1 \end{vmatrix}$.

Volume V of the tetrahedron with vertices P_1, P_2, P_3, and P_4.

G43: $\dfrac{x^2}{a^2} + \dfrac{y^2}{b^2} + \dfrac{z^2}{c^2} = 1$.

Ellipsoid with center at the origin and axes $2a$, $2b$, $2c$.

G44: $\dfrac{x^2}{a^2} + \dfrac{y^2}{b^2} + \dfrac{z^2}{b^2} = 1$.

Ellipsoid of revolution with center at the origin and axes $2a$, $2b$, $2b$.

G45: $x^2 + y^2 + z^2 = r^2$.

Sphere with center at the origin and radius r.

G46: $x^2 + y^2 + z^2 + u^2 + v^2 = r^2$.

Typical hyper-sphere, five-dimensional case.

ANSWERS

Exercise No. 1

A. R1; 5

R2; 9

R3; 1

R4; 5

R5; 6

R6; 24

R7; −6

R8; 6

R9; 25

R10; −1

R11; 13

R12; $\frac{3}{2} = 1.5$

R13; $\frac{3}{2} = \frac{6}{4} = \frac{3/2}{2/2}$

R14; $\frac{1}{3/2} = \frac{2}{3}$

R15; $\frac{3}{5} + \frac{2}{5} = \frac{3+2}{5} = \frac{5}{5}$

R16; $\frac{3}{5} + \frac{2}{6} = \frac{18}{30} + \frac{10}{30} = \frac{28}{30}$

R17; $(\frac{3}{5})(\frac{2}{6}) = \frac{6}{30}$

R18; $(6)^2 = 3^2 \cdot 2^2 = 36$

R19; $(\frac{3}{2})^2 = \frac{9}{4}$

R20; $3^3 \cdot 3^2 = 3^5 = 243$

R21; $\frac{3^3}{3^2} = 3^1$

R25; $(3^3)^2 = 3^6 = 729$

E. R26; $4^{\frac{1}{2}} = \sqrt[2]{4} = 2$

R27; $4^{\frac{3}{2}} = (4^{\frac{1}{2}})^3 = 2^3 = 8$

F. R26; $27^{\frac{1}{3}} = 3$

R27; $27^{\frac{2}{3}} = 9$

G. R24; $3^{-2} = \frac{1}{9}$

R25; $3^6 = 729$

R26; $3^{\frac{1}{2}} = \sqrt{3} = 1.732$

R27; $3^{\frac{3}{2}} = \sqrt{27} = 5.196$

R28; 25

R29; 5

R30; 25

R31; 19

R32; 35

H. R33; 7   R34; 1   R35; 8   R36; 2

I. R37; 20

R38; 16

R39; 10

R40; 25

R41; 18π

R42; 81π

R43; 60

R44; 64

R45; 35

R46; $\frac{35}{3}$

R47; 324π

R48; 972π

J. R40; $c = 2, b = \sqrt{3}$

R49; $\frac{1}{2}$

R50; $\frac{\sqrt{3}}{2}$

R51; $\frac{1}{\sqrt{3}}$

R52; $\frac{1}{\sqrt{3}}$

R53; $\frac{1}{4} + \frac{3}{4} = 1$

R54; $\frac{1}{\frac{1}{2}} = \frac{\sqrt{3}}{\sqrt{3}/2} = \frac{2}{1}$

R55; $1 = 3 + 4 - 2(\sqrt{3} \times 2)\cos 30°$

$$= 7 - 4\sqrt{3}\left(\frac{\sqrt{3}}{2}\right) = 7 - 6 = 1$$

Exercise No. 2

(1) $x = 2 \quad y = 5$

(2) $x = 2 \quad y = 5$

(3) $x = -4 \quad y = 2$

(4) $x = -1 \quad y = 8$

(5) $x = -\frac{28}{3} \quad y = -\frac{29}{3}$

(6) $x = 2 \quad y = 3$

(7) $x = \frac{1}{2} \quad y = -1$

(8) $x = 7 \quad y = 8$

(9) $x = 9 \quad y = -5$

(10) $x = 6 \quad y = -7$

Exercise No. 3

(1) $x = \frac{19}{11} \quad y = \frac{7}{11}$

(2) $x = -\frac{37}{31} \quad y = -\frac{15}{31}$

(3) Inconsistent

(4) $x = \frac{34}{3} \quad y = -\frac{19}{3}$

(5) Inconsistent

(6) $x = \frac{4}{3} \quad y = \frac{1}{3}$

(7) 50×100, 5000 sq. ft.

(8) In 1 hr; 4 hrs later

Exercise No. 4

(1) Indeterminate $x = 0, y = -\frac{7}{6}$ $x = 1, y = -\frac{4}{6}$

$x = 10, y = \frac{23}{15}$

(2) Inconsistent

(3) $x = \frac{7}{6} \quad y = \frac{7}{10}$

(4) Inconsistent

(5) Head wind speed 50 m.p.h. original plane speed 150 m.p.h.

(6) Conditions give indeterminate problem.

Exercise No. 5

1. (b) & (c) are dependent. Answer to (a) and (b) (or (c))

$x = \frac{15}{11}, y = \frac{1}{11}$

2. (b) & (c) are dependent. Answer to (a) and (b) (or (c))

$x = \frac{32}{29}, y = \frac{39}{29}$

3. Redundant (a) & (b): $x = -\frac{5}{19}, y = \frac{41}{19}$

(a) & (c): $x = \frac{9}{11}, y = \frac{16}{3}$

(b) & (c): $x = \frac{13}{13}, y = \frac{2}{13}$

4. (a) & (b) are dependent; solution to (b) & (c) (or (a) and (c)) $x = \frac{202}{23}, y = \frac{80}{23}$

5. (c) depends on (a) & (b), $x = 1, y = 2$.

6. (a), (b), and (c) are all dependent; infinitely many solutions.

7. (a) & (b) $x = 1, y = 2$ (b) & (c) $x = 3, y = -1$
(a) & (c) are inconsistent.

8. (a), (b), and (c) have a common solution $x = 1, y = 3$.

9. (a) & (b) are *dependent*, (c) is inconsistent.

10. (a), (b), & (c) have a common solution $x = 2, y = -1$

Exercise No. 6

A. (1) 0   (2) $\frac{1}{2}$ (3) 1

B. (1) 2   (2) 4   (3) 5

C.

$x =$	0	1	2	3	4	5	6	7	8	9	10
(1) $y =$	2	5	8	11	14	17	20	23	26	29	32
(2) $y =$	45	42	39	36	33	30	27	24	21	18	15
(3) $y =$	7	7.2	7.4	7.6	7.8	8	8.2	8.4	8.6	8.8	9
(4) $y =$	2	2	2	2	2	2	2	2	2	2	2
(5) $y =$	−6	−4	−2	0	2	4	6	8	10	12	14
(6) $y =$	7	5	3	1	−1	−3	−5	−7	−9	−11	−13

Exercise No. 7

$x =$	0	1	2	3	4	5	6	7	8	9	10
(1) $y = \frac{1}{2}x + \frac{8}{7}$	$\frac{8}{7}$	$\frac{9}{7}$	$\frac{11}{7}$	$\frac{14}{7}$	$\frac{17}{7}$	$\frac{20}{7}$	$\frac{23}{7}$	$\frac{26}{7}$	$\frac{29}{7}$	$\frac{32}{7}$	$\frac{35}{7}$
(2) $y = -\frac{5}{7}x + 1$	1	$\frac{2}{7}$	$-\frac{3}{7}$	$-\frac{9}{7}$	$-\frac{13}{7}$	$-\frac{18}{7}$	$-\frac{23}{7}$	$-\frac{28}{7}$	$-\frac{33}{7}$	$-\frac{38}{7}$	$-\frac{43}{7}$
(3) $y = \frac{1}{2}x + \frac{8}{7}$					Same as 1						
(4) $y = -\frac{2}{3}x + \frac{8}{3}$	$\frac{8}{3}$	$\frac{4}{3}$	$\frac{2}{3}$	0	$-\frac{2}{3}$	$-\frac{4}{3}$	$-\frac{6}{3}$	$-\frac{8}{3}$	$-\frac{10}{3}$	$-\frac{12}{3}$	$-\frac{14}{3}$

Exercise No. 8

	$x = 4\frac{1}{2}$	$6\frac{1}{2}$	$8\frac{1}{2}$	$y = 1$	2	3	$4\frac{1}{2}$
(1)	$\frac{31}{7}$	$\frac{43}{7}$	$\frac{55}{7}$	$-\frac{1}{3}$	0	$\frac{1}{3}$	$\frac{5}{6}$
(2)	$\frac{63}{2}$	$\frac{51}{2}$	$\frac{39}{2}$	$\frac{44}{3}$	$\frac{43}{3}$	$\frac{42}{3}$	$\frac{81}{6}$
(3)	$\frac{79}{10}$	$\frac{88}{10}$	$\frac{97}{10}$	-30	-25	-20	$-\frac{25}{2}$
(4)	2	2	2	meaningless			
(5)	3	7	11	$\frac{7}{2}$	4	$\frac{9}{2}$	$\frac{21}{4}$
(6)	-2	-6	-10	3	$\frac{5}{2}$	2	$\frac{5}{4}$

Exercise No. 9

	$x = -7$	-3	5	$y = -6$	-2	4
(1)	$-\frac{5}{3}$	$-\frac{2}{3}$	$\frac{10}{3}$	$-\frac{41}{3}$	$-\frac{17}{3}$	$\frac{19}{3}$
(2)	6	$\frac{22}{5}$	$-\frac{18}{5}$	$\frac{49}{7}$	$\frac{21}{7}$	$-\frac{21}{7}$
(3)	$-\frac{5}{3}$	$-\frac{2}{3}$	$\frac{10}{3}$	$-\frac{41}{3}$	$-\frac{17}{3}$	$\frac{19}{3}$
(4)	4	$\frac{12}{5}$	$-\frac{4}{5}$	18	8	-7

Exercise No. 10

A. (3) The lines are the same.
   (4) The three lines meet in a common point.
   (5) The first two lines are the same line.
   (6) The first and third lines are parallel.
   (7) The three lines meet in three points.

B. (1) b and c are the same line. a and b meet at $(\frac{16}{11}, \frac{1}{11})$.
   (2) b and c are the same line. a and b meet at $(\frac{32}{29}, \frac{39}{29})$.
   (3) The three lines form a triangle with vertices

   $(-\frac{5}{19}, \frac{41}{19})$, $(\frac{9}{7}, \frac{16}{7})$, $(\frac{13}{27}, \frac{2}{27})$.

   (4) a and b are the same line. b and c meet at $(\frac{202}{23}, \frac{80}{23})$.
   (5) a and b meet at $(1, 2)$. c depends on them and goes through the same point.
   (6) a, b, and c are the same line.
   (7) a and c are parallel. a and b meet at $(1, 2)$, b and c at $(3, -1)$.
   (8) a, b, and c meet at a common point $(1, 3)$.
   (9) All three are parallel.
   (10) a, b, and c meet at a common point $(2, -1)$.

Exercise No. 11

(1) $x = \pm\frac{5}{7}$
(2) $x = \pm 3$
(3) $x = \pm 6$
(4) $x = \pm 11$
(5) $x = \pm 5$

Exercise No. 12

(1) $x = 3, 4$
(2) $x = -6, -1$
(3) $x = 10, -2$
(4) $x = 2, 4$
(5) $x = 2, -5$

Exercise No. 14

B.	Discriminant	Roots
(1)	1	$x = -2, -1$
(2)	17	$x = \dfrac{-3 \pm \sqrt{17}}{2}$
(3)	1	$x = 2, 1$
(4)	49	$x = 1, -\frac{2}{7}$
(5)	1	$x = \frac{4}{3}, 1$
(6)	-4	no real roots

Exercise No. 15

(1) $x = \dfrac{7 \pm \sqrt{89}}{4}$ (3) $x = \dfrac{2 \pm \sqrt{88}}{14}$

(2) $x = \dfrac{-5 \pm \sqrt{59}\,i}{6}$ (4) $x = \dfrac{1 \pm \sqrt{31}\,i}{8}$

(5) $x = 38.6$, the other root is negative.

Exercise No. 16

(1) $\dfrac{19 \pm \sqrt{89}}{8}$

(2) $x^2 = \dfrac{5 \pm \sqrt{13}}{2}$, $x = \pm\dfrac{\sqrt{5 \pm \sqrt{13}}}{2}$

(3) $x^2 = \dfrac{26 \pm 2\sqrt{144}}{2} = 13 \pm 12 = 25, 1 = \sqrt{25}, 1$

(4) $x = 2, \frac{1}{3}$

Exercise No. 17

Exercise No. 11

(1) — The parabola hits the x-axis at $\pm\frac{5}{7}$.

Exercise No. 14

(3) — The parabola hits the x-axis at 2 and 1.
(5) — The parabola hits the x-axis at $\frac{4}{3}$ and 1.
(6) — The parabola doesn't cross the x-axis.

Exercise No. 15

(2) — The parabola doesn't cross the x-axis.

(3) — The parabola crosses the x-axis at $\dfrac{1 \pm \sqrt{22}}{7}$.

Exercise No. 18

1. $1.6 = T$. The other answer is negative, *hence meaningless*.
2. $R = 8.385$ or -2.385. The negative answer is meaningless.
3. (a) $T = 0$, $T = 62.5$ sec.
   (b) 64.5 sec or -1.94 sec which can be discarded.

Exercise No. 19

(1) $y = \pm \sqrt{9 - x^2}$

(2) $y = \pm \sqrt{x - 1}$

(3) $y = \pm \sqrt{\dfrac{x^2 - 1}{2}}$

(4) $y = \dfrac{5}{x}$

(5) $y = \dfrac{-4 \pm \sqrt{4(x - 2)}}{2} = -2 \pm \sqrt{x - 2}$

(6) $y = \pm \sqrt{25 - (x - 1)^2} + 2$

(7) $y = \pm \sqrt{3 - \dfrac{3x^2}{4}}$

(8) $y = \pm x$

Exercise No. 20

(1) $y = \dfrac{-1 \pm \sqrt{5}}{2}, \; x = \dfrac{5 \pm \sqrt{5}}{2}$ They meet at two points.

(2) $x = \dfrac{-24 \pm \sqrt{26}}{10}, \; y = \dfrac{8 \pm 3\sqrt{26}}{10}$ They meet at two points.

(3) $(y = 0, x = 2), \; (y = -\tfrac{12}{7}, \; x = \tfrac{2}{7})$ The line crosses the curve at two points.

(4) The roots are imaginary, the curves do not meet.

Roots: $y = \dfrac{1 \pm \sqrt{39}\,i}{4}, \; x = \dfrac{-3 \pm \sqrt{39}\,i}{8}$

(5) $y = \dfrac{2 \pm \sqrt{14}}{2}, \; x = \dfrac{4 \pm \sqrt{14}}{2}$ The curves meet in two points.

Exercise No. 21

(1) $y = \pm 3\sqrt{7}\,i, \; x = \pm 2\sqrt{22}$ The curves do not meet.

(2) $x = \dfrac{1 \pm \sqrt{89}}{2}, \; y = \pm\sqrt{\dfrac{-1 - \sqrt{89}}{2}}, \; y = \pm\sqrt{\dfrac{-1 + \sqrt{89}}{2}}$

Curves meet at two points.

(3) $x = \tfrac{1}{2}, \; y = \dfrac{\pm \sqrt{35}}{2}$ The curves meet at two points.

(4) $(y = 0, x = 0), \; (y = 1, x = 1)$ The curves meet at two points.

(5) $x = \pm \dfrac{1}{\sqrt{2}}, \; y = \tfrac{1}{2}$ The curves meet at two points.

(6) $x = \pm \sqrt{\dfrac{9 \pm \sqrt{65}}{2}}, \; y = \dfrac{2}{x}$ The curves meet in four points.

Exercise No. 22

(1) $y = \pm \dfrac{8}{\sqrt{21}}, \; x = \pm 5\sqrt{\dfrac{5}{21}}$

(2) $x^2 = \tfrac{5200}{289}, y^2 = \tfrac{324}{289} \therefore x = \pm \sqrt{\tfrac{5200}{289}}, y = \pm \sqrt{\tfrac{324}{289}}$

(3) $x = \pm \sqrt{1 \pm \sqrt{15}}\,i, \; y = \dfrac{4}{\pm\sqrt{1 \pm\sqrt{15}\,i}}$

(4) $y = -\dfrac{1 \pm \sqrt{10001}}{50}$ Note two roots are imaginary.

Exercise No. 23

(2) $x = \pm 1, y = \pm 1$ Two points

(3) $y = \pm\tfrac{7}{3}, x = \pm\tfrac{2}{3}$

(4) $x = \dfrac{3 \pm \sqrt{5}}{2}, y = \dfrac{3 \mp \sqrt{5}}{4}$

(5) $x = \pm 2, y = \pm 3$

(6) $y = -1; \; y = \tfrac{3}{7};$
$x = -5; \; x = \tfrac{15}{7}; \; y = \pm 3\sqrt{3}, x = \pm 2\sqrt{3}$

(7) $x = -1; \; x = 4$
$y = -4; \; y = 1$

Exercise No. 24

(1) $x = \pm 2, y = \pm\tfrac{1}{3}$

(2) $x = \pm i\sqrt{5}; \; x = \pm 1$
$y = -4; \quad y = 2$

(3) $y = 3; \; y = -2 \left[y = \dfrac{-1 \pm \sqrt{21}}{2}; \; x = \dfrac{-1 \mp \sqrt{21}}{2} \right]$
$x = 3; \; x = -2$

(4) $x = 0, y = 0; \; x = 1, y = 1$

(5) $y = \pm 2, x = \mp 2$

Exercise No. 25

B.

	$w = 0$	-1	5
$z = \dfrac{19w - 25}{20}$	$-\tfrac{5}{4}$	$-\tfrac{11}{5}$	$\tfrac{7}{2}$
$x = \dfrac{75 - 21w}{20}$	$\tfrac{15}{4}$	$\tfrac{24}{5}$	$-\tfrac{3}{2}$
$y = \dfrac{175 + 27w}{20}$	$\tfrac{35}{4}$	$\tfrac{37}{5}$	$\tfrac{31}{2}$

C. (1) $x = \tfrac{4}{11}, y = \tfrac{9}{5}, z = -\tfrac{40}{9}, w = \tfrac{51}{5}$

(2) $x = \tfrac{23}{12}, y = -\tfrac{21}{2}, z = \tfrac{11}{4}, w = \tfrac{13}{6}$

D. The problem is indeterminate. If the answers are integers one of the motors is 3 H.P. The others are 2 and 1 H.P.

Exercise No. 26

A.

z	x	y
0	5	$\tfrac{65}{4}$
-5	0	$\tfrac{35}{2}$
25	30	10

B. $y = \dfrac{z}{5} + 10; \; x = \dfrac{2z}{5}$

z	x	y
0	0	10
20	8	14
50	20	20

C. (1) $x = 2, y = -1, z = 0, w = -3$

(2) $x = 1, y = -2, z = 1, w = 1$

(3) $x = 2, y = -1, z = 5, w = -1$

(4) $x = 1, y = -2, w = 4, z = -3$

Exercise No. 27

(1) Dependent Equations   (4) and (5) $y = 3, z = 5, x = 2$

(2) $x = -\tfrac{53}{7}, y = 0, z = -\tfrac{41}{7}, w = -\tfrac{11}{7}$ The same number as the number of resistances.

(3) Linearly inconsistent.

Exercise No. 30

(1) -5 (2) -36 (3) 15 (4) abc (5) 30 (6) $4(x - 1)(x + 2)$

Exercise No. 31

(1) 0 (2) 0 (3) 0

Exercise No. 33

C. (1) 0 (2) $x - 117$

Exercise No. 34

(1) The plane flies 350 m.p.h. The wind is 25 m.p.h.
(2) 400 lb. lead, 50 lb. zinc, 20 lb. tin and 30 lb. antimony.
(3) $x = 1, y = -2, w = -4, z = 3$
(4) The equations were

$$x + 2y - z + w = 1$$
$$3x + 3y + 2z + 5w = 5$$
$$5x + 4y + 3z - 6w = 7$$
$$-7x + 6y - 2z + 7w = 2$$

Exercise No. 35

D. $M = \frac{1}{2}, N = \frac{2}{3}$

Exercise No. 37

(1) $\sin x = 1 \quad \sin x = 0$
$\cos x = 0, \cos x = 1$

(2) $\sin x = 1 \quad \sin x = 0$
$\cos x = 0 \quad \cos x = -1$

(3) $\sin x = \pm i \dfrac{1}{\sqrt{2}}$
$\cos x = \pm \sqrt{\frac{3}{2}}$

(4) $\cot x = -1 \pm \sqrt{2}$
$\tan x = 1 \pm \sqrt{2}$

(5) $\cot x = \pm 1$
$\tan x = \pm 1$

(6) $\sin x = \dfrac{1 \pm \sqrt{7}\,i}{4}$

(7) $\sin x = \pm 2$

(8) $\tan x = \dfrac{1 \pm i\sqrt{3}}{2}$

Exercise No. 38

(1) $90°, 0°$
(2) $90°, -180°$
(3) Impossible
(4) $67° 30'$
(5) $45°$
(6) Impossible
(7) Impossible
(8) Impossible

Exercise No. 39

(1) $\pm 180°, \pm 540°, \pm 900°$
(2) $-90°, -450°, -810°, +270°, +630°, +990°$
(3) $0°, \pm 360°, \pm 720°$
(4) $45°, 405°, 765°, -315°, -675°$
(5) $225°, 585°, 945°, -135°, -495°, -855°$

Exercise No. 40

A. (1) $\dfrac{\pi}{6}$ (3) $\frac{41}{10}\pi$ (5) $\frac{5}{2}\pi$ (7) $\dfrac{\pi}{18}$

(2) $-\dfrac{\pi}{3}$ (4) $-\frac{5}{4}\pi$ (6) -4π (8) $-\dfrac{\pi}{90}$

B. (9) $45°$ (11) $30°$ (13) $720°$ (15) $5.7°$
(10) $-60°$ (12) $-540°$ (14) $-1620°$ (16) $-11.4°$

Exercise No. 41

B.

Quadrant	I	II	III	IV
Sin	+	+	−	−
Cos	+	−	−	+
Tan	+	−	+	−
Cot	+	−	+	−
Sec	+	−	−	+
Csc	+	+	−	−

C. (sin, csc), (cos, sec), (tan, cot), since they are reciprocals.
D. $\tan \theta_1 = -\tan(\pi - \theta_1) = \tan(\pi + \theta_1) = -\tan(2\pi - \theta_1)$
E. (1) $\sqrt{\frac{3}{2}} = .866$ (3) 1   (5) $\sqrt{\frac{3}{2}} = .866$ (7) -2
(2) -1 (4) 2   (6) $-\frac{1}{2}$ (8) 2

Exercise No. 42

A.

	OP_3	OP_4
x component	$r\cos\theta_3 = -r\cos\theta_1$	$r\cos\theta_4 = r\cos\theta_1$
y component	$r\sin\theta_3 = -r\sin\theta_1$	$r\sin\theta_4 = -r\sin\theta_1$
slope	$\tan\theta_3 = \tan\theta_1$	$\tan\theta_4 = -\tan\theta_1$

B.

	x component	y component	Slope
5	-7.07	-7.07	1
6	-10	0	0
7	-8.66	-5.00	$\dfrac{1}{\sqrt{3}} = .577$
8	0	-10	doesn't exist
9	7.07	-7.07	-1
10	-8.66	$+5.00$	$-\dfrac{1}{\sqrt{3}} = -.577$

C. 17.32 lbs. up, 10 lbs. out of the wall; if the pole were parallel to the window, or 0°.

Exercise No. 43

A. (1) $-x = 90°, x = 0°,$ etc.
(2) $x = 90°, x = 180°,$ etc.
(3) Impossible
(4) $67° 30', 247° 30', -22° 30', 337° 30', 157° 30'$
(5) $x = 45°, 135°, 225°, 315°,$ etc.
(6) Impossible
(7) Impossible
(8) Impossible

B. $x = \sin^{-1} \frac{1}{2}, x = \sin^{-1} -2$ (impossible), $y = \sin^{-1}\frac{1}{2} = 30°$, 150°, etc.

Exercise No. 44

Employ diagrams in work.
A. $38° 10' = .213\pi$
B. $42° = .23\pi$

Exercise No. 45

(1) $0°, 180°, 360°, 60°, 300°$
(2) $45°, 225°$
(3) $220°$
(4) $0°, 270°$
(5) $90°, 270°$
(6) $90°, 270°$

(1)

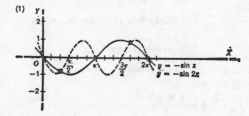

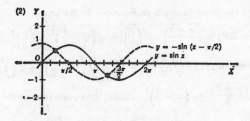

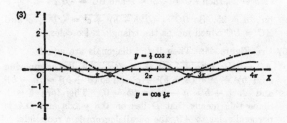

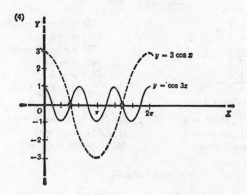

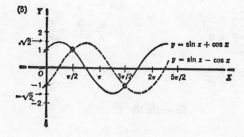

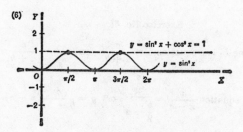

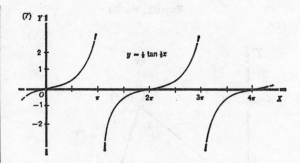

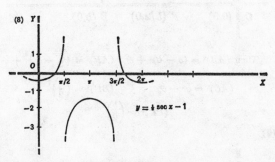

The graphic solutions of Examples 1 to 6 are the intersection points of each pair of curves.

Exercise No. 46

Midpoint	Tri-section Point
1. $(-31.5, -11)$	$(-27, -9), (-36, -13)$
2. $(-22.5, 1)$	$(-21, -1), (-24, 3)$
3. $(-9, 7)$	$(-12, 3), (-6, 11)$
4. $(-13.5, 13)$	$(-18, 11), (-9, 15)$
5. $(9, 7)$	$(6, 11), (12, 3)$
6. $(22.5, 1)$	$(27, -1), (36, 3)$
7. $(31.5, -11)$	$(27, -9), (36, -13)$
8. $(36, -5)$	$(33, -1), (39, -9)$
9. $(8, 12.5)$	

Exercise No. 47

A. (1) 13   (2) 6   (3) $\sqrt{205}$

B. (4) $3\sqrt{97}$ (5) 30 (6) 30 (7) 15 (8) 15 (9) $3\sqrt{97}$ (10) 30

Exercise No. 48

(1) $\frac{1}{2}$ (3) doesn't exist   (5) $1.732 = \sqrt{3}$ (7) .14

(2) $-\frac{2}{7}$ (4) 0   (6) $\frac{40}{9}$

Exercise No. 49

A. (1) $\tan \theta = .595$; $\theta = 30.8°$ (2) $90°$

B. (3) $\tan \theta = +\frac{40}{11}$ (4) $\tan \theta = .5586$; $\theta = 29°10'$

     (5) $\tan \theta = .5586$; $\theta = 29.2°$

Exercise No. 50

(4)

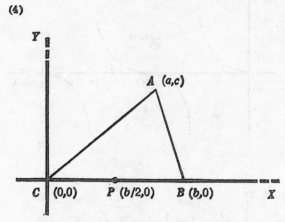

Then $(AB)^2 = (a-b)^2 + c^2$ $(AP)^2 = \left(a - \dfrac{b}{2}\right)^2 + c^2$

$(AC)^2 = a^2 + c^2$ $(BP)^2 = \left(\dfrac{b}{2}\right)^2$

$(PC)^2 = \left(\dfrac{b}{2}\right)^2$

(5)

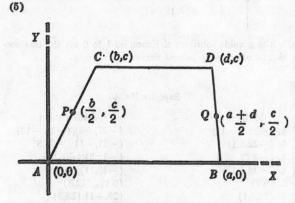

Then PQ is parallel to AB

and $(PQ)^2 = \left(\dfrac{a+d-b}{2}\right)^2$; $PQ = \dfrac{a+d-b}{2}$

$AB = a$, $CD = d - b$

(6) See diagram 47 with A: $(0,0)$; B: $(a,0)$; C: (b,c)

$$P: \left(\dfrac{b}{2}, \dfrac{c}{2}\right) \quad Q: \left(\dfrac{a+b}{2}, \dfrac{c}{2}\right)$$

If $AC = BC$; then $\sqrt{b^2 + c^2} = \sqrt{(b-a)^2 + c^2}$ and $b^2 + c^2 = (b-a)^2 + c^2$. Therefore $b^2 + c^2 = b^2 - 2ab + a^2 + c^2$ and subtracting gives $0 = a^2 - 2ab = a(a - 2b)$; so either $a = 0$ or $a = 2b$. Since a is not zero the second holds. BP and AQ are the medians. Then $BP =$

$$\sqrt{\left(a - \dfrac{b}{2}\right)^2 + \left(\dfrac{c}{2}\right)^2} \text{ and } AQ = \sqrt{\left(\dfrac{a+b}{2}\right)^2 + \left(\dfrac{c}{2}\right)^2}.$$ But

$a = 2b$ and substituting we have:

$$BP = \sqrt{\left(2b - \dfrac{b}{2}\right)^2 + \left(\dfrac{c}{2}\right)^2} = \sqrt{\left(\dfrac{3b}{2}\right)^2 + \left(\dfrac{c}{2}\right)^2} \text{ and}$$

$$AQ = \sqrt{\left(\dfrac{2b+b}{2}\right)^2 + \left(\dfrac{c}{2}\right)^2} = \sqrt{\left(\dfrac{3b}{2}\right)^2 + \left(\dfrac{c}{2}\right)^2}$$

(7) Using the same diagram as Example 6: we have

$$AQ = BP; \sqrt{\left(\dfrac{a+b}{2}\right)^2 + \left(\dfrac{c}{2}\right)^2} = \sqrt{\left(a - \dfrac{b}{2}\right)^2 + \left(\dfrac{c}{2}\right)^2}.$$

Squaring we have $\left(\dfrac{a+b}{2}\right)^2 + \left(\dfrac{c}{2}\right)^2 = \left(a - \dfrac{b}{2}\right)^2 + \left(\dfrac{c}{2}\right)^2$

and $\left(\dfrac{a+b}{2}\right)^2 = \left(a - \dfrac{b}{2}\right)^2$. Taking the square root we get

$\dfrac{a+b}{2} = a - \dfrac{b}{2}$, multiplying by 2 we get $a + b = 2a - b$

or $a = 2b$, then $AC = \sqrt{b^2 + c^2}$ and $BC = \sqrt{(a-b)^2 + c^2}$

but $a = 2b$. So $BC = \sqrt{(2b-b)^2 + c^2} = \sqrt{b^2 + c^2}$ and $AC = BC$ which means the triangle is isosceles.

(8) Use Figure 45: Then if the diagonals are equal $AC = BD$: $\sqrt{(a+b)^2 + c^2} = \sqrt{(a-b)^2 + c^2}$ or squaring $(a+b)^2 + c^2 = (a-b)^2 + c^2$ and so $(a+b)^2 = (a-b)^2$ and $\therefore$ $a+b = a-b$ or $2b = 0$. Therefore $b = 0$. Since this means that D lies on the y-axis and DA is perpendicular to AB, the parallelogram is a rectangle.

Exercise No. 51

A. (1) $\dfrac{y-2}{x-1} = 2$ (5) $y = -2x - 14$

     (2) $\dfrac{y-6}{x+4} = -1$ (6) $y = x$

                                   (7) $y = -3x + 15$

     (3) $\dfrac{y+4}{x-5} = 3$ (8) $y = \frac{1}{2}x - \frac{7}{2}$

                                   (9) $y = 5$

     (4) $y = 2x - 4$ (10) $\dfrac{y+2}{x+1} = \dfrac{1}{4}$

B. (1) $(0,3)$, slope $= 0$ (4) $(0,-2)$, slope $= -\frac{2}{3}$

     (2) $(0,\frac{7}{3})$, slope $= \frac{2}{3}$ (5) $(0,\frac{9}{2})$, slope $= \frac{1}{2}$

     (3) doesn't exist   (6) $(0,-10)$, slope $= 3$

Exercise No. 52

A. (1) $y = -\frac{1}{3}x + 1$

     (2) $y = -\frac{1}{2}x - 2$ (6) $\dfrac{y+7}{x-4} = 2$

     (3) $y = -x - 5$

     (4) $\dfrac{y-7}{x-4} = -2$ (7) $\dfrac{y-4}{x+7} = -\dfrac{7}{13}$

     (5) $\dfrac{x-7}{x+4} = 2$ (8) $\dfrac{y+5}{x+18} = \dfrac{4}{9}$

B. (9) $\dfrac{4}{9} = \dfrac{y+5}{x+18}$ (10) $\dfrac{y-19}{x} = \dfrac{4}{9}$

Exercise No. 53

(a) a' (d) b'

(b) c', d', e' (e) c'

(c) None

Exercise No. 54

(1) The line L_1 has equation $\dfrac{y}{x} = \dfrac{c}{a+b}$ or $y = \dfrac{c}{a+b} x$

L_2 has equation $\dfrac{y}{x-a} = \dfrac{\frac{c}{2}}{\frac{b}{2} - a}$ or $y = \dfrac{c}{b - 2a}(x - a)$.

The intersection of L_1 and L_2 is obtained from $\frac{c}{a+b}x =$ $\frac{c}{b-2a}(x-a)$. Dividing by c and multiplying out $(b-2a)x = (a+b)(x-a)$ we get that $bx - 2ax = ax + bx - a^2 - ab - 3ax = -a^2 - ab$, or dividing by $-3a$, $x = \frac{a+b}{3}$. Therefore $y = \frac{c}{a+b}x = \frac{c}{a+b} \cdot \frac{a+b}{3}$ $= \frac{c}{3}$. This divides L_1 from $(0,0)$ to $\left(\frac{a+b}{2}, \frac{c}{2}\right)$ in the ratio of $2:1$. And so for all the medians.

(2) We need to find both points and show they are the same.

The diagonal AC has equation $\frac{y}{x} = \frac{c}{d}$ or $y = \frac{c}{d}x$. The diagonal BD has the equation $\frac{y}{x-a} = \frac{c}{b-a}$ or $y = \frac{c}{b-a}(x-a)$. They meet when $\frac{c}{d}x = \frac{c}{b-a}(x-a)$. Dividing and clearing fractions $(b-a)x = (x-a)d$ or $bx - ax = xd - ad$ or $ad = x(d+a-b)$ and $x = \frac{ad}{d+a-b}$. Then $y = \frac{c}{d}x = \frac{c}{d}\frac{ad}{d+a-b} = \frac{ac}{d+a-b}$.

If this point lies on the line joining the mid-points, the proposition is proved. The line joining the mid-points of the parallel sides, $\left(\frac{d+b}{2}, c\right)$ and $\left(\frac{a}{2}, 0\right)$ has equation $\frac{y}{x-\frac{a}{2}} = \frac{c}{\frac{d+b-a}{2}}$ or $y = \frac{2c}{d+b-a}\left(x - \frac{a}{2}\right)$: when $x = \frac{ad}{d+a-b}$ we get $y = \frac{2c}{d+b-a}\left[\frac{ad}{d+a-b} - \frac{a}{2}\right]$ $= \frac{c}{d+b-a}\left[\frac{ad - a^2 + ab}{(d+a-b)}\right] = \frac{ac}{d+a-b}$.

(3) See 2 above: $\left(\frac{ad}{d+a-b}, \frac{ac}{d+a-b}\right)$ is the common intersection point.

(4)

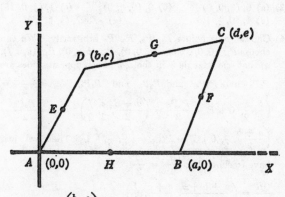

Then E: $\left(\frac{b}{2}, \frac{c}{2}\right)$

F: $\left(\frac{d+a}{2}, \frac{e}{2}\right)$

G: $\left(\frac{d+b}{2}, \frac{c+e}{2}\right)$

H: $\left(\frac{a}{2}, 0\right)$

The line connecting EF has equation

$\frac{y - \frac{c}{2}}{x - \frac{b}{2}} = \frac{\frac{c-e}{2}}{\frac{b-(d+a)}{2}}$ or $y - \frac{c}{2} = \frac{c-e}{b-(d+a)}\left(x - \frac{b}{2}\right)$

The line connecting GH

$\frac{y}{x - \frac{a}{2}} = \frac{\frac{c+e}{2}}{\frac{d+b-a}{2}}$ or $y = \frac{c+e}{d+b-a}\left(x - \frac{a}{2}\right)$

When they meet

$\frac{c+e}{d+b-a}\left(x - \frac{a}{2}\right) - \frac{c}{2} = \frac{c-e}{d+b-(a)}\left(x - \frac{b}{2}\right)$

Multiplying by $2(d+b-a)(b-d-a)$ we get

$2(b-d-a)(c+e)\left(x - \frac{a}{2}\right) - c(d+b-a)(b-d-a)$

$= 2(d+b-a)(c-e)\left(x - \frac{b}{2}\right)$ or $2x(b-d-a)(c+e)$

$- a(b-d-a)(c+e) - c(d+b-a)(b-d-a)$

$= 2x(d+b-a)(c-e) - b(d+b-a)(c-e)$.

Clearing parenthesis and collecting terms we get:

$$x = \frac{a+b+d}{4}, \quad y = \frac{c+e}{4}$$

Exercise No. 55

A. The denominators are zero.

B. They have the same slope and same y-intercept.

C. (1) parallel   (2) parallel   (3) parallel   (4) not parallel

D. (1) $2x - 3y + 13 = 0$ (4) $y = 3$
   (2) $3x - 2y + 5 = 0$ (5) $y = x$
   (3) $4x - 7y - 53 = 0$

Exercise No. 56

C. Only the lines of problem 4 are perpendicular.

D. (1) $\frac{y-3}{x+2} = -\frac{3}{2}$ (3) $\frac{y+3}{x-8} = -\frac{7}{4}$ (5) $\frac{y+2}{x+2} = -1$
   (2) $\frac{y-1}{x+1} = -\frac{2}{3}$ (4) $x = -1$

Exercise No. 57

A. (1) $\frac{5}{\sqrt{13}}$ (3) 7 (5) $\frac{7}{\sqrt{5}}$

   (2) 5 (4) $\frac{3}{\sqrt{5}}$ (6) $\sqrt{5}$

B. (1) $\frac{1}{\sqrt{13}}$ (2) 0 (3) $\frac{5}{7}$

C. (1) $\frac{5}{\sqrt{13}}$ (3) $\frac{9}{\sqrt{10}}$ (5) $\frac{3}{\sqrt{5}}$

   (2) $\frac{4}{\sqrt{41}}$ (4) 5

Exercise No. 58

(1) $\left(b, \dfrac{b(a-b)}{c}\right)$

(2) (3) The lines go through $\left(\dfrac{b-a}{2}, \dfrac{c}{2}\right)$ with slope $\left(\dfrac{a}{2}, 0\right)$

no slope, and $\left(\dfrac{b}{2}, \dfrac{c}{2}\right)$ with slope $-\dfrac{b}{c}$. They intersect

at $\left(\dfrac{a}{2}, \dfrac{b^2 - ab + c^2}{2c}\right)$.

(4) Expand the determinant.

(5) $P_1P_2 = \sqrt{(x_1 - x_2)^2 + (y_1 - y_2)^2}$ The line through P_1P_2

has equation $\dfrac{y - y_1}{x - x_1} = \dfrac{y_1 - y_2}{x_1 - x_2}$ or $y - y_1 = \dfrac{y_1 - y_2}{x_1 - x_2}(x - x_1)$

or $(x_1 - x_2)(y - y_1) = (y_1 - y_2)(x - x_1)$.   Collecting

terms $(y_1 - y_2)x - (x_1 - x_2)y - x_1(y_1 - y_2) + y_1(x_1 - x_2)$.

By formula 615: the distance from (x_3, y_3) to this is

$h = \dfrac{(y_1 - y_2)x_3 - (x_1 - x_2)y_3 - x_1(y_1 - y_2) + y_1(x_1 - x_2)}{\sqrt{(y_1 - y_2)^2 + (x - x_2)^2}}$.

The area of a triangle is $\frac{1}{2}h \times P_1P_2$

$= \frac{1}{2}[(y_1 - y_2)x_3 - (x_1 - x_2)y_3 - x_1(y_1 - y_2)$

$+ y_1(x_1 - x_2)]$

$= \frac{1}{2}[(y_1 - y_2)x_3 - (x_1 - x_2)y_3 - x_1y_1 + x_1y_2 + x_1y_1$

$- y_1x_2] = \frac{1}{2}\left[x_3\begin{vmatrix} y_1 & 1 \\ y_2 & 1 \end{vmatrix} - y_3\begin{vmatrix} x_1 & 1 \\ x_2 & 1 \end{vmatrix} + 1\begin{vmatrix} x_1 & y_1 \\ x_2 & y_2 \end{vmatrix}\right]$

$= \frac{1}{2}\begin{vmatrix} x_1 & y_1 & 1 \\ x_2 & y_2 & 1 \\ x_3 & y_3 & 1 \end{vmatrix}$

(6) $P_1P_2P_3 : 216$
$P_4P_2P_2 : 216$
$P_3P_4P_5 : 432$
$P_5P_6P_4 : 216$
$P_6P_5P_7 : 216$

Exercise No. 59

A. (1) $x = -\frac{7}{4}$, $(\frac{7}{4}, 0)$ (3) $x = +\frac{21}{4}$ $(-\frac{21}{4}, 0)$

    (2) $y = -\frac{9}{2}$ $(0, \frac{9}{2})$ (4) $y = \frac{73}{4}$ $(0, -\frac{73}{4})$

B. (1) $y^2 = 20x$ (3) $y^2 = 2x - 3$

    (2) $x^2 = -12y$

       (4) $x^2 + y^2 - 4x - 44y + 244 + 2xy = 0$

Exercise No. 60

(1) (a) $6\sqrt{5}$ (b) $6\sqrt{15}$

(2) (a) $\frac{100}{9}$; $\frac{481}{36} = 13.35''$ (b) $\frac{225}{36} = 6.25$; $8.5''$

(3) (a) $.04'$ (b) $36'$ (c) $64'$

(4) (a) $50\sqrt{10} = 158'$ (b) 3541 (c) $474'$

(5) (a) 3.6 (b) 1.6 (c) 0.016

(6) (a) 2.5 (b) 17.7 (c) $15\sqrt{2.5} = 23.75$

(7) (a) $x^2 = 64,000,000y$ (b) $32,000,000$ miles

Exercise No. 61

B. (1) $\dfrac{x^2}{100} + \dfrac{y^2}{96} = 1$

    x intercepts y intercepts
      ± 10 $\pm 4\sqrt{6}$

    (2) $\dfrac{y^2}{64} + \dfrac{x^2}{60} = 1$

    x intercepts y intercepts
   $\pm 2\sqrt{15}$ ± 8

(3) $\dfrac{(y - \frac{5}{2})^2}{\frac{25}{16}} + \dfrac{(x - 1)^2}{\frac{21}{16}} = 1$

Center at $(1, \frac{5}{2})$. Long axis parallel to the y-axis

length $\frac{5}{2}$, short axis parallel to the x-axis, length $\dfrac{\sqrt{21}}{2}$

(4) $\dfrac{x^2}{16} + \dfrac{y^2}{9} = 1$

    x intercepts y intercepts
     ± 4 ± 3

(5) $\dfrac{y^2}{9} + \dfrac{x^2}{18} = 1$

    x intercepts y intercepts
   $\pm 3\sqrt{2}$ ± 3

Exercise No. 62

(1) $19\sqrt{3}$ (3) $91\frac{1}{2}$ million miles, $94\frac{1}{2}$ million miles

(2) $19\sqrt{15}$

Exercise No. 63

(1) 24 (2) 20 (5) $x^2 + y^2 = k^2$ a circle of radius k.

Exercise No. 64

(1) $c = \dfrac{\sqrt{41}}{4}$, $e = \dfrac{\sqrt{41}}{5}$ foci: $\left(\dfrac{\sqrt{41}}{4}, 0\right)$ & $\left(-\dfrac{\sqrt{41}}{4}, 0\right)$ and

$c = \dfrac{\sqrt{41}}{4}$, $e = \dfrac{\sqrt{41}}{5}$ foci: $\left(0, \dfrac{\sqrt{41}}{4}\right)$ & $\left(0, -\dfrac{\sqrt{41}}{4}\right)$

(2) 2 by $2\frac{1}{2}$; same shape as large rectangle.

(3) $\frac{5}{2}\sqrt{35} = 14.8$ ft.

Exercise No. 65

(1) A, O, B, D (4) $Q(-x_1, -y_1, z_1)$; $R(x_1, -y_1, 0)$;

(2) C, E, B, O $S(x_1, -y_1, -z_1)$; $T(x_1, 0, -z_1)$

(3) C, F, A, O (5) $(-x_1, -y_1, -z_1)$

Exercise No. 66

(1) $B: (x_2\ y_2, 0)$, $C: (x_1, y_2, 0)$, $D: (x_1, y_1, 0)$, $E: (x_2, y_1, z_1)$,
$F: (x_1, y_2, z_1)$, $Q: (x_2, y_2, z_1)$, $P_1: (x_1, y_1, z_1)$, $P_2: (x_2, y_2, z_2)$,
$I: (x_1, y_2, z_2)$, $J: (x_1, y_1, z_2)$, $H: (x_2, y_1, z_2)$, $K: (x_2, 0, z_2)$,
$L: (x_2, 0, z_1)$, $M: (x_1, 0, z_1)$, $N: (x_1, 0, z_2)$, $R: (0, y_2, z_1)$,
$S: (0, y_2, z_2)$.

(2) (a) $9, (\frac{9}{2}, 0, 0)$ (b) $9, (0, \frac{9}{2}, 0)$ (c) $1, (\frac{1}{2}, 0, 15)$
(d) $1, (0, \frac{1}{2}, 15)$ (e) $\sqrt{307}, (\frac{1}{2}, \frac{9}{2}, \frac{15}{2})$

(4) Choose your points P_1, P_2, P_3, P_4 arbitrarily. We can choose $P_1: (0, 0, 0)$, $P_2: (a, 0, 0)$, $P_3: (b, c, 0)$, $P_4: (d, e, f)$, so that one triangle is in the x-y plane, opposite sides are

P_1P_3 and P_2P_4, and P_2P_3 and P_1P_4. $Q = \dfrac{P_1P_3}{2} =$

$\left(\dfrac{b}{2}, \dfrac{c}{2}, 0\right)$, $R = \dfrac{P_2P_4}{2} = \left(\dfrac{a + d}{2}, \dfrac{e}{2}, \dfrac{f}{2}\right)$. $S = \dfrac{P_2P_3}{2} =$

$\left(\dfrac{a + b}{2}, \dfrac{c}{2}, 0\right)$, $T = P_1P_4 = \left(\dfrac{d}{2}, \dfrac{e}{2}, \dfrac{f}{2}\right)$. Then we just

have to show that $\dfrac{QR}{2} = \dfrac{ST}{2}$.

$\dfrac{QR}{2} = \left(\dfrac{a + b + d}{4}, \dfrac{e + c}{4}, \dfrac{f}{4}\right)$

$\dfrac{ST}{2} = \left(\dfrac{a + b + d}{4}, \dfrac{c + e}{4}, \dfrac{f}{4}\right)$

and so they meet at their mid-points.

(5) We can choose as our vertices P_1: $(0, 0, 0)$, P_2: $(a, 0, 0)$, P_3: $(0, b, 0)$, P_4: $(0, 0, c)$, P_5: $(a, b, 0)$, P_6: (a, b, c), P_7: $(0, b, c)$, P_8: $(a, 0, c)$. The diagonals are P_1P_6, P_2P_7, P_4P_5, and P_3P_8. $\dfrac{P_1P_6}{2} = \left(\dfrac{a}{2}, \dfrac{b}{2}, \dfrac{c}{2}\right)$, $\dfrac{P_3P_8}{2} = \left(\dfrac{a}{2}, \dfrac{b}{2}, \dfrac{c}{2}\right)$, $\dfrac{P_4P_5}{2} = \left(\dfrac{a}{2}, \dfrac{b}{2}, \dfrac{c}{2}\right)$ and therefore they all have a common mid-point.

(6) Choose coordinates P_1: $(0, 0, 0)$, P_2: $(a, 0, 0)$, P_3: $(b, c, 0)$, P_4: (d, e, f). This will include the two-dimensional case where $f = 0$. Therefore we only have to prove the three-dimensional case. (See example 4). There the lines joining mid-points of opposite sides are QR and ST.

$$(QR)^2 = \left(\frac{a+d-b}{2}\right)^2 + \left(\frac{e-c}{2}\right)^2 + \left(\frac{f}{2}\right)^2$$

$$(ST)^2 = \left(\frac{a+b-d}{2}\right)^2 + \left(\frac{e-c}{2}\right)^2 + \left(\frac{f}{2}\right)^2$$

The diagonals are P_1P_3 and P_2P_4
$(P_1P_3)^2 = a^2$
$(P_2P_4)^2 = (b-d)^2 + (c-e)^2 + f^2$

$$(QR)^2 + (ST)^2 = \frac{a^2 + d^2 + b^2 + 2ad - 2ab - 2bd}{4} +$$

$$\frac{a^2 + b^2 + d^2 - 2ad - 2bd + 2ab}{4} + \frac{2(e-c)^2}{4} + \frac{2f^2}{4}$$

$$= \frac{2a^2 + 2d^2 + 2b^2 - 4bd}{4} + 2\frac{(e-c)^2}{4} + \frac{2f^2}{4} \quad \text{so}$$

$2[(QR)^2 + (ST)^2] = a^2 + d^2 + b^2 - 2bd + (e-c)^2 + f^2$ and $(P_1P_3)^2 + (P_2P_4)^2 = a^2 + (b^2 - 2bd + d^2) + (c-e)^2 + f^2$ and so the statement is true.

Exercise No. 67

B. (1) $\cos\alpha = \dfrac{1}{\sqrt{307}}$

$\cos\beta = \dfrac{-9}{\sqrt{307}}$

$\cos\gamma = \dfrac{15}{\sqrt{307}}$

(2) $\cos\alpha = \dfrac{-9}{\sqrt{307}}$

$\cos\beta = \dfrac{1}{\sqrt{307}}$

$\cos\gamma = \dfrac{15}{\sqrt{307}}$

(3) $\cos\alpha = -\frac{8}{17}$
$\cos\beta = 0$
$\cos\gamma = \frac{15}{17}$

(4) $\cos\alpha = \dfrac{-1}{\sqrt{2}}$ $\quad \alpha = \dfrac{3\pi}{4} = 135°$

$\cos\beta = \dfrac{1}{\sqrt{2}}$ $\quad \beta = \dfrac{\pi}{4} = 45°$

$\cos\gamma = 0$ $\quad \gamma = \dfrac{\pi}{2} = 90°$

(5) $\cos\alpha = \dfrac{-1}{\sqrt{2}}$ $\quad \alpha = 135°$

$\cos\beta = \dfrac{1}{\sqrt{2}}$ $\quad \beta = 45°$

$\cos\gamma = 0$ $\quad \gamma = 90°$

(6) $\cos\alpha = 1$ $\quad \alpha = 0°$
$\cos\beta = 0$ $\quad \beta = 90°$
$\cos\gamma = 0$ $\quad \gamma = 90°$

(7) $\cos\alpha = 0$
$\cos\beta = 1$
$\cos\gamma = 0$

(8) $\cos\alpha = 1$
$\cos\beta = 0$
$\cos\gamma = 0$

C. (1) $\cos w = \dfrac{297}{17\sqrt{307}} = .995;\ w = 5°45'$

(2) $\cos w = .995;\ w = 5°45'$

(3) $\cos w = \dfrac{8}{17\sqrt{2}} = 0.333;\ w = 70°33'$

(4) $\cos w = \frac{207}{307} = 0.675;\ w = 47°32'$

D. Lines must be perpendicular to one of the axes.

E. When they are parallel to an axis.

F. No, the squares must add up to 1.

G. By definition of cosine.

H. $\cos w = \dfrac{1}{\sqrt{3}};\ \therefore\ w = 54°42'$

Exercise No. 68

A. (1) $3(x-2) - (y-1) + 4(z+2) = 0$ or
$3x - y + 4z = -3$

(2) $3(x+3) - z = 0$ or $3x - z = -9$

B. (3) $2(x+1) - 3(y-2) + (z+3) = 0$ or
$2x - 3y + z = -11$

(4) $y + 1 = 0$

(5) $2(x+1) - (y+2) + (z+1) = 0$ or
$2x - y + z + 1 = 0$

C. If we form a parallelopiped from the tetrahedron $P_1 P_2 P_3 P_4$, by making parallelograms out of each face, then the tetrahedron has $\frac{1}{6}$ the volume of the parallelopiped. It remains only to find the volume of this. The volume of a parallelopiped is the area of the base B times the altitude H. If we let $P_1 P_2 P_3$ form the base and P_4 be the opposite vertex, then it will be the area of the parallelogram formed on $P_1 P_2 P_3$ multiplied by the distance of P_4 to the plane of $P_1 P_2 P_3$. The equation of the plane is:

$\begin{vmatrix} x & y & z & 1 \\ x_1 & y_1 & z_1 & 1 \\ x_2 & y_2 & z_2 & 1 \\ x_3 & y_3 & z_3 & 1 \end{vmatrix} = 0.$ This can be rewritten as $Ax + By + Cz + D = 0$ where

$A = \begin{vmatrix} y_1 & z_1 & 1 \\ y_2 & z_2 & 1 \\ y_3 & z_3 & 1 \end{vmatrix}$, $B = \begin{vmatrix} x_1 & z_1 & 1 \\ x_2 & z_2 & 1 \\ x_3 & z_3 & 1 \end{vmatrix}$, $C = \begin{vmatrix} x_1 & y_1 & 1 \\ x_2 & y_2 & 1 \\ x_3 & y_3 & 1 \end{vmatrix}$

and $D = \begin{vmatrix} x_1 & y_1 & z_1 \\ x_2 & y_2 & z_2 \\ x_3 & y_3 & z_3 \end{vmatrix}$. The height H is given by

$$\pm \frac{\begin{vmatrix} x_4 & y_4 & z_4 & 1 \\ x_1 & y_1 & z_1 & 1 \\ x_2 & y_2 & z_2 & 1 \\ x_3 & y_3 & z_3 & 1 \end{vmatrix}}{\sqrt{A^2 + B^2 + C^2}} = H$$

We now have to find the area of B. B is a parallelogram, and the area of a parallelogram is given by the height h times the base b. Consider $P_1 P_2$ as the base and the distance from P_1 to $P_2 P_3$ as the height h. However h is equal to the length $P_1 P_3$ times the $\sin\theta$ where θ is the angle between $P_1 P_2$ and $P_2 P_3$. Therefore:

$B = P_1P_2 \times P_2P_3 \times \sin\theta =$

$\sqrt{(x_1 - x_2)^2 + (y_1 - y_2)^2 + (z_1 - z_2)^2}$

$\times \sqrt{(x_2 - x_3)^2 + (y_2 - y_3)^2 + (z_2 - z_3)^2} \times \sin\theta.$

We know that $\cos\theta =$

$$\frac{(x_1 - x_2)(x_2 - x_3) + (y_1 - y_2)(y_2 - y_3) + (z_1 - z_2)(z_2 - z_3)}{\sqrt{(x_1-x_2)^2(y_1-y_2)^2+(z_1-z_2)^2} \cdot \sqrt{(x_2-x_3)^2+(y_2-y_3)^2+(z_2-z_3)^2}}$$

But $\sin\theta =$

$$\sqrt{1 - \cos^2\theta} = \frac{\sqrt{A^2 + B^2 + C^2}}{P_1P_2 \times P_2P_3} \qquad \text{and so}$$

$$B = \sqrt{A^2 + B^2 + C^2}$$

Therefore $H \cdot B = \pm \begin{vmatrix} x_4 & y_4 & z_4 & 1 \\ x_1 & y_1 & z_1 & 1 \\ x_2 & y_2 & z_2 & 1 \\ x_3 & y_3 & z_3 & 1 \end{vmatrix}$,   with the $\pm$ sign,

to indicate that we want a positive number as the answer. The determinant changes sign if we interchange two rows which means when we change in what order the points are taken.

Exercise No. 69

A. $(2,3,0)$ $(0,1,\tfrac{4}{3})$ $(-1,0,2)$

B. $\cos A_1 = \dfrac{127}{3\sqrt{5}\sqrt{411}}$ $\cos A_2 = \dfrac{-79}{\sqrt{411}\sqrt{118}}$

$\cos A_3 = \dfrac{452}{27\sqrt{5}\sqrt{118}}$

INDEX

Wesley Gibson leads the life of quiet desperation. His dead-end job numbs him, his girlfriend cheats on him and his hypochondria convinces him he has every disease from cancer to the common cold. His world is the depth of dismay.

But there is another world, one buried just inches below what Wesley thinks is his life. What happens when the scales fall from your eyes and the real clockwork of the world is laid bare? What happens when you're Wesley Gibson, one minute the most downtrodden wretch the world has ever seen, and the next... you're Wanted?

WANTED created by **Mark Millar and J.G. Jones**

WANTED issues #1-#6
Written by **Mark Millar**
Pencils and Inks by **J.G. Jones**
Colors by **Bongotone's Paul Mounts**
Flashback sequences pgs. 6-10 of **WANTED #6** Penciled and Inked by **Dick Giordano**
For the original editions: **WANTED issues #1-6** lettering by
Dreamer Design's Robin Spehar, Dennis Heisler and Mark Roslan

For this edition
    Book Design by **Jason Medley**
    Cover Art by **J. G. Jones**

ISBN # 978-1-58240-497-4
Published by Image Comics®
Wanted Vol.1 2007 Third Printing. Office of Publication: 1942 University Ave., Suite 305, Berkeley, CA 94704. Originally published as WANTED issues #1-6 and WANTED: Dossier. WANTED is ™
and © 2007 Mark Millar and J.G. Jones. "WANTED," its logos, all related characters and their likenesses are registered trademarks of Mark Millar and J.G. Jones. The entire contents of this
book are © 2007 Top Cow Productions, Inc. The characters, events and stories in this publication are entirely fictional. With the exception of art used for review purposes, none of the contents
of this book may be reprinted in any form without the express written consent of Top Cow Productions, Inc.
Printed in Canada

Special Thanks to Wizard for permission to reprint Materials from the **WANTED #1 Wizard Ace Edition**™ variant.

What did you think of this book? We love to hear from our readers. Please e-mail us at: **fanmail@topcowent.com**

or write us at: **WANTED Letters**
c/o Top Cow Productions, Inc.
10350 Santa Monica Blvd., Suite #100
Los Angeles, CA 90025

for **Top Cow Productions, Inc.:**
Marc Silvestri_Chief Executive Officer
Matt Hawkins_President and Chief Operating Officer
Rob Levin_VP - Editorial
Filip Sablik_VP - Marketing & Sales
Chaz Riggs_Graphic Design
Phil Smith_Managing Editor
Joshua Cozine_Assistant Editor
Alyssa Phung_Controller
Adrian Nicita_Webmaster

for Image Comics
Erik Larsen
publisher
Eric Stephenson
creative director

visit us on the web at **www.topcow.com** and **www.topcowstore.com**

and for more on Mark Millar go to **www.millarworld.tv**

To find the comics shop nearest you call **1-888-COMICBOOK**

TABLE OF CONTENTS

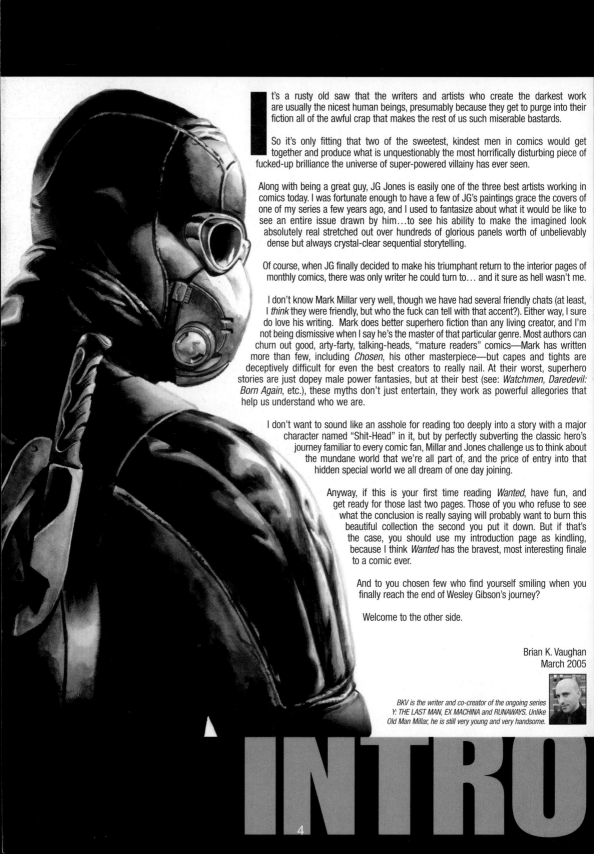

t's a rusty old saw that the writers and artists who create the darkest work are usually the nicest human beings, presumably because they get to purge into their fiction all of the awful crap that makes the rest of us such miserable bastards.

So it's only fitting that two of the sweetest, kindest men in comics would get together and produce what is unquestionably the most horrifically disturbing piece of fucked-up brilliance the universe of super-powered villainy has ever seen.

Along with being a great guy, JG Jones is easily one of the three best artists working in comics today. I was fortunate enough to have a few of JG's paintings grace the covers of one of my series a few years ago, and I used to fantasize about what it would be like to see an entire issue drawn by him...to see his ability to make the imagined look absolutely real stretched out over hundreds of glorious panels worth of unbelievably dense but always crystal-clear sequential storytelling.

Of course, when JG finally decided to make his triumphant return to the interior pages of monthly comics, there was only writer he could turn to... and it sure as hell wasn't me.

I don't know Mark Millar very well, though we have had several friendly chats (at least, I *think* they were friendly, but who the fuck can tell with that accent?). Either way, I sure do love his writing. Mark does better superhero fiction than any living creator, and I'm not being dismissive when I say he's the master of that particular genre. Most authors can churn out good, arty-farty, talking-heads, "mature readers" comics—Mark has written more than few, including *Chosen*, his other masterpiece—but capes and tights are deceptively difficult for even the best creators to really nail. At their worst, superhero stories are just dopey male power fantasies, but at their best (see: *Watchmen, Daredevil: Born Again*, etc.), these myths don't just entertain, they work as powerful allegories that help us understand who we are.

I don't want to sound like an asshole for reading too deeply into a story with a major character named "Shit-Head" in it, but by perfectly subverting the classic hero's journey familiar to every comic fan, Millar and Jones challenge us to think about the mundane world that we're all part of, and the price of entry into that hidden special world we all dream of one day joining.

Anyway, if this is your first time reading *Wanted*, have fun, and get ready for those last two pages. Those of you who refuse to see what the conclusion is really saying will probably want to burn this beautiful collection the second you put it down. But if that's the case, you should use my introduction page as kindling, because I think *Wanted* has the bravest, most interesting finale to a comic ever.

And to you chosen few who find yourself smiling when you finally reach the end of Wesley Gibson's journey?

Welcome to the other side.

Brian K. Vaughan
March 2005

BKV is the writer and co-creator of the ongoing series Y: THE LAST MAN, EX MACHINA and RUNAWAYS. Unlike Old Man Millar, he is still very young and very handsome.

INTRO

BRING ON
THE BAD GUYS

Written by: Mark Millar

Penciled and Inked by: J.G. Jones

Colored by: Paul Mounts

Lettered by: Dreamer Design's
Robin Spehar
Dennis Heisler
Mark Roslan

THIS IS ME MEETING HIM FOR **DINNER** TWO DAYS LATER AND PRETENDING NOT TO KNOW ABOUT IT AS WE ENJOY SOME REALLY NICE **KOREAN FOOD** TOGETHER.

THIS IS THE OFFICE WHERE I WORK AS AN ASSISTANT TO THE ASSOCIATE EDITOR ON **HYPOTHYROIDISM TODAY,** THE THIRD-BIGGEST **AUTO-IMMUNE** PERIODICAL ON THE EASTERN SEABOARD.

THIS IS ME TAKING SHIT FROM MY AFRICAN-AMERICAN **BOSS.**

AS YOU CAN SEE, I'M **SMILING** AS SHE INSULTS ME, BUT IT'S ONLY BECAUSE I'M **EMBARRASSED** BY THE SITUATION AND MORE THAN A LITTLE **AFRAID** OF THE SCARY FUCKING BITCH.

THIS IS THE SESAME-CRUSTED SALMON OVER SOURDOUGH WITH MUSTARD GREENS AND WASABI MAYONNAISE I LIKE TO HAVE FOR LUNCH JUST TO PROVE I'M DIFFERENT FROM THE **HERD.**

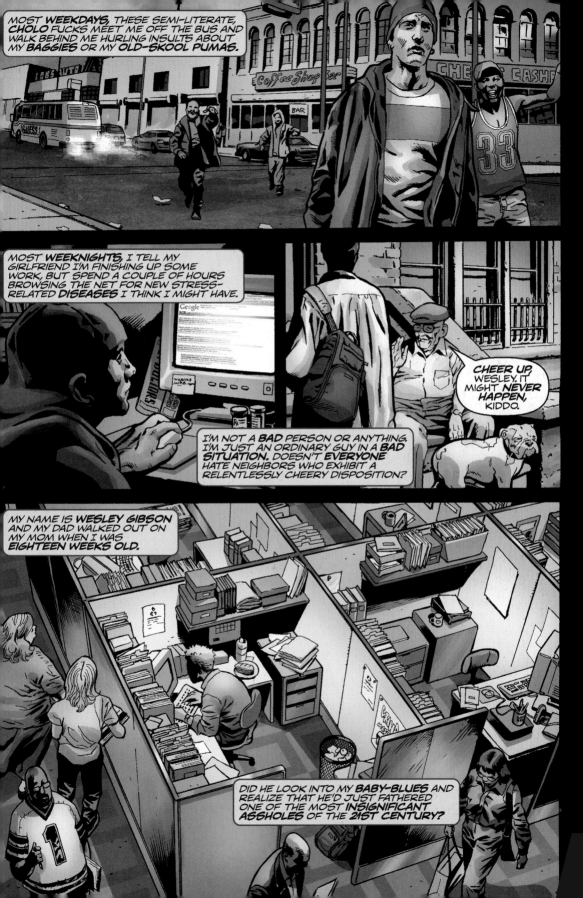

GET YOUR
ASSES BACK
HERE, YOU
LITTLE
SNOTS!

TH-THIS WASN'T *OUR* IDEA, MAN. I SWEAR, I DON'T EVEN KNOW WHO YOU *ARE.* YOU WERE JUST A NAME IN AN ENVELOPE...

OH, *REALLY?* AND HERE I THOUGHT YOU BOYS WERE CRIMINAL FUCKING *MASTER-MINDS!* WHO *SENT* YOU, DICK-WAD? WHO THE FUCK *ARE* YOU PEOPLE?

JUST THE *DECOYS,* MAN.

JUST THE MOTHER-FUCKING *DECOYS.*

WHAT?

"YOU *KIDDING?* THAT'S HOW THEY WHACKED *THE KILLER?*"

I *DUNNO*, SUCKER. I MEAN, HIT-MEN CAN CLOCK UP A *SHIT-LOAD* OF *ENEMIES*, MY FRIEND...

YEAH, BUT BEING THE *WORLD'S GREATEST* HIT-MAN MEANS NONE OF THEM ARE PACKING A PULSE BESIDES HIS *OLD BOSS*.

YOU THINK WHAT YOU *WANT* TO THINK, IMP, BUT I KNOW THIS WAS JUST A *SETTLING* OF *OLD SCORES* AND IF I WAS THE FOX, I'D BE HAVING TROUBLE *SLEEPING* RIGHT NOW.

SAME HERE, BUT ONLY BECAUSE I'D BE DREAMING ABOUT SPENDING MY OLD *BOYFRIEND'S* MULTI-MILLION DOLLAR *ESTATE*.

DOES THE FACT THEY WEREN'T TECHNICALLY *MARRIED* MEAN SHE HAS TO SPLIT THE LOOT WITH THE FIVE *SUPER-FAMILIES* TOO?

ACTUALLY, POOR *FOX* AIN'T IN LINE TO INHERIT A *DIME*, LITTLE MAN. DIDN'T YOU *HEAR*?

TURNS OUT THE KILLER WAS MARRIED FOR TEN MINUTES TO SOME *SOCIAL WORKER* BITCH, TWENTY FIVE YEARS AGO.

I HEARD SHE DIED OF AN *EMBOLISM* A WHILE BACK, BUT THE PROFESSOR'S TRYING TO TRACK DOWN THEIR KID AND LET HIM KNOW ABOUT THIS LITTLE *CASH WINDFALL* HE'S GOT COMING.

LUCKY LITTLE BASTARD.

TALK ABOUT THE RIGHT *WOMB* AT THE RIGHT *TIME*.

REMEMBER ME? WESLEY GIBSON?

IF I LOOK **TIRED** IT'S BECAUSE I WAS UP ALL NIGHT CONSOLING LISA AFTER SHE BROKE DOWN AND CONFESSED TO FUCKING ELEVEN **CO-WORKERS** IN THE TWENTY-TWO MONTHS WE'VE BEEN TOGETHER.

I WANT US TO TRY FOR A **BABY**, WESLEY. I'D FEEL MORE **ATTRACTIVE** IF YOU MADE A COMMITMENT AND WOULDN'T NEED THE **VALIDATION** OF THESE **CONSTANT AFFAIRS.**

BUT HOW CAN WE BRING A **BABY** INT THIS? SHE HASN'T EVEN SHAVED HE LEGS SINCE LAST CHRISTMAS.

HOW CAN I GET OUT OF THIS **JOB?** HOW CAN I GET OUT OF THIS **RELATIONSHIP?** I SHOULD HAVE BEEN A FUCKING **MILLIONAIRE** BY TWENTY-FOUR.

BY **10 A.M.,** I'M ALREADY BROWSING THE **INTERNET** AGAIN TRYING TO FIND AN ANSWER TO THE WORRYING LEVELS OF THE **CHRONIC FATIGUE** I'VE BEEN FEELING A LOT LATELY.

HEY, WHAT'S **THIS** YOU'RE LOOKING UP, WESLEY? **KU-KLUX-KLAN.COM,** OR **WWW.SMALL-WHITE-DICKS** AGAIN?

AT **NOON,** MY BEST FRIEND CALLS AND CANCELS OUR KOREAN LUNCH APPOINTMENT BECAUSE SOMETHING **UNEXPECTED** CAME UP

IS **LISA** WITH HIM, I WONDER? IS SHE SLIDING HER CHUBBY HAND INTO HIS BRAND NEW **SEAN PAULS** TO GET OVER THIS STRING OF GODDAMN FIGHTS WE JOKINGLY CALL A **RELATIONSHIP?**

I CAN'T BELIEVE YOU JUST *DID* THAT. THE *COPS* ARE GOING TO BE *ALL OVER* YOU FOR THIS. YOU'RE GONNA GO TO *JAIL*...

JESUS, WESLEY. WOULD YOU *LIGHTEN UP?* I KNOW YOU BEEN SCARED YOUR *ENTIRE LIFE,* BUT I BEEN SENT HERE TO TELL YOU THAT THOSE DAYS ARE *BEHIND* YOU, MAN.

AS LONG AS ONE OF US IS WEARING THIS *PIN,* OR DRIVING A CAR WITH THESE *NUMBER-PLATES,* WE CAN DO WHATEVER WE *WANT.*

YOU CAN SHOOT, KILL, RAPE OR DESTROY ANYONE YOU *LIKE* NOW, BABY. CONSEQUENCES ARE FOR THE *LITTLE PEOPLE* WHEN YOU GOT A SEAT IN *THE FRATERNITY.*

ONLY DIFFERENCE BETWEEN A *DREAM* AND A *NIGHTMARE* IS HOW BIG YOUR *BALLS* ARE, BITCH.

OH, GOD. THIS IS A *NIGHTMARE.* THIS IS A FUCKING *NIGHTMARE* I'M HAVING...

THIS IS ME AND LISA LYING IN BED AND NOT HAVING SEX. THIS IS HER NOT CALLING IN TO WORK AND PRETENDING TO BE SICK. THIS IS ME **NOT** GETTING A BLOW-JOB.

FUCK YOU

Written by: Mark Millar

Penciled and Inked by: J.G. Jones

Colored by: Paul Mounts

Lettered by: Dreamer Design's
Robin Spehar
Dennis Heisler
Mark Roslan

NOW BEFORE ANYONE COMPLAINS, I JUST WANT TO STRESS THAT THESE INNOCENT PEOPLE WERE DEAD AND BURIED LONG BEFORE I STARTED PUMPING HOT LEAD INTO THEIR LIFELESS CADAVERS.

THE FOX JUST THOUGHT I SHOULD GET USED TO THE SIGHT OF **FLESH SPLINTERING** AND **BONE FLYING** BEFORE SHE MOVED ME ON TO ANY LIVE HUMAN TARGETS.

THAT'S THE WAY, BABY. DON'T THINK OF THESE PEOPLE AS LOVEABLE GRANDMAS AND GRANDPAS NO MORE! JUST THINK OF THIS AS HOUSE OF MOTHER-FUCKING *DEAD* OR SOMETHING!

LIKEWISE, THE REASON I'M DOING FOURTEEN DAYS IN A **SLAUGHTERHOUSE** HERE IS TO GET ME AS NUMB AND DESENSITIZED AS YOUR AVERAGE **EIGHT-YEAR-OLD.**

I'M A **FRIEND OF THE EARTH,** A **GREEN-PEACE** CAMPAIGNER, AND A **VEGETARIAN** OF SOME ELEVEN YEARS STANDING, YOU UNDERSTAND.

THAT'S A SHIT-LOAD OF *EMPATHY* I NEED TO GET RID OF, BUT THREE CALVES A MINUTE SEEMS A PRETTY GOOD PLACE TO *START*.

FUCK 'EM, WESLEY! IF THEY WAS SMART, THEY WOULDN'T *BE* IN THIS SITUATION, RIGHT? FUCK 'EM HARD WITH YOUR BIG, STEEL GUN!

I SPEND THE MORNINGS WORKING *OUT* AND THE AFTERNOONS GETTING *WORKED OVER* BY THE BIGGEST BASTARD I'VE EVER LAID EYES ON.

LIKE ME, YOU MIGHT BE WONDERING WHAT KIND OF TRAINING INVOLVES BEING TIED TO A CHAIR AND HAVING YOUR FACE REARRANGED, BUT IT'S ALL PART OF THE PLAN, THE FOX ASSURES ME...

HOW *ELSE* YOU GONNA LOSE THIS FAGGOTY FEAR OF GETTING PUNCHED IN THE FACE, HUH?

AFTER THREE WEEKS, I'VE LOST FOUR TEETH, FRACTURED MY JAW, BROKEN THREE RIBS AND PERSONALLY KILLED TWELVE THOUSAND, EIGHT HUNDRED AND SEVENTEEN *FARM ANIMALS.*

WEEK FIVE THEY FINALLY TAKE THE HANDCUFFS OFF...

GOOD MORNING, WESLEY. YOU READY FOR YOUR DAILY *SHIT-KICKING,* SHORT-ASS?

AS LONG AS I CAN REMEMBER, I'VE BEEN CURIOUS WHAT MY FATHER MUST HAVE BEEN LIKE.

EVERY ONCE IN A WHILE I'D SEE A GUY IN THE STREET JUST THE RIGHT AGE AND JUST THE RIGHT BUILD AND MY HEART WOULD SKIP A BEAT. JUST FOR A FRACTION OF A SECOND.

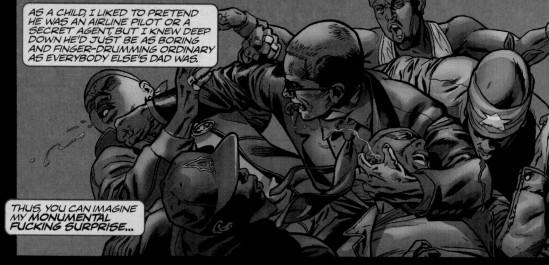

AS A CHILD, I LIKED TO PRETEND HE WAS AN AIRLINE PILOT OR A SECRET AGENT, BUT I KNEW DEEP DOWN HE'D JUST BE AS BORING AND FINGER-DRUMMING ORDINARY AS EVERYBODY ELSE'S DAD WAS.

THUS, YOU CAN IMAGINE MY MONUMENTAL FUCKING SURPRISE...

YOU'VE JOINED A SECRET FRATERNITY OF SUPER-CRIMINALS?

THAT'S RIGHT, LISA. MY DAD WAS A COMIC-BOOK SUPER-VILLAIN AND WHEN HE GOT BUMPED OFF, I AUTOMATICALLY GOT HIS PLACE IN THIS *SECRET ORGANIZATION* NOBODY KNOWS ABOUT.

OH, JESUS, WESLEY. OH, JESUS, JESUS, JESUS H. *CHRIST!* HAVE YOU ANY IDEA HOW FUCKING *LOW* THIS IS?

WHAT?

I KNOW YOU'VE GOT MAJOR *CONFRONTATION ISSUES*, BUT THIS IS *TOO MUCH.* COULD YOU PLEASE JUST ADMIT THAT YOU'VE MET SOMEONE ELSE AND ACT LIKE A *MAN* FOR A CHANGE?

BUT I *HAVEN'T* MET SOMEONE ELSE. OKAY, I'M GETTING SEX-LESSONS FROM A JEWEL THIEF WHO'S TEACHING ME HOW TO *FUCK* PROPERLY, BUT I'M NOT ACTUALLY *SEEING* ANYBODY.

YOU'RE HAVING SEX WITH SOMEBODY ELSE? YOU'RE ACTUALLY *ADMITTING* THIS?

OH, PLEASE. SPARE ME THE INDIGNATION, LISA. LIKE YOU HAVEN'T BEEN FUCKING MY BEST FRIEND FOR THE LAST YEAR AND A HALF?

WHAT?

IT'S OKAY. YOU DON'T HAVE TO LIE ANYMORE. I SHOT HIM IN THE FACE AND DROPPED HIS BODY IN A DUMPSTER LAST NIGHT ANYWAY.

CHRIST, WESLEY. CAN'T YOU SEE HOW *DELUSIONAL* YOU'RE BECOMING? CAN'T YOU SEE YOU'RE IN THE MIDDLE OF A *NERVOUS BREAKDOWN* HERE?

NO, I'M NOT. I'M HAVING THE *OPPOSITE* OF A BREAKDOWN. I'M HAVING A GODDAMN *KICK-START*, LISA, AND I'VE NEVER FELT BETTER IN MY *LIFE*.

YOU'RE REALLY GONNA THROW IT ALL AWAY? JUST LIKE THAT?

THROW *WHAT* AWAY? WE NEVER REALLY *HAD* ANYTHING. WE JUST SAT HOME WATCHING TELEVISION EVERY NIGHT AND FUCKED LIKE *OLD PEOPLE* ONCE A MONTH.

"NOT JUST THE TEN OR TWELVE SUPERVILLAINS THAT MADE UP EACH OF THESE ROGUE GALLERIES AND SUCH, BUT THE HUNDREDS AND THOUSANDS OF SUPER-CRIMINALS ALL ACROSS THE PLANET.

"INDIVIDUALLY, WE'D ALWAYS FAILED TO MAKE MUCH IMPACT, BUT AS AN ARMY I HYPOTHESIZED WE'D BE PRETTY MUCH UNBEATABLE.

"THE FINAL BATTLE TOOK PLACE IN 1986. IT LASTED ALMOST THREE MONTHS AND WE LOST A GREAT MANY FRIENDS DURING THAT ENCOUNTER, BUT WE **BEAT** THEM IN THE END.

"BY THE MIDDLE OF AUGUST, THERE WASN'T A SUPERHERO LEFT STANDING FROM ONE END OF THIS GLOBE TO THE OTHER."

I DON'T UNDERSTAND. HOW COME THIS ISN'T IN THE HISTORY BOOKS? EVEN IF THERE'D BEEN **ONE** SUPERHERO, WOULDN'T THAT HAVE BEEN ALL OVER THE **NEWS** AND STUFF?

AH, BUT IT WASN'T ENOUGH JUST TO **BEAT** THEM, WESLEY. WE HAD TO STRIP THEM OF THEIR MEMORIES AND MAKE SURE THAT EVEN THEIR **GREATEST FANS** DIDN'T REMEMBER THEM.

SUCH SCIENCE MIGHT SEEM COMICAL IN THIS NEW WORLD THAT WE **MOLDED** FOR YOU, BUT BELIEVE ME WHEN I SAY THAT **REALITY ITSELF** CAN BE REWRITTEN IF WE DESIRE IT, BOY.

SEVEN-DIMENSIONAL IMPS AND ALIEN SUPER-COMPUTERS ARE AMONG OUR RANKS, YOU KNOW. THERE'S REALLY **NOTHING** WE CAN'T DO IF WE ALWAYS STAND **UNITED.**

SUPERGANGBANG

Written by: Mark Millar

Penciled and Inked by: J.G. Jones

Colored by: Paul Mounts

Lettered by: Dreamer Design's
Robin Spehar
Dennis Heisler
Mark Roslan

DON'T LET FUCKWIT GET ANY CLOSER, BOYS. THIS IS THE SAME RADIATION THAT CRIPPLED THE FELLOW I *CLONED* HIM FROM AND WE DON'T WANT OUR BACKWARDS FRIEND GETTING *HURT* NOW, *DO* WE?

WHAT DO YOU WANT IT FOR ANYWAY, SIR? I MEAN THE LIBERTY UNION DON'T HAVE *COUNTERPARTS* HERE ANYMORE. THERE'S NO SUPERHEROES LEFT TO *USE* IT AGAINST.

NO, BUT IT'S THE PERFECT GIFT FOR A DEAR FRIEND I'LL BE SEEING TONIGHT AT THE *COUNCIL OF FIVE,* WESLEY. JUST A SMALL REMINDER OF OUR DIM-AND-DISTANT *YOUTH.*

SPEAKING OF WHICH, YOU THINK I COULD BE EXCUSED FROM *BODY-GUARD DUTY* AT THIS THING, PROFESSOR?

THE LAST THREE TIMES I JUMPED REALITIES I WAS THROWING UP FOR A DAY AND A HALF AND I'M STARTING TO FEEL QUEASY *ALREADY.*

NOT A PROBLEM, FOX. WESLEY CAN HELP FUCKWIT WITH SECURITY FOR THE EVENING AND GET A LITTLE TASTE OF THE *HIGH-LIFE* FOR ALL THAT *FINE WORK* HE'S DONE LATELY.

COOL BY ME.

ALL PERSONNEL PLEASE CLEAR THE AREA! BRIDGE OPENING UP TO NORTH AFRICA, PEOPLE! CLEAR SOME SPACE FOR THE NEXT BIG BATCH OF HUNGRY, GRATEFUL *HOOKERS!*

LIKE I SAID, IT'S BEEN THREE MONTHS. THREE MONTHS SINCE I WAS LOCKED IN THAT DEAD-END JOB, THAT DEAD-END RELATIONSHIP AND SURFING FOR INTERNET PORN EVERY NIGHT.

NOW I'M MURDERING PEOPLE ALL OVER TIME AND SPACE AND DRIVING A THREE HUNDRED THOUSAND DOLLAR SPORTS CAR.

FUCK YOU, MOM. FUCK YOU, ALL THOSE TEACHERS WHO SAID I WAS TOO LAZY TO EVER *AMOUNT* TO ANYTHING.

THAT NERVOUS YOUNG FELLOW WITH THE SODA AND LIME IS ANOTHER MEMBER OF THE RICTUS FAMILY. HAVE YOU EVER HEARD OF JOHNNY TWO-DICKS, THE GRAND MEMBER OF CRIME?

NO? WELL, HE'S A VERY, VERY DECENT YOUNG PHARMACIST WHO'S EVERY DECISION IS QUITE LITERALLY MADE BY THAT THIRTEEN INCH CRIMINAL MASTERMIND HE'S PACKING IN HIS UNDERWEAR.

WHO'S THE FREAK SPIKING THE PUNCH?

OH, THAT'LL BE THE FRIGHTENER. HE SPECIALIZES IN PSYCHIC VIRUSES. AND THOSE TWO MEN BESIDE HIM ARE THE PUZZLER AND THE AVIAN.

I HAVEN'T MET THE FELLOW DRESSED AS THE MAD MARCH HARE BEFORE, SO I CAN ONLY ASSUME HE'S THE YOUNG BOY WHO MURDERED HIS PREDECESSOR AND BLACKMAILED HIS WAY INTO THE GANG.

SOUNDS LIKE A NICE, DECENT, CHURCHGOING CROWD.

ACTUALLY, MISTER RICTUS WAS A DEVOUT CHRISTIAN MANY YEARS AGO. ACCORDING TO THE NEWSPAPERS, HE WAS QUITE THE PILLAR OF SOCIETY UNTIL HIS NASTY INDUSTRIAL ACCIDENT.

"IT'S NOT THAT HE BECAME EMBITTERED AFTER BEING SO HORRIBLY DISFIGURED. HE JUST DIED FOR A MOMENT ON THE OPERATING TABLE AND RECOGNIZED THE POINTLESSNESS OF EXISTENCE."

"HE DISCOVERED THAT THERE WAS NO GOD OR HEAVEN. IN FACT, NO ANYTHING THAT HE'D BEEN PROMISED FOR HIS LIFETIME OF GROVELING SERVITUDE."

THIS MIND-SHATTERING NEWS, I'M AFRAID, REMOVED WHATEVER MORAL COMPASS HAD BEEN GUIDING HIS LIFE UNTIL THAT POINT AND ESSENTIALLY CREATED A MAN WITHOUT A CONSCIENCE.

I WOULDN'T MIND GIVING HIM ANOTHER NEAR-DEATH EXPERIENCE AFTER WHAT THE FUCKER DID TO MY DAD.

IT WAS THE TROPHY I BROUGHT BACK FROM THE PARALLEL WORLD, RIGHT?

EXCUSE ME?

THAT'S WHAT YOU USED TO GET THE EMPEROR BACK ONSIDE. YOU TAPPED INTO A LITTLE SHARED HISTORY AND GOT HIM ALL MISTY-EYED ABOUT THE GOOD OLD DAYS. THAT WAS BRILLIANT.

OH, THE TROPHY WOULDN'T HAVE DONE IT BY ITSELF, WESLEY, BUT IT WAS A POTENT COMBINATION WITH THE ANGLE I WAS SITTING AT AND THE SUBLIMINAL CODE I WAS TAPPING WITH MY FINGERS.

NOT TO MENTION, OF COURSE, THE COLOGNE I WAS WEARING WHICH I BELIEVE WAS A FAVORITE OF HIS LATE, BELOVED FATHER.

UNBELIEVABLE.

ACTUALLY, IT'S ALL QUITE SIMPLE WHEN YOU BREAK IT DOWN AND EXAMINE THE BASIC COMPONENTS, BUT EVEN I MUST ADMIT I'M FEELING PLEASED WITH MYSELF AFTER A SOLID NIGHT'S WORK.

I THINK I MIGHT HEAD BACK TO THE LAB AND TREAT MYSELF TO A FEW HOURS EXTRA TIME ON THIS MAP OF THE HUMAN SOUL I'VE BEEN PIECING TOGETHER IN MY RECREATION TIME.

FUCKWIT, WOULD YOU MIND PULLING OVER AND LETTING MISTER GIBSON OUT HERE, PLEASE?

OR RATHER, WOULD YOU MIND NOT PULLING OVER AND NOT LETTING MISTER GIBSON OUT OF THE CAR?

"YOU KNOW THE FIRST SUPERVILLAINS WERE COMPLETELY NAKED?"

CRIME PAYS

Written by: Mark Millar

Penciled and Inked by: J.G. Jones

Colored by: Paul Mounts

Lettered by: Dreamer Design's
Robin Spehar
Dennis Heisler
Mark Roslan

"BUT THEY WAS JUST THE FIRST OF MANY: WHEN THINGS WAS AT THEIR PEAK, YOUR DADDY AND I COUNTED TWENTY-TWO SUPERVILLAINS FOR EVERY SINGLE HERO OUT THERE AND THEY WAS JUST GETTING WACKIER BY THE MINUTE.

"IF A SUPERHERO HAD SUPER-STRENGTH, THE VILLAINS HAD TO BE ABLE TO FLY. IF A SUPERHERO HAD A MAGIC RING, THE VILLAINS HAD TO BE INVULNERABLE TO THAT RING AND MAYBE EVEN TURN THEIR ASS *INVISIBLE.*

"THE NEWSPAPERS WERE JUST FULL'A FREAKY SHIT BACK IN THOSE DAYS, MAN; VILLAINS THAT COULD WALK THROUGH WALLS, VILLAINS DRESSED LIKE CROSSWORD PUZZLES, VILLAINS WHO COULD TRAVEL THROUGH CLOCKS AND PAINTINGS.

"THE PAPERS JUST LOOK SO GODDAMN *BORING* NOW WITH THEIR PRESIDENTIAL ELECTIONS AND DOW JONES INDEXES."

SO HOW DID *YOU* GET INTO ALL THAT SHIT?

WHAT? THE GAME?

GROWING UP IN THE PROJECTS YOUR ONLY CHANCE OF MAKING A BUCK WAS EITHER SPORT OR SUPER-CRIME, AND I WASN'T *TALL* ENOUGH TO DUNK A BASKETBALL.

I BEEN DOING THIS JOB SINCE I WAS FOURTEEN YEARS OLD, HONEY, AND I DON'T REGRET A SINGLE THING. HOW MANY PEOPLE GET TO THIRTY-FIVE AND STILL BE ABLE TO SAY THAT, HUH?

I WAS **WONDERING** WHEN YOU WERE FINALLY GOING TO MAKE YOUR MOVE. I TAKE IT THIS ISN'T JUST *ME* YOU'RE GUNNING FOR HERE?

THIS ISN'T A *PERSONAL ATTACK* OR ANYTHING, IS IT?

I'M AFRAID THEY NEVER *MADE* IT TO THEIR GRANDMA'S HOUSE, LITTLE MAN. THE BAD NEWS IS THAT THE BOYS AND I CAUGHT UP WITH THEM SHORTLY AFTER THEY TOOK OFF IN THEIR PEOPLE CARRIER AND, WELL...

THEY SAY A *PICTURE'S* WORTH A *THOUSAND WORDS.*

SOMETHING YOU'D LIKE TO *SAY*, OLD FRIEND?

SOMETHING *DIGNIFIED*? SOMETHING *CLASSY* AND *FORGIVING*?

BOYS, I WANT YOU TO *KILL* THESE MOTHER-FUCKERS...

JESUS CHRIST!

THE SHIT LIST

Written by: Mark Millar

Penciled and Inked by: J.G. Jones

Colored by: Paul Mounts

Lettered by: Dreamer Design's
Robin Spehar &
Dennis Heisler

LAGOS, NIGERIA.

VIDEO-CALL FROM *MANHATTAN*, ADAM ONE.

TELL THE PROFESSOR HE'LL JUST HAVE TO WAIT, HAKEEM. MY SON IS VERY WEAK AND I WANT TO BE WITH BOTH HE AND HIS GRANDCHILDREN IN THESE FINAL HOURS. TELL SOLOMON I SHALL CALL HIM BACK IN THE MORNING.

BUT IT ISN'T *THE PROFESSOR*, YOUR HIGHNESS. IT'S THE PROFESSOR'S CODES, YES, BUT THE GENTLEMAN IN QUESTION APPEARS TO BE MISTER *RICTUS* OF THE *AUSTRALIAN* FAMILY.

TELL HIM I'M ON MY WAY.

WELL, NOW. HERE'S A SURPRISE, EH? *YOURS TRULY* CALLING ON *THE PROFESSOR'S* PRIVATE LINE? "THAT CAN ONLY MEAN ONE THING," YOU MUST BE THINKING TO YOURSELF.

AND YOU *KNOW* SOMETHING, MY WISE AND ANCIENT FRIEND...?

--ERRRSSSS!

DEAD OR ALIVE

Written by: Mark Millar

Penciled and Inked by: J.G. Jones

Colored by: Paul Mounts

Flashback sequences pgs 6-10
Penciled and Inked by: Dick Giordano

Lettered by: Dreamer Design's
Robin Spehar &
Dennis Heisler

"YOUR MOTHER WAS STILL DATING THE AVIAN BACK IN THOSE DAYS. SHE'D BEEN HIS SOCIAL WORKER WHEN HE WAS IN PRISON AND, WELL, ONE THING LED TO ANOTHER AND HE BROUGHT HER HERE ONE TIME.

"I'LL NEVER FORGET HOW BEAUTIFUL SHE LOOKED THAT NIGHT. THE SHININESS OF HER HAIR, THE SPIKINESS OF HER BOOTS, THE AVIAN'S HAND SQUEEZING HER LATEX ASS AS THEY TALKED TO THEIR FRIENDS...

"I WANTED TO HAVE HER OVER THAT BAR RIGHT THEN AND THERE.

"FORTUNATELY, EVEN WHAT STARTED AS DRINKS WOULD ALWAYS END UP AS AN ORGY BACK IN THOSE DAYS AND IT WASN'T LONG BEFORE THE HAT-MAKER OR KING-BEE WOULD SUGGEST WE DO A *KEY-SWAP*--

"--AS IN EVERYONE DROPPING THE KEYS OF THEIR RAT-MOBILES OR SPECTRO-COPTERS INTO A HAT AND WHICHEVER GIRLFRIEND PICKED THEM OUT WOULD GO BACK TO THEIR LAIRS AND FUCK THEM SENSELESS."

IS THIS WHAT HAPPENED WITH YOU AND MOM? SHE PICKED OUT YOUR *CAR KEYS?* THAT'S *DISGUSTING...*

SON, WE NEVER EVEN *GOT* HOME THAT NIGHT. WE FUCKED IN MY FLAME-CHASER FOR TWO DAYS SOLID AND, MY GOD, SHE WAS HOT. THE THINGS THAT WOMAN COULD DO WITH HER FINGERS.

UH, COULD WE CHANGE THE SUBJECT, PLEASE?

WHY? I THOUGHT THE FOX HAD DEPROGRAMMED ALL THOSE WEIRD SEXUAL HANG-UPS YOU'D BEEN CARRYING AROUND?

A GIRL CAN ONLY DO SO MUCH, BABY. TOOK ALMOST TWO ENTIRE WEEKS TO GET HIM OVER HIS FEAR OF CUNNILINGUS.

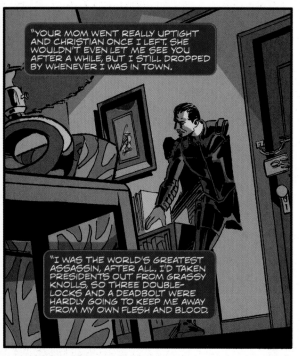

"YOUR MOM WENT REALLY UPTIGHT AND CHRISTIAN ONCE I LEFT. SHE WOULDN'T EVEN LET ME SEE YOU AFTER A WHILE, BUT I STILL DROPPED BY WHENEVER I WAS IN TOWN.

"I WAS THE WORLD'S GREATEST ASSASSIN, AFTER ALL. I'D TAKEN PRESIDENTS OUT FROM GRASSY KNOLLS, SO THREE DOUBLE-LOCKS AND A DEADBOLT WERE HARDLY GOING TO KEEP ME AWAY FROM MY OWN FLESH AND BLOOD."

"YOU WERE ALWAYS ASLEEP WHEN I APPEARED. SOMETIMES I'D LEAVE PRESENTS LIKE SOME HALF-ASSED SANTA, BUT MOSTLY I JUST SAT THERE STROKING YOUR HAIR UNTIL THE SUN CAME UP."

"THE SKY WAS SO BLUE IN THOSE DAYS, WESLEY. THE TREES WERE A DEEPER GREEN THAN YOU CAN POSSIBLY IMAGINE AND THE FOOD WAS SO RICH AND TASTY COMPARED TO THAT SHIT YOU EAT TODAY.

"THERE WAS A MOMENT WE ALMOST DIDN'T GO AHEAD WITH THAT REVOLUTION WE'D BEEN PLANNING. A MOMENT WE DIDN'T WANT TO LET THINGS GO ALL GRIM AND GRITTY...

"...BUT IT WAS ONLY FOR A MOMENT.

"SOME OF US HAD SPENT HALF OF OUR ADULT LIVES IN JAIL AND THE HEROES WERE GETTING SO GOOD AT WHAT THEY DID. WE *HAD* TO STRIKE. WHAT IF THIS WAS OUR *FINAL OPPORTUNITY?*"

"TWO MILLION PEOPLE DIED AROUND THE WORLD IN THOSE TERRIBLE WEEKS, BUT I MADE SURE YOU AND YOUR MOM WERE SAFE.

"EVEN WHEN THE OTHERS WERE BUILDING THAT ENORMOUS MACHINE PEOPLE REFER TO AS THE EMPIRE STATE BUILDING, I WAS BACK AT YOUR BUILDING WATCHING OVER YOU BOTH.

"IT WAS THE PERFECT PLAN. WHAT BETTER WAY TO KEEP CONTROL THAN MAKING EVERYONE FORGET THERE HAD EVER *BEEN* A REVOLUTION?

"THAT MACHINE REARRANGED THE VERY ATOMS OF THE UNIVERSE THAT HOT, DAMP SUMMER IN 1986 AND THIS BEAUTIFUL LITTLE WORLD OF OURS WOULD NEVER BE THE SAME AGAIN.

"THE HEROES WERE GONE, THE VILLAINS WERE THE ONLY ONES THAT WOULD *REMEMBER* THEM AND I HELD YOUR HAND TIGHT AS THIS *NEW* EARTH FORMED ITSELF AROUND US.

"BY MORNING, ALL THE MAGIC IN THE WORLD WAS GONE AND YOUR MOTHER THOUGHT YOUR FATHER HAD BEEN AN *AIRLINE PILOT.*"

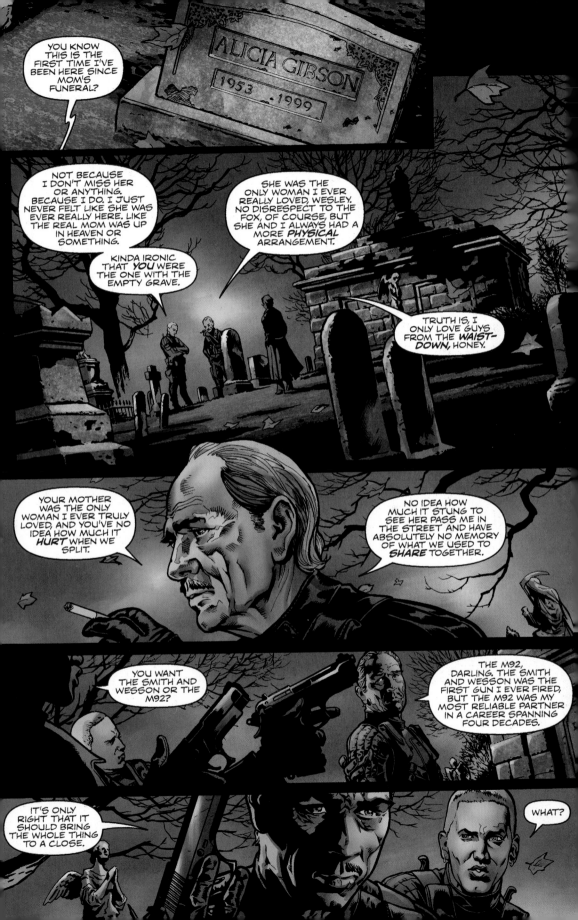

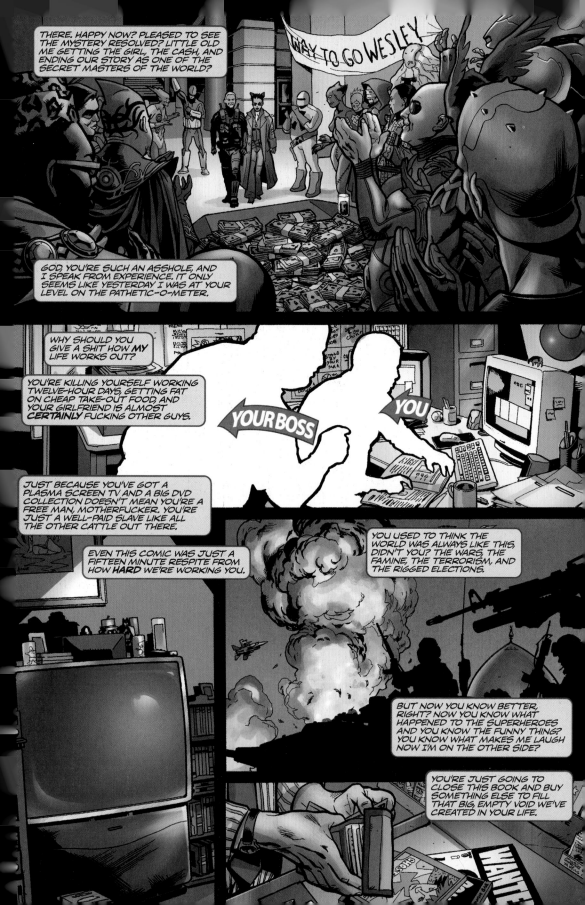

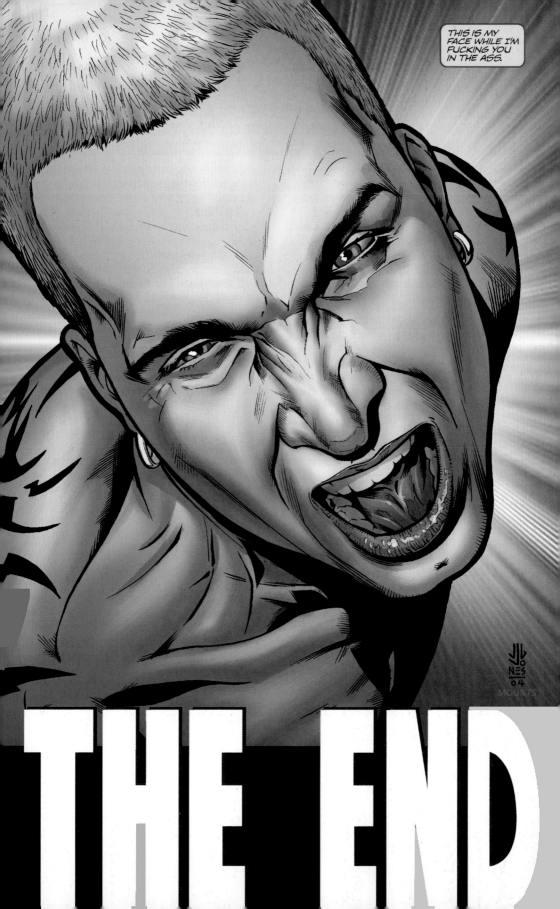

In some ways *Wanted* was the first thing I ever wrote.

I'm sitting here typing these words at the age of thirty-four, but the idea first came to me almost thirty years ago when one of my brothers pulled a great, big scam and outwitted their tiny sibling. Bear in mind that I was maybe five years old at the time and the brother in question was already halfway through a microbiology degree at university and you'll have some insight into my traumatic and often painful childhood. Anyway, here's how it happened.

When I was in what you people probably refer to as first grade, my classroom had a little library at the back where we could all sit and read some books once we'd finished our work for the day. I picked up a hardback with an image of the Statue of Liberty on the cover, sounded out the word "America" and thought it seemed extremely cool. When you're living in the ass-end of nowhere in one of the poorest countries in Western Europe (i.e. Scotland), America really does sound very cool indeed and thus I was drawn to all those images of hot dogs, Mount Rushmore and Jimmy Carter inside. I can remember the book in incredible detail because it all seemed so bright and glitzy and expensive compared to the low-budget, Braveheart-meets-Trainspotting kind of life going on outside and that was BEFORE I even saw that picture of the real-life Superman standing there with a gangster pointing a gun at his chest.

This wasn't a comic. This wasn't Superman as channeled by Curt Swan and Tex Blaidsell with dialogue by Cary Bates and Curt Swan. This was the real thing. I looked at the picture and read the caption over and over again and all it said was "Superman: The Great American Hero." To me, this was as big and as life changing as Moses getting a call from that burning bush when he was on the way to Wherever-The-Hell in the desert (comics always seemed much more interesting than the Old Testament). This was Superman and he was as real as Jimmy Carter, Mount Rushmore or any of those hot dogs. Just as I was starting to have my doubts about Santa Claus, here was a whole new preposterous character to think about and I couldn't bear to part with this crappy book. I wanted to tell the world what I'd found and so I did something I've never done before or since; I slipped it into my bag and sneaked it home to pore over the picture at some considerable length.

An important thing to realize is that we didn't have cable TV in the UK back in those days. We had a mere three channels and,

quite interestingly, one of the lowest obesity rates in the western world (think there might be a connection, fan-boy?). We paid a LICENSE for our television sets every year and the notion of paying money for re-runs was decidedly un-British. Sure, Superman appeared here in the late fifties when it first aired in the States, but it had never been shown again after cancellation and I didn't even know the show had existed. George Reeves was as unknown to me as Linda Lovelace when I was six years old. When I saw this picture of that middle-aged, slightly dumpy-looking Superman standing there it was Curt Swan's art incarnate. This was Superman and the one thing I dearly wanted to know was what the fuck had happened to him.

I can remember the scene with absolute clarity. I was sitting in my shared bedroom (there were eight of us in the house) re-reading the book again and again by the glow from by two-bar electric fire and I asked my brother Bobby what had happened to Superman. Why were there plane crashes on the news? Why were there earthquakes? Why didn't Superman help people in real life like he did in the comics if he was this great American hero? Bobby grinned the same grin he pulled when he told me my Dad killed Hitler during the war, that he could read my mind and that I'd inherit superhuman powers on my seventh birthday. "What happened to Superman? Didn't you hear?" he asked. "Superman disappeared during a big war with all the super-villains. Superman, Batman, Spider-Man, Captain America; they all disappeared during this enormous battle and they've never been seen again."

Gutted, I sloped off to bed. The greatest man in human history was gone forever and, with him, my hopes of becoming some kind of wisecracking boy sidekick. All my training had been for nothing. Superman was dead, the other heroes were dead and, as Bobby explained, the villains had made us forget they ever existed. All we had in their place were comic books. Crude, four-color approximations of their adventures by stoned, slightly crazy misfits who could somehow half-remember their adventures and get them down on the printed page. But who even read comics anymore? And what happened to the villains?

Read on.

Mark Millar
18th March 2004
The Ass-End of Nowhere

Excerpted from the WANTED: Dossier May 2004

WANTED: DOSSIER

The **WANTED: Dossier** was a printed supplement to the **WANTED** Series. In it, the heroes and villains of Wesley Gibson's world were illustrated by some of comics' brightest stars. The following characters were illustrated by the artists listed below, turn the page to see their unique visions of Mark Millar's characters from **WANTED**.

THE KILLER
Pencils: **John Romita Jr.**
Inks: **Scott Hanna**
Colors: **Paul Mounts**

FOX
Pencils: **Marc Silvestri**
Inks: **Joe Weems V**
Colors: **Steve Firchow**

PROFESSOR SOLOMON SELTZER
Pencils, inks and colors:
Dave Johnson

ORIGINAL KILLER
Pencils and Inks: **Tim Bradstreet**
Colors: **Paul Mounts**

THE COUNCIL OF FIVE
Pencils, inks and tones: **Ashley Wood**
Colors: **Paul Mounts**

THE HEROES
Pencils, inks and colors:
Ty Templeton

DOLL-MASTER
Pencils and inks: **Brian Michael Bendis**
Colors: **Paul Mounts**

IMP and DEADLY NIGHTSHADE
Pencils, inks and colors:
Chris Bachalo

SHIT-HEAD
Pencils and inks: **Bill Sienkiewicz**
Colors: **Paul Mounts**

FUCKWIT
Pencils and Inks: **Frank Quitely**
Colors: **Paul Mounts**

SUCKER
Pencils: **Joe Quesada**
Inks: **Mark Millar**
Colors: **Paul Mounts**

MISTER RICTUS
Pencils and inks: **Jae Lee**
Colors: **June Chung**

The Killer

Wesley Gibson's life spun out of control one deadly day in a wash of blood and a hail of gunfire. Or did it finally spin into control?

Gibson had been leading the life of quiet desperation—boring girlfriend, dead-end job, surfing for Internet porn on a daily basis. But when the Fox plucked him out of his humdrum existence, she showed him possibilities unimagined: He was The Killer, heir to a vast fortune and superpowers he didn't even know he had. The scales finally dropped from Wesley Gibson's eyes, and he saw the hidden clockwork of the real power behind the throne. A vast supervillain Fraternity was running the world, and had been since 1986, when they teamed up to destroy all the superheroes.

Wesley was taken under the wing of The Fox and Professor Solomon Seltzer, and embraced his life as the new Killer. He could steal, kill, or fuck whoever he wanted, all without repercussion. He was born again, baptized in blood and fire, as The Killer.

Pencils: John Romita Jr.
Inks: Scott Hanna
Colors: Paul Mounts

FOX

She shoots from the hip—and what hips they are.

The woman who would become The Fox grew up poor, in the kind of neighborhood where the only ways up and out are sports, hip-hop, or crime.

Guess what? She can't rap and she can't dunk abasketball.

The Fox used her natural gifts of agility and strength, and combined them with a crass disdain of humanity to become one of the deadliest killers the world has ever seen. She kills, because she does not care. Stopping someone else's breathing is as natural as breathing to her.

The Fox has never forgotten where she came from. She may lack formal education, but she has street smarts in spades—and she knows how to hitch her wagon to the right star. She unleashed her insatiable appetite for sex on the original Killer, becoming his consort, and cementing her position in the supervillain Fraternity. When The Killer died, she turned her vulva on his son—Wesley Gibson, the new Killer.

What's she want? Money? Sex? Power? All of the above. And Wesley Gibson just might be nothing more than the latest tool in her arsenal.

Pencils: Marc Silvestri
Inks: Joe Weems V
Colors: Steve Firchow

Professor Solomon Seltzer

Professor Solomon Seltzer is a schemer who maps the human soul...and fucks 18-year-old prostitutes.

He started talking when he was still in the womb, was reading and solving math problems before he could walk, and had graduated magna cum laude with a bio-mechanics degree before his friends had reached kindergarten.

The Professor is the smartest man who's ever lived, a billionaire by ten years old. Rumor has it—and there's not much doubt—that this Level Nine intelligence put together the plan to vanquish all the superheroes in 1986. The villainous Mister Rictus likes to take credit for that act, but he knows in his blackest of hearts that he can't. Rictus is insanely jealous of The Professor for his vast intelligence, and the two are adversaries of the highest order.

Pencils, inks, and colors: Dave Johnson

Original Killer

The world's deadliest assassin is dead, the victim of assassination himself. But while he was alive, The Killer lived up to his name—frequently, and without remorse.

The Killer killed for money, for sport, to advance his own cause. He killed for fun. The riches he gained bought him the Epicurean lifestyle he so greatly desired. He bathed in the finest champagne. He devoured the finest caviar. He bedded the finest women—and sometimes, the finest men.

So dangerous a foe was he, that when The Killer met his end, he was shot through the head with a gun from two cities away. No one would dare get closer. But The Killer passed on his legacy—and his unerring aim and killer instinct—to his illegitimate son, Wesley Gibson.

Pencils and inks: Tim Bradstreet
Colors: Paul Mounts

DOSSIER

The Council of Five

The Illuminati is real—and God help us, they have superpowers.

When the supervillains finally destroyed all the heroes in 1986, they carved the world up five ways. Five "heads" of the supervillain families hold sway over their territories:

• Professor Solomon Seltzer got North and South America
• Adam-One, the world's oldest man, took Africa
• The Future, a savage Nazi bastard, got Europe
• The Emperor, a Chinese crimelord, took control of Asia
• Mister Rictus got stuck with Australia

From behind a veil of secrecy, the Council of Five act as the real clockwork of the world, pulling the strings of global society. Not a legbreaking gets done, not an illegal act takes place, without their knowledge and consent.

But that veil of secrecy has been called into question. The Professor and Adam-One like things to remain behind the scenes—when the underworld stays underground, there's more loot for everyone to divvy up.

The Future and Rictus, on the other hand, tire of remaining in shadow. They want the world to piss its pants when it hears the mere letters in their names.

The Emperor? He's the swing vote, constantly in play. Meetings of the Council of Five have taken on the air of an armed camp. Daggers are drawn, searching for a back to be planted in.

The Illuminati is real—and God help them, they may turn against each other.

Pencils, inks and tones: Ashley Wood
Colors: Paul Mounts

The Heroes

The bad guys won. Led by Professor Solomon Seltzer, ALL the supervillains finally teamed up and destroyed all the superheroes in a cataclysmic battle back in 1986.

But it wasn't enough just to kill them—they had to destroy them MORE completely. Reality was folded and unfolded by the Fraternity's cognoscenti, seven-dimensional imps, and alien super-computers. All vestiges that the heroes had ever existed were destroyed. We were left with a world where humanity, at best, has vague, Alzheimer's-esque memories of superheroes. That, and the comic books.

But who reads comic books anymore?

Pencils, inks and colors: Ty Templeton

Doll-Master

He has a wife, two beautiful young daughters, a house in the suburbs, and wears a bow tie. He's the most docile, kind-hearted, well-spoken gentleman you're likely to meet. You'd be proud to have him as your neighbor. Too bad he's also one of the most nefarious super-criminals the world has ever seen.

The Doll-Master combines an uncanny knack for micro-mechanics with a Gepetto-like love for his "babies" that's definitely three steps over the line of creepy, and borders on the insane. His smile is halfway between "kindly old uncle" and "pedophile." He is indeed the master of his doll-like automatons...but these babies are killing machines that can rend the flesh from your bones.

He'll order his dolls to fillet a man, but he never swears in front of children. He'll blow up a bank, killing dozens...but always leaves the toilet seat down for his wife. And if so much as a tiny tear appears in his dolls' suits, he'll lovingly stitch it up himself. He is the master of his puppets...but they pull the strings of his heart.

Pencils and inks: Brian Michael Bendis
Colors: Paul Mounts

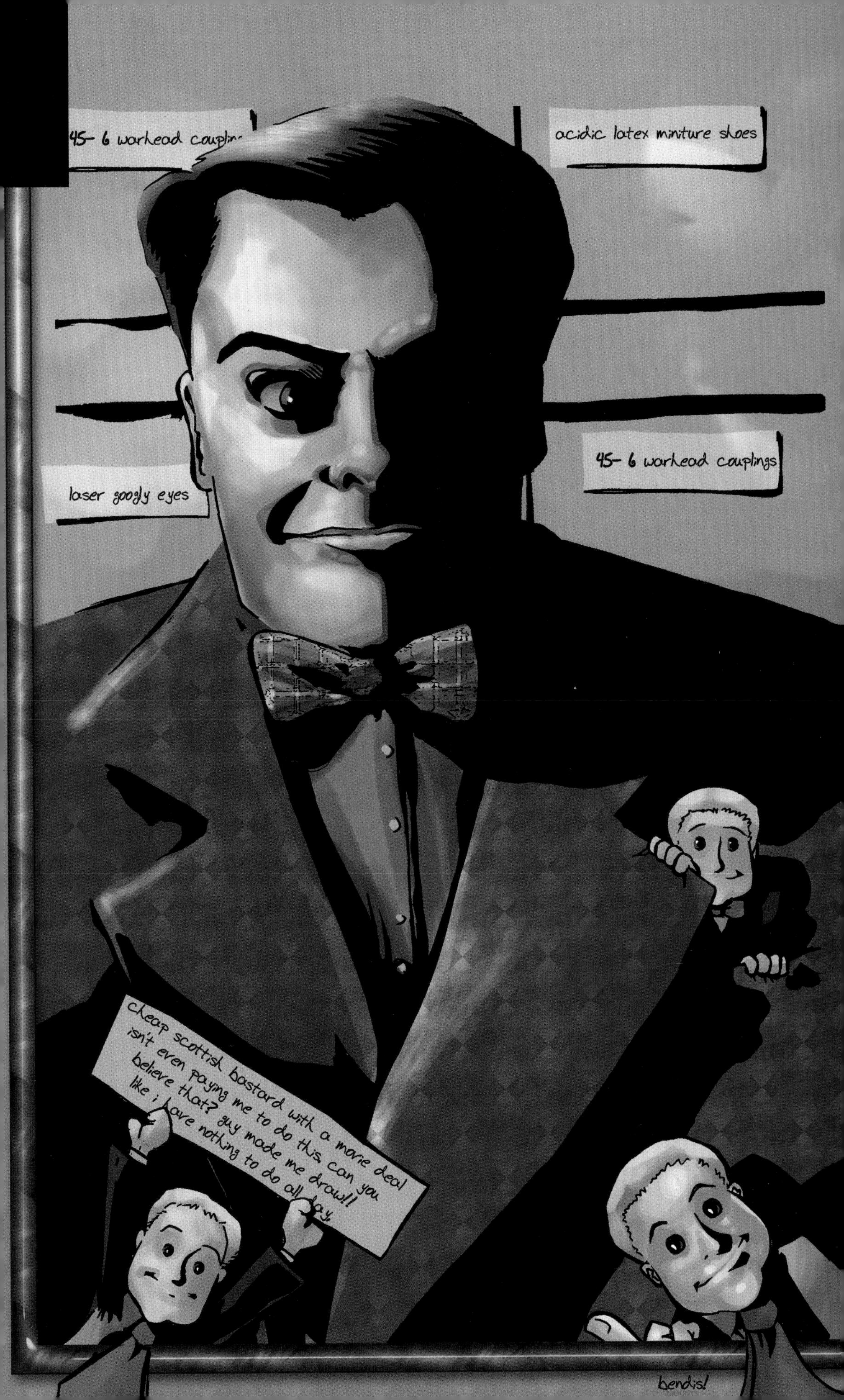

IMP

The power to fold and unfold reality is in the hands of a child.

Despite the fact that he's thousands of years old, Imp is considered an infant in the seven-dimensional reality from which he hails. He visits us when his parents aren't looking. What we see when we look at him is just the three-dimensional aspect—but there are four more levels you and I can't even dream of.

With his higher consciousness, Imp has the power to shape reality in our dimension as easily as an artist can shape a world with a pencil and eraser. He's limited only by his imagination. In the past, he turned America into a marshmallow-land for 12 hours. He made buildings come alive and slug it out with buildings from other major cities. He turned his greatest super-foe into ice cream on a lark. Part of the Professor's crew, he has been convinced to tone things down a tad when he visits our realm, as he is so powerful, he could accidentally unmake reality.

Deadly Nightshade

She is the deadliest of blossoms. She can envelop you in her petals, and give you the most exquisite of deaths as her pollens fill your lungs, choking the life out of you.

Deadly Nightshade is an assassin in Rictus' crew with a secret. She's so enamored of living on the edge, that she's having an affair with Imp. The cross-camp romance is the most dangerous of games, as tensions between Rictus and the Professor always threaten to boil over. Couple that with the fact that she knows Imp could unmake reality in an orgasm-induced loss of control, and Deadly Nightshade takes all of our lives into her petals every time she slips between the sheets.

Pencils, inks and colors: Chris Bachalo

SHIT-HEAD

The collected feces of the 666 most evil beings ever to walk the earth have taken on sentience. There's a little Hitler in there, a touch of Ed Gein, half a pound of Jeffrey Dahmer.

No one knows how this walking, steaming shit-pile came to life. Many say it formed in the world's sewers, the result of a mystic spell, or perhaps science gone horribly awry. Suffice it to say that Shit-Head is part of Mister Rictus' army, perhaps the most vile creature among a horrifically vile crew.

Shit-Head is a mere footsoldier in Rictus' crew, used as muscle. He can make his body diarrhea-soft, bloody constipation-hard, or any consistency in-between. As such, he can pound an enemy with the force of a freight train, then slip away, liquid-smooth, through a sewer grate.

Shit happens. Just pray he never happens to you.

Pencils and inks: Bill Sienkiewicz
Colors: Paul Mounts

FUCKWIT

Meet the Down's Syndrome copy of the world's (former) greatest hero.

Fuckwit is the product of an experiment to create a superman. Part of the experiment worked: The body is there. The brain...not so much so.

Fuckwit is massively powered, but dumb as a bag of hammers. He is the concept of the child with the gun, writ large. He'll yank your arm off trying to shake your hand, and not even realize he's done anything wrong...or was that right?

Fuckwit has all the endearing qualities of a new puppy—he's friendly, loyal, and loves you unconditionally. Unfortunately, he also has all the toilet training of a new puppy.

What if the most powerful man in the universe had brain damage? He'd be a Fuckwit, now wouldn't he?

Pencils and inks: Frank Quitely
Colors: Paul Mounts

FRANK
QUITELY
MOUNTS

SUCKER

Sucker came to Earth thousands of years ago, a parasitical alien organism. He forms a symbiotic bond with a host and, to stay alive, has to feed it with the life-force of other living creatures. The second he stops delivering, the host dies, and Sucker moves on to a new one.

This is especially effective when Sucker drains the life-force of a superpowered being—Sucker has access to the powers himself, and a superpowered pawn to put into play.

Sucker enjoys a good cigar—Cubans, of course. He leaves those like he leaves his parasitical victims—drained, dead, and dusty.

Pencils: Joe Quesada
Inks: Mark Millar
Colors: Paul Mounts

Mister Rictus

Once, he saved souls. Now he damns them.

The man who would become Mister Rictus was once a devout Christian, the most pious of men. But he died briefly on the operating table after a horrible accident. He expected Heaven, but found...

Nothing.

No God. No Heaven. No afterlife. None of what he had been promised for living his devout life. Just a void. And the void left a void in his heart.

The incident created a man without a conscience. Rictus now lives completely without moral compass. He knows there is no eternal consequence to his action. He lives for the day, and that is all.

Rictus does everything you've ever thought about in your darkest moments. Every whim he comes up with, he caters to. And his whims tend toward the dark end of the spectrum.

If he wants to eat, fuck, or kill something, he does it, without thought. He is the creature of pure id. Rictus spent years embracing the light, a light that was never there. Now, there is only darkness.

Pencils and inks: Jae Lee
Colors: June Chung

COVER GALLERY

"Now get in the fucking car while I still got this Little Miss Patient smile on my face, asshole!"

—THE FOX

"Only difference between a dream and a nightmare is how big your balls are, bitch."

—THE FOX

"That's exactly what you think it is. My own little 'fuck you' to the world."

—PROFESSOR SOLOMON SELTZER

3

171

172

4

5

6

DOSSIER

The Death Row editions were second printings with variant covers and contained bonus materials which are

SEX

MONEY

SUPER-POWERS

COSTUME

YOU KNOW
YOU WANT IT!

WANTED #6 VARIANTS

...k behind the scenes of Wanted with these initial character designs from artist extraordinaire **J.G. Jones**

...CTER DESIGN_
KILLER, A.K.A. WESLEY GIBSON

...s nailed the design of Wesley Gibson in the "Killer" garb pretty much right out of the box. The costume i...
...tside your window," all real-world stuff that the stylish killer–on–the–go wouldn't be caught dead witho...

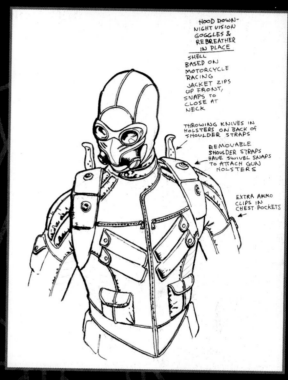

HOOD DOWN-
NIGHT VISION
GOGGLES &
REBREATHER
IN PLACE

SHELL
BASED ON
MOTORCYCLE
RACING
JACKET ZIPS
UP FRONT,
SNAPS TO
CLOSE AT
NECK

THROWING KNIVES IN
HOLSTERS ON BACK OF
SHOULDER STRAPS

REMOVABLE
SHOULDER STRAPS
HAVE SWIVEL SNAPS
TO ATTACH GUN
HOLSTERS

EXTRA AMMO
CLIPS IN
CHEST POCKETS

...CHARACTER DESIGN_
IMP

...he Imp gets his in Wanted #4. But before he ...could die, he had to come to life in the design ...of J.G. Jones. J.G.'s design was pretty much ...spot-on, and no real changes were made from ...he initial take. "I was sorry to see Imp go." said ...Jones. "He and Sucker were two of my favorite ...characters, both to design and to draw."

WHITE
HAIR

Dark Purple-ish

SPITE

Purple or
Brown leather
Collar

Yellow Ochre
Cloth costume

The Original KILLER has lots of Nasty scars on his old, bald head

CHARACTER DESIGN
THE ORIGINAL KILLER

Wesley's papa, on the other hand, went through a bit of revision. The original guy seemed a bit too chunky for a man of such power. The revised Original Killer was sleek and sophisticated.

♪ KILLING ME SOFTLY WITH HIS SONG... ♪ KILLING ME Softly... ♪

The Martyr

CHARACTER DESIGN
MISTER RICTUS

The haunting presence of Mister Rictus appears in WANTED #2. Different names were bandied about and at one point the character was going to be called *The Martyr*.

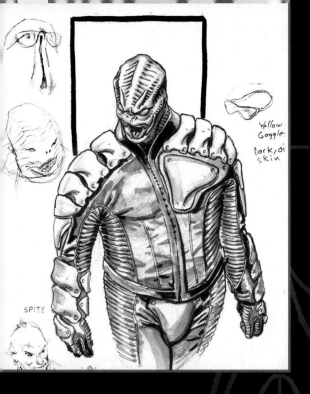

SPITE

Yellow Goggle,
Dark/oi
skin

CHARACTER DESIGN__
SUCKER

Like most of the villains, Sucker shows up for the firs[t]
time in costume in WANTED #2. J.G. Jones' first tr[y]
at Sucker's look and costume was the only try needed[.]
The design was perfect right off the bat!

CHARACTER DESIGN__
FOX

[N]ot a single accessory went overlooked in J.G. Jones' design for the Fox!
[C]uffs, collar, and even the cut of her pants are all addressed, along with
[a] provision for costume changes.

Fox Fur
Cuffs &
Collar on
coat

Amber
Goggles

Fox
Ears
head-
band

The Fox is
a chic, so there[f]
NO WAY she only
has one costume
like most
superheroes.
I think she
she have
variations
on the ox-
blood leather
outfit. This
seems more
like a woman
to me!

Fox has
pointed Boots
with bell-bottom
leather pants
over them.
she wears Reddish
Brown leather
head to toe.

WANTED #1_ PG. 8

Man that he is, artist J.G. Jones hoes to the end of the row! J.G. did two distinct versions of The Killer getting his head blown up, and went with the "giant hole" version. "Drawing all the bits of the brain made it look a little slow and fussed over," J.G. says. "I wanted to go with something that had the impact of a bullet through the head."

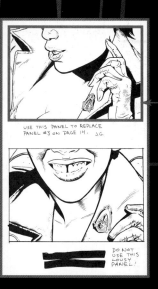

USE THIS PANEL TO REPLACE
PANEL #3 ON PAGE 14. J.G.

DO NOT
USE THIS
LOUSY
PANEL!

WANTED #1_PG. 14

The all important pin that symbolizes The Fraternity went through three separate takes before J.G. settled on winner. "I was just monkeying around with Masonic-type symbols. *The Man Who Would Be King* is one of my favorite movies after all," says J.G.

WANTED #1_PG. 18

As if this massive splash page could get any more elaborate, J.G. added additional backgrounds and cosmic goo-gob to page 18. "I drew this page early on so that we could have a teaser to release online. After drawing the rest of the book, the lab didn't match what I needed for the rest of the scene, so I changed part of the page to make the whole thing work," says J.G.

PAGE 21, PANEL 3   Replacement Panel

WANTED #1_ PG. 21

The scared to death expression of Wesley on page 21 went under the knife before J.G. settled on the final version. "I had to get the face just right to show Wesley starting to freak out," says J.G.

WANTED #2_ PG. 4

Writer Mark Millar changed the pace of the story just a tad on pages 4 and 5. Millar felt that having the "Wesley quits" scene extended over two pages didn' set up the next scene well enough. Millar offered to pay artist J.G. Jones out of his own pocket to make the changes to tighten up the "Wesley quits" scene to end on pg. 4.   (And cheap Scotsman that he is, Millar is yet to pay

WANTED #2_PG. 5

Similarly a rewrite on page 5 led to a redraw of page 5. J.G. Jones worked like a demon to get the revisions done, finishing on Christmas Eve, then retiring for the holidays with a tumbler of Scotch. "It was Macallan," Jones says. "A single malt, rich and smoky, with a good dose of that peat-y flavor."

WANTED #2_PG. 20

Artist J.G. Jones was moved to re-draw a panel on page 20 when he really "got" the impact of what was meant here. "I think you'll note a pasing resemblance to a certain well-known strange visitor," Jones notes. "But only a passing resemblance. We'd never venture into the arena of identity theft."

Replacement Panel

WANTED #3_PG. 6

Artist J.G. Jones realized he was running into an angle
problem on page 6, the handoff of the radioactive condor
from Doll Master to the professor. He redrew the shot for a
better reveal, and even a nice warm glow from the
radiation on the Prof's face.

ISSUE 3
Page 16
Panel 4
Replacent Panel

WANTED #3_PG. 16

WANTED #4_ PG. 6

Artist J.G. Jones did a bit of a flip-flop on "generic patriot good guy" on page 3, panel 1. "The figure looked a little flat to me, so I went for something with more energy and impact."

WANTED #4_ PG. 4

Similarly, the Fox and Wesley were reversed on page 4, panel 3. "Here I wanted to emphasize the dim, moody lighting, so I switched to a shot that would backlight the figures and give me more shadows."

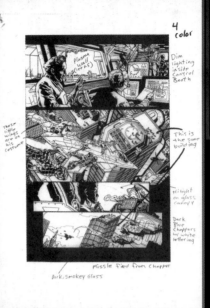

COLOR NOTES_
WANTED #3_PGS. 3 AND 4

Perfectionist that he is, Dr. Jones supplies copius color notes to huemaster Paul Mounts. J.G.'s notes include tone and feel ("Keep it moody and grungy-think Blade Runner"), specific color direction ("Dark blue choppers w/white lettering"), and even head off questions ("This is the same building").

COLOR NOTES_
WANTED #4_PG. 21

J.G. also supplies balloon placements to the letterer, so as to not cover important elements of the art. Compare this sample to the actual page 21, and you'll see that letterer Robin Spehar follows these placements…to the letter.

WANTED
NEW BONUS MATERIAL, FROM SCRIPT TO PRESS

Here is a quick look at the evolution of writer Mark Millar's story from the original script to the finished page, using the classic final fight scenes between Mr. Rictus and Wesley from *Wanted* #5 pages 19 and 20.

Page Nineteen

1/ Pull back and the most astonishing, Kill Bill-kinda overhead where we see the smallish figure of Wesley standing here with the circle of dead bodies around him. Rictus is standing here alone and is very obviously the next target. Wesley stands here with his head down and a smoking gun in each hand, The Fox still down on the ground and surrounding by all these bloodied corpses.

NO DIALOGUE

2/ Close on Wesley as we go all John Woo and have him raise his head and look right into our eyes with an I'm-gonna-fuck-you-up intensity. He's covered in blood and dripping with intense, but restrained emotion here as he takes out his knife.

1 WESLEY CAPTION : Act tough. Act tough. Act tough.
2 WESLEY GIBSON : I'll give you one shot.

3/ Mister Rictus reacts by firing off a shot towards us, slightly panicked looking. He's a dangerous mother-fucker, but even HE knows his time is up.

3 SOUND F/X   : BLAM!

4/ Super-cool shot as the intense Wesley just smacks (yes, smacks) the bullet aside with the blade of his knife like he's playing ping-pong.

4 SOUND F/X   : SMACK!

Page Twenty

1/ Reaction shot as this bullet fires back towards it's intended target and just shoots Mister Rictus right through the throat, coming out the back off his neck. His hat comes off for the first time and we're quite surprised by the suddenness of all this.

1 RICTUS (big)   : GRARGH!
2 WESLEY CAPTION : Holy shit.

2/ Reaction shot from The Fox and even SHE'S surprised by the suddenness of all this, hardly able to take in just how good this off-panel kid has become.

3 THE FOX   : That was GREAT.

3/ Wesley squats over the wounded body of Mister Rictus as he lies here bleeding on the floor. He's got his hunting knife out and looks cold and menacing as he basically says goodbye. The Fox is watching the whole thing with some interest.

4 WESLEY GIBSON : Now answer the question, Rictus: I don't care about the peace treaty. I don't care about Goddamn The Professor. I just wanna know who killed my Dad and if you had anything to do with it!
5 WESLEY GIBSON : Who killed him, you old bastard? I just wanna know who put that fucking BULLET in his head!

4/ Close on Mister Rictus as he lies back here, the bullet-wound in his throat, as he coughs through blood to hiss his defiant last words.

6 RICTUS (huge)   : HOW THE FUCK SHOULD I KNOW?

5/ Reaction shot from Wesley as he grits his teeth and looks quite feral. Off-panel, he's clearly plunging the knife in here and we get a big spurt of blood right across his face.

7 WESLEY GIBSON : Wrong ANSWER, cock-sucker.

Also available from Top Cow Productions, In

The Agency
(ISBN 13: 978-1-58240-776-0)

Arcanum: Milleniums End
(ISBN 13: 978-1-58240-538-4)

Aphrodite IX: Time Out of Mind
(ISBN: 1-58240-372-4)

Common Grounds
(ISBN 13: 978-1-58240-436-3)

Compendium Editions
The Darkness Compendium, vol. 1
(ISBN 13: 978-1-58240-643-5)
Witchblade Compendium, vol. 1
(ISBN 13: 978-1-58240-634-3)
Tomb Raider Compendium, vol. 1
(ISBN: 13 978-1-58240-637-4)

Cyberforce: Assault with a Deadly Woman
(ISBN 13: 978-1-887279-04-8)
Cyberforce: Tin Men of War
(ISBN 13: 978-1-58240-190-4)
Cyberforce: From the Ashes, vol. 1
(ISBN 13: 978-1-58240-708-1)

The Darkness: Ultimate Collection
(ISBN 13: 978-1-58240-780-7)
The Darkness: Spear of Destiny
(ISBN 13: 978-1-58240-147-8)
The Darkness: Heart of Darkness , vol. 2
(ISBN 13: 978-1-58240-205-5)
The Darkness: Original Sin, vol. 3
(ISBN 13: 978-1-58240-459-2)
The Darkness: Flesh and Blood, vol. 3.5
(ISBN 13: 978-1-58240-538-4)
The Darkness: Resurrection, vol. 4
(ISBN 13: 978-1-58240-349-6)
The Darkness, vol. 5
(ISBN: 978-1-58240-646-6)
The Darkness, vol. 6
(ISBN: 978-1-58240-795-1)
The Darkness: Levels vol. 1
(ISBN: 978-1-58240-797-5)
Art of the Darkness, art collection
(ISBN 13: 978-1-58240-649-7)

Delicate Creatures
(ISBN 13: 978-1-58240-225-3)

Freshmen, vol. 1
(ISBN 13: 978-1-58240-593-3)
Freshmen, vol. 2
(ISBN 13: 978-1-58240-732-6)

Hunter-Killer, vol. 1 slipcase edition
(ISBN 13: 978-1-58240-647-3)
Hunter-Killer, vol. 1 hardcover edition
(ISBN 13: 978-1-58240-644-2)
Hunter-Killer, vol. 1 softcover
(ISBN 13: 978-1-58240-647-3)

Kin: Descent of Man
(ISBN 13: 978-1-58240-224-6)

Magdalena: Blood Divine
(ISBN 13: 978-1-58240-215-4)
The Magdalena vol. 1
(ISBN 13: 978-1-58240-645-9)

Midnight Nation
(ISBN 13: 978-1-58240-460-8

Myth Warriors
(ISBN 13: 978-1-58240-427-1

The Necromancer
(ISBN 13: 978-1-58240-648-0

No Honor
(ISBN 13: 978-1-58240-321-2

Obergeist
(ISBN 13: 978-1-58240-243-7

Rising Stars: Born in Fire, vol. 1
(ISBN 13: 978-1-58240-172-0
Rising Stars: Power, vol. 2
(ISBN 13: 978-1-58240-226-0
Rising Stars: Fire and Ash, vol. 3
(ISBN 13: 978-1-58240-491-2
Rising Stars: Visitations
(ISBN 13: 978-1-58240-268-0

Top Cow's Best of: Dave Finch
(ISBN 13: 978-1-58240-638-1
Top Cow's Best of: Michael Turn
(ISBN 13: 978-1-58240-544-5
Top Cow's Best of: Warren Ellis
Down/Tales of the Witchblade
(ISBN 13: 978-1-58240-623-7

Proximity Effect
(ISBN 13: 978-1-58240-375-5

Tom Judge: The Rapture
(ISBN 13: 978-1-58240-389-2

Strykeforce vol. 1
(ISBN 13: 978-1-58240-471-4

Tokyo Knights
(ISBN 13: 978-1-58240-428-8

Tomb Raider: Saga of the Medus
(ISBN 13: 978-1-58240-164-5
Tomb Raider: Mystic Artifacts, vo
(ISBN 13: 978-1-58240-202-4
Tomb Raider: Chasing Shangri-
(ISBN 13: 978-1-58240-267-3
Tomb Raider/Witchblade: Troub
(ISBN 13: 978-1-58240-279-6

Wanted softcover edition
(ISBN 13: 978-1-58240-497-4

Witchblade: Revelations, vol. 2
(ISBN 13: 978-1-58240-458-5
Witchblade/Darkness: Family Ti
(ISBN 13: 978-1-58240-030-3

Witchblade: Obakemono, vol. 6
(ISBN 13: 978-1-58240-259-8
Witchblade: Blood Relations , vo
(ISBN 13: 978-1-58240-315-1
Witchblade: Witchhunt, vol. 10
(ISBN 13: 978-1-58240-590-2
Witchblade: Awakenings, vol. 1
(ISBN 13: 978-1-58240-635-0
Witchblade/Darkminds: The Ret